MODERN PROTECTIVE STRUCTURES

Theodor Krauthammer

CRC Press
Taylor & Francis Group
Boca Raton   London   New York

CRC Press is an imprint of the
Taylor & Francis Group, an informa business

Civil and Environmental Engineering

A Series of Reference Books and Textbooks

Editor

Michael D. Meyer

Department of Civil and Environmental Engineering
Georgia Institute of Technology
Atlanta, Georgia

1. *Preliminary Design of Bridges for Architects and Engineers,* Michele Melaragno
2. *Concrete Formwork Systems,* Awad S. Hanna
3. *Multilayered Aquifer Systems: Fundamentals and Applications,* Alexander H.-D. Cheng
4. *Matrix Analysis of Structural Dynamics: Applications and Earthquake Engineering,* Franklin Y. Cheng
5. *Hazardous Gases Underground: Applications to Tunnel Engineering,* Barry R. Doyle
6. *Cold-Formed Steel Structures to the AISI Specification,* Gregory J. Hancock, Thomas M. Murray, and Duane S. Ellifritt
7. *Fundamentals of Infrastructure Engineering: Civil Engineering Systems: Second Edition, Revised and Expanded,* Patrick H. McDonald
8. *Handbook of Pollution Control and Waste Minimization,* Abbas Ghassemi
9. *Introduction to Approximate Solution Techniques, Numerical Modeling, and Finite Element Methods,* Victor N. Kaliakin
10. *Geotechnical Engineering: Principles and Practices of Soil Mechanics and Foundation Engineering,* V. N. S. Murthy
11. *Estimating Building Costs,* Calin M. Popescu, Kan Phaobunjong, and Nuntapong Ovararin
12. *Chemical Grouting and Soil Stabilization, Third Edition,* Reuben H. Karol
13. *Multifunctional Cement-Based Materials,* Deborah D. L. Chung

14. *Reinforced Soil Engineering: Advances in Research and Practice,* edited by Hoe I. Ling, Dov Leshchinsky, and Fumio Tatsuoka
15. *Project Scheduling Handbook,* Jonathan F. Hutchings
16. *Environmental Pollution Control Microbiology: A Fifty-Year Perspective,* Ross E. McKinney
17. *Hydraulics of Spillways and Energy Dissipators,* Rajnikant M. Khatsuria
18. *Wind and Earthquake Resistant Buildings: Structural Analysis and Design,* Bungale S. Taranath
19. *Natural Wastewater Treatment Systems,* Ronald W. Crites, E. Joe Middlebrooks, Sherwood C. Reed
20. *Water Treatment Unit Processes: Physical and Chemical,* David Hendricks
21. *Construction Equipment Management for Engineers, Estimators, and Owners,* Douglas D. Gransberg, Calin M. Popescu, and Richard C. Ryan
22. *Modern Protective Structures,* Theodor Krauthammer

CRC Press
Taylor & Francis Group
6000 Broken Sound Parkway NW, Suite 300
Boca Raton, FL 33487-2742

© 2008 by Taylor & Francis Group, LLC
CRC Press is an imprint of Taylor & Francis Group, an Informa business

No claim to original U.S. Government works
Printed in the United States of America on acid-free paper
10 9 8 7 6 5 4 3 2 1

International Standard Book Number-13: 978-0-8247-2526-6 (Hardcover)

This book contains information obtained from authentic and highly regarded sources Reasonable efforts have been made to publish reliable data and information, but the author and publisher cannot assume responsibility for the validity of all materials or the consequences of their use. The Authors and Publishers have attempted to trace the copyright holders of all material reproduced in this publication and apologize to copyright holders if permission to publish in this form has not been obtained. If any copyright material has not been acknowledged please write and let us know so we may rectify in any future reprint

Except as permitted under U.S. Copyright Law, no part of this book may be reprinted, reproduced, transmitted, or utilized in any form by any electronic, mechanical, or other means, now known or hereafter invented, including photocopying, microfilming, and recording, or in any information storage or retrieval system, without written permission from the publishers.

For permission to photocopy or use material electronically from this work, please access www.copyright.com (http://www.copyright.com/) or contact the Copyright Clearance Center, Inc. (CCC) 222 Rosewood Drive, Danvers, MA 01923, 978-750-8400. CCC is a not-for-profit organization that provides licenses and registration for a variety of users. For organizations that have been granted a photocopy license by the CCC, a separate system of payment has been arranged.

Trademark Notice: Product or corporate names may be trademarks or registered trademarks, and are used only for identification and explanation without intent to infringe.

Library of Congress Cataloging-in-Publication Data

Krauthammer, Theodor.
  Modern protective structures / Theodor Krauthammer.
    p. cm.
  Includes bibliographical references and index.
  ISBN 0-8247-2526-3 (alk. paper)
  1. Buildings--Protection. 2. Structural analysis (Engineering) I. Title.

TH9025.K73 2007
693.8--dc22   2006051078

Visit the Taylor & Francis Web site at
http://www.taylorandfrancis.com

and the CRC Press Web site at
http://www.crcpress.com

Dedication

To my family, teachers, students, and sponsors.

Preface

Abnormal loading on a facility or vehicle during a terrorist attack has become quite common in many parts of the world, and it directly impacts public safety. We have observed an increasing need to protect critical infrastructure facilities and systems against industrial explosive accidents, criminal activities, and social/subversive unrest. This problem may, in fact, exceed the reasons for hardening developments (i.e., military-sponsored work on fortifications). Careful attention must be devoted to typical modern civilian and military facilities and systems whose failure could severely disrupt the social and economic infrastructures of nations. Addressing such challenges requires special knowledge about existing capabilities and how to treat a wide range of threats. Homeland security and protective technology engineering, and staff at various R&D and safety and security organizations need to:

- know how to assess the risk associated with threats, hazards, and various explosive incidents
- have access to knowledge on how critical facilities behave under blast, shock, and impact loads
- know how to analyze and design various facilities to protect lives and property
- implement such knowledge for conducting effective rescue and recovery operations, and forensic investigations

This book addresses a broad range of scientific and technical issues involved in mitigating the severe loading effects associated with blast, shock, and impact. The content is based on the author's 30-year involvement in cutting-edge research in this general field, and on his experience gained from teaching this subject in graduate courses and special training programs. It combines theoretical, numerical, and valuable practical information that can be incorporated effectively and immediately into a broad range of essential activities.

Ted Krauthammer, Ph.D., FACI
Goldsby Professor of Civil Engineering
Director, Center for Infrastructure Protection and Physical Security
University of Florida

Author

Dr. Theodor Krauthammer is currently the Goldsby Professor of Civil Engineering at the University of Florida and director of the Center for Infrastructure Protection and Physical Security. Until the end of 2006, he was a professor of civil engineering at The Pennsylvania State University, and director of the Protective Technology Center. He earned his Ph.D. in civil engineering from the University of Illinois at Urbana–Champaign.

Dr. Krauthammer's main research and technical activities are directed at structural behavior under severe dynamic loads, including considerations of both survivability and fragility aspects of facilities subjected to blast, shock impact, and severe vibrations. His work has included the development of design recommendations for enhancing structural performance, physical security, and safety of buildings, facilities, and systems. He has conducted both numerical simulations and experimental studies of structures subjected to a broad range of loading conditions. His research has been supported by many government agencies in the United States and abroad.

Dr. Krauthammer is a fellow of the American Concrete Institute (ACI), a member of the American Society of Civil Engineers (ASCE), and a member of the American Institute of Steel Construction (AISC), and is also involved with several other national and international professional organizations. He serves on 12 technical committees of ASCE, ACI, and AISC. Dr. Krauthammer is chair of the ASCE Committee on Blast Shock and Vibratory Effects and its Task Committee on Structural Design for Physical Security. He also chairs the Joint ACI–ASCE Committee 421 on Design of Reinforced Concrete Slabs and is a member of the ASCE–SEI Bridge and Tunnel Security Committee. He was a member of the National Research Council Defensive Architecture Committee.

The professor has written more than 350 research publications, and has been invited to lecture throughout the United States and abroad. He has served as a consultant to industry and governments in the United States and overseas. Dr. Krauthammer's teaching background includes courses on structural design and behavior, structural analysis, advanced dynamics, protective structures, and numerical methods. He is very active in organizing cooperative international scientific, technical, and engineering education activities.

Table of Contents

Chapter 1 Introduction 1
1.1 General Background 1
1.2 Protective Planning and Design Philosophy 5
1.3 Protection Methodology, Threat, and Risk Assessment 8
    1.3.1 Threat, Hazard, and Vulnerability Assessments 10
    1.3.2 Risk Assessment 11
1.4 From Threat and Hazard Environment to Load Definition 16
    1.4.1 Military Threat Assessment 19
        1.4.1.1 Tactics and Strategy 19
        1.4.1.2 Weapons Systems 21
        1.4.1.3 Summary 24
    1.4.2 Terrorism and Insurgency Threat Assessment 24
1.5 Technical Resources and Blast Mitigation Capabilities 24
    1.5.1 Design Manuals 26
    1.5.2 Design and Construction Considerations 28
1.6 Analysis Requirements and Capabilities 30
    1.6.1 Computational Analysis 30
    1.6.2 Experimental Analysis 30
1.7 Protective Technology: Current State and Future Needs 31
    1.7.1 Relationships to R&D 33
    1.7.2 Policy and Technology Needs 35
    1.7.3 Development and Implementation of Protective Design Methodology 37
        1.7.3.1 Basic and Preliminary Applied Research 38
        1.7.3.2 Applied Research, Advanced Technology Development, and Demonstration and Validation Tests 38
        1.7.3.3 Demonstration, Validation, and Implementation 38
1.8 Summary 39

Chapter 2 Explosive Devices and Explosions 41
2.1 Introduction 41
2.2 Characteristics of Explosive Processes, Devices, and Environments 41
    2.2.1 Nuclear Weapons 41
    2.2.2 Conventional Weapons 41

		2.2.2.1	Small Arms and Aircraft Cannon Projectiles 42
		2.2.2.2	Direct and Indirect Fire Weapon Projectiles 42
		2.2.2.3	Grenades 44
		2.2.2.4	Bombs 46
		2.2.2.5	Rockets and Missiles 47
		2.2.2.6	Special-Purpose Weapons 48
	2.2.3	Conventional Weapons Summary 49
2.3	Explosives, Explosions, Effects and Their Mitigation 50
	2.3.1	Blast Effects and Mitigation 50
	2.3.2	Explosives and Explosions 53
	2.3.3	Explosive Types and Properties 58
	2.3.4	Combustion Phenomena and Processes 59
	2.3.5	Detonation Process and Shock Waves 60

Chapter 3 Conventional and Nuclear Environments 65

3.1	Introduction 65
3.2	Airblast 68
	3.2.1 HE Devices and Conventional Weapons 68
	3.2.1.1 External Explosions 68
	3.2.1.2 Internal Explosions 75
	3.2.1.3 Leakage Blast Pressure 77
	3.2.2 Nuclear Devices 79
3.3	Penetration 83
	3.3.1 Concrete Penetration 87
	3.3.2 Rock Penetration 92
	3.3.3 Soil and Other Granular Material Penetration 96
	3.3.4 Armor Penetration 98
	3.3.5 Penetration of Other Materials 101
	3.3.6 Shaped Charges 103
3.4	HE-Induced Ground Shock, Cratering, and Ejecta 106
	3.4.1 HE Charges and Conventional Weapons 106
	3.4.2 Three-Dimensional Stress Wave Propagation 106
	3.4.3 One-Dimensional Elastic Wave Propagation 110
	3.4.4 Reflection and Transmission of One-Dimensional D-Waves between Two Media 114
	3.4.5 Computational Aspects 116
	3.4.6 Shock Waves in One-Dimensional Solids 117
	3.4.7 Stress Wave Propagation in Soils 119
	3.4.7.1 Numerical Evaluation 125
	3.4.8 Application to Protective Design 129
	3.4.9 Cratering 135
	3.4.10 Ejecta 137
3.5	Cratering, Ejecta, and Ground Shock from Nuclear Devices 141
	3.5.1 Cratering 141

		3.5.2	Ejecta 145
		3.5.3	Ground Shock 147
3.6	Fragmentation 153
		3.6.1	Fragment Penetration 158
3.7	Fire, Chemical, Bacteriological, and Radiological Environments 167

Chapter 4 Conventional and Nuclear Loads on Structures 169

4.1	Conventional Loads on Structures 169
	4.1.1 Buried Structures 169
	4.1.2 Above-Ground Structures 170
	4.1.3 Mounded Structures 175
	4.1.4 Surface-Flush Structures 175
	4.1.5 Blast Fragment Coupling 175
4.2	Nuclear Loads on Structures 177
	4.2.1 Above-Ground Structures 177
	4.2.2 Buried Structures 179
	4.2.3 Soil Arching 185
4.3	Step-by-Step Procedures for Deriving Blast Design Loads 187
	4.3.1 Constructing Load–Time History on Buried Wall or Roof 187
	4.3.2 Computing Pressure–Time Curve on Front Wall from External Explosion (Surface Burst) 187
	4.3.3 Computing Pressure–Time Curve on Roof or Sidewall (Span Perpendicular to Shock Front) 190
	4.3.4 Computing Pressure–Time Curve on Roof or Sidewall (Span Parallel to Shock Front) 190
	4.3.5 Blast Load on Rear Wall 191

Chapter 5 Behaviors of Structural Elements 195

5.1	Introduction 195
5.2	Government and Non-Government Manuals and Criteria 196
	5.2.1 Tri-Service Manual TM 5-1300 (Department of the Army, 1990) 196
	5.2.2 Army Technical Manuals 5-855-1 (Department of the Army, 1986 and 1998), and UFC 3-340-01 (Department of Defense, June 2002) 197
	5.2.3 ASCE Manual 42 (ASCE, 1985) 197
	5.2.3.1 Design of Blast Resistant Buildings in Petrochemical Facilities (ASCE, 1997) 197
	5.2.3.2 Structural Design for Physical Security (ASCE, 1999) 197
	5.2.4 DoD and GSA Criteria 198
5.3	Distances from Explosion and Dynamic Loads 198

5.4 Material Properties of Steel and Concrete 198
    5.4.1 Steel 198
    5.4.2 Concrete 199
    5.4.3 Dynamic Effects 202
5.5 Flexural Resistance 202
5.6 Shear Resistance 205
5.7 Tensile and Compressive Members 206
5.8 Principal Reinforcement 207
5.9 Cylinders, Arches, and Domes 208
5.10 Shear Walls 212
5.11 Frames 214
5.12 Natural Periods of Vibration 215
5.13 Advanced Considerations 218
    5.13.1 Membrane Behavior 218
        5.13.1.1 Membrane Application for Analysis and Design 225
    5.13.2 Direct Shear 226
    5.13.3 Diagonal Shear Effects 228
    5.13.4 Size Effects and Combined Size and Rate Effects 230
    5.13.5 Connections and Support Conditions 231
5.14 Application to Structural Design 233
5.15 Practical Damage and Response Limits 234

Chapter 6 Dynamic Response and Analysis 237

6.1 Introduction 237
6.2 Simple Single Degree of Freedom (SDOF) Analysis 240
    6.2.1 Theoretical Solution for SDOF Systems: Duhamel's Integral 242
    6.2.2 Graphical Presentations of Solutions for SDOF Systems 247
    6.2.3 Numerical Solutions of SDOF Systems 247
    6.2.4 Advanced SDOF Approaches 250
6.3 Multi-Degree-of-Freedom (MDOF) Systems 251
    6.3.1 Introduction to MDOF 251
    6.3.2 Numerical Methods for MDOF Transient Responses Analysis ...255
6.4 Continuous Systems 258
    6.4.1 General Continuous Systems 259
6.5 Intermediate and Advanced Computational Approaches 262
    6.5.1 Intermediate Approximate Methods 263
    6.5.2 Advanced Approximate Computational Methods 264
    6.5.3 Material Models 265
6.6 Validation Requirements for Computational Capabilities 267

	6.7	Practical Computation Support for Protective Analysis and Design Activities..272	
		6.7.1	Equivalent SDOF System Approach.....................................274
		6.7.2	Applications to Support Analysis and Design.....................276
		6.7.3	Approximate Procedure for Multi-Segmental Forcing Functions...281

Chapter 7 Connections, Openings, Interfaces, and Internal Shock............291

7.1	Connections 291	
	7.1.1	Introduction 291
	7.1.2	Background 292
	7.1.3	Studies on Behavior of Structural Concrete Detailing..........296
	7.1.4	Studies on Behavior of Structural Steel Detailing 308
	7.1.5	Summary 309
7.2	Openings and Interfaces 310	
	7.2.1	Entrance Tunnels..310
	7.2.2	Blast Doors...310
	7.2.3	Blast Valves 313
	7.2.4	Cable and Conduit Penetrations...315
	7.2.5	Emergency Exits 317
7.3	Internal Shock and Its Isolation 317	
7.4	Internal Pressure 319	
	7.4.1	Internal Pressure Increases...319
	7.4.2	Airblast Transmission through Tunnels and Ducts 319

Chapter 8 Pressure–Impulse Diagrams and Their Applications...................325

8.1	Introduction...325		
8.2	Background 326		
8.3	Characteristics of P-I Diagrams 327		
	8.3.1	Loading Regimes..328	
	8.3.2	Influence of System and Loading Parameters 330	
8.4	Analytical Solutions of P-I Diagrams 331		
	8.4.1	Closed-Form Solutions 331	
		8.4.1.1	Response to Rectangular Load Pulse 331
		8.4.1.2	Response to Triangular Load Pulse....................334
	8.4.2	Energy Balance Method...335	
		8.4.2.1	Approximating Dynamic Regions 336
		8.4.2.2	Continuous Structural Elements 338
8.5	Numerical Approach to P-I Curves..339		
	8.5.1	P-I Curves for Multiple Failure Modes 341	
	8.5.2	Summary 341	
8.6	Dynamic Analysis Approach 343		
	8.6.1	Introduction 343	

	8.6.2	Dynamic Material and Constitutive Models..........343
	8.6.3	Flexural Behavior..........344
		8.6.3.1 Dynamic Resistance Function..........346
	8.6.4	Direct Shear Behavior..........347
8.7	Dynamic Structural Model..........348	
	8.7.1	Flexural Response..........349
		8.7.1.1 Equation of Motion..........349
		8.7.1.2 Transformation Factors..........350
	8.7.2	Direct Shear Response..........351
		8.7.2.1 Equation of Motion..........351
		8.7.2.2 Shear Mass..........351
		8.7.2.3 Dynamic Shear Force..........352
	8.7.3	Summary..........357
8.8	Application Examples for SDOF and P-I Computational Approaches..........357	
	8.8.1	SDOF and P-I Computations for Reinforced Concrete Beams..........357
	8.8.2	SDOF and P-I Computations for Reinforced Concrete Slabs..........363
	8.8.3	Summary..........366

Chapter 9 Progressive Collapse..........373

9.1	Introduction..........373	
	9.1.1	Progressive Collapse Phenomena..........373
9.2	Background..........374	
	9.2.1	Abnormal Loadings..........374
	9.2.2	Observations..........375
9.3	Progressive Collapses of Different Types of Structures..........377	
	9.3.1	Precast Concrete Structures..........378
	9.3.2	Monolithic Concrete Structures..........378
	9.3.3	Truss Structures..........379
	9.3.4	Steel Frame Buildings..........379
9.4	Department of Defense and General Services Administration Guidelines..........380	
	9.4.1	Department of Defense (DoD) Guidelines..........380
		9.4.1.1 Design Requirements for New and Existing Construction..........380
		9.4.1.2 Design Approaches and Strategies..........381
		9.4.1.3 Damage Limits..........384
		9.4.1.4 Other Topics..........384
	9.4.2	GSA Guidelines..........384
		9.4.2.1 Potential for Progressive Collapse Assessment of Existing Facilities..........385

		9.4.2.2	Analysis and Design Guidelines for Mitigating Progressive Collapse in New Facilities...385
		9.4.2.3	Analysis and Acceptance Criteria....................387
		9.4.2.4	Material Properties and Structural Modeling.....389
		9.4.2.5	Redesigns of Structural Elements....................389
9.5	Advanced Frame Structure Analysis...390
	9.5.1	Background 390
	9.5.2	Semi-Rigid Connections 392
	9.5.3	Computer Code Requirements...394
	9.5.4	Examples of Progressive Collapse Analysis........................395
		9.5.4.1	Semi-Rigid Connections 395
		9.5.4.2	Analyses 397
		9.5.4.3	Results 400
		9.5.4.4	Conclusions 400
9.6	Summary..404

Chapter 10 A Comprehensive Protective Design Approach.........................405

10.1	Introduction...405
10.2	Background..405
10.3	Protection Approaches and Measures 407
10.4	Planning and Design Assumptions..410
10.5	Siting, Architectural, and Functional Considerations 411
	10.5.1	Perimeter Line 413
	10.5.2	Access and Approach Control 413
	10.5.3	Building Exterior...414
	10.5.4	Building Interior..416
	10.5.5	Vital Nonstructural Systems 417
	10.5.6	Post-Incident Conditions...417
10.6	Load Considerations..417
10.7	Structural Behavior and Performance 422
10.8	Structural System Behavior...424
10.9	Structural System and Component Selection....................................430
10.10	Multi-Hazard Protective Design..433
10.11	Other Safety Considerations..434
10.12	Development and Implementation of Effective Protective Technology 434
	10.12.1	Recommended Actions...435
		10.12.1.1	Basic and Preliminary Applied Research...........435
		10.12.1.2	Applied Research, Advanced Technology Development, Demonstrations, and Validation Tests 436
		10.12.1.3	Demonstration, Validation, and Implementation 436

        10.12.2 Education, Training, and Technology Transfer Needs 437
        10.12.3 Recommended Long-Term Research and
                 Development Activities 438
10.13 Summary 441

References 443

Index 459

1 Introduction

1.1 GENERAL BACKGROUND

One of the basic needs of all living creatures is to have safe and secure shelter. Throughout history, humans have demonstrated a remarkable ability to address this need. They have developed capabilities to protect themselves against both natural disasters and hazards associated with man-made activities. In 1989, the world viewed a watershed event with the demise of the Cold War and the former Soviet Union. The so-called "Doomsday Clock" that signified the time remaining for civilization was ritualistically set back. Nations across the globe began reducing their armed forces in response to the peace dividend achieved with increased world stability. Unfortunately, the euphoria did not exist for long. Besides the proliferation of advanced weapon systems (including long range ballistic missiles that can carry nuclear, biological, and chemical devices), one notes a dramatic increase in international terrorism activities.

Terrorism is not a new phenomenon, and one can find historic references that such activities existed for more than 2000 years. More recently, the use of terrorism as a means to achieve national objectives was a primary cause for World War I. During the Cold War, terrorism was related to the struggle between the superpowers. In most cases, this tactic seemed restricted to internal disputes by rival political factions within well defined nation states. Industrialized countries viewed the annoyance of terrorism as a third world phenomenon and, with perhaps the exceptions of a few countries, paid such events little notice. World opinion quickly condemned the more spectacular events (e.g., Munich in 1972, Beirut in 1981, etc.) but then just as quickly dismissed them as anomalistic behavior. The application of terror as a global political tactic, associated mainly with the Middle East since the early 1950s, has increased since the early 1980s, and escalated after the end of the Cold War. Many regions around the world have been increasingly burdened by this phenomenon during the last quarter century (e.g., various regions in Africa and the Middle East, Afghanistan, France, Germany, India, Iraq, Italy, Japan, Russia, Spain, Sri Lanka, and United Kingdom). The locations, causes, participants, intensities, and means have been changing rapidly. Vehicle bombs and other types of improvised explosive devices (IEDs) have become the preferred mechanisms for terrorist attacks, followed by the use of homicide bombers and the renewed threats from weapons of mass destruction (WMDs).

Prior to 1993, the United States had been relatively unaffected by terrorism within its borders. Then, in February 1993, the U.S. was attacked by externally supported terrorists who targeted the Word Trade Center. In April 1995, the U.S. was shocked by the devastating home-grown terrorist attack against the Alfred P. Murrah Federal Building in Oklahoma City. The events of September 11, 2001, and subsequent incidents in Indonesia, Spain, and the United Kingdom demonstrated the ability of terrorists to cause civilian deaths and property damage at levels not seen since the waning days of World War II. These recent horrific terrorist attacks changed forever the way various federal, state, and local government agencies in the United States and many other organizations around the world look at national security and the need for protection from terrorism.

Clearly, in today's geopolitical environment, the need to protect both military facilities and civilian populations from enemy attack has not diminished. Furthermore, we noted an increasing need to protect civilian populations against terrorism and social and subversive unrest. This situation is true for many parts of the world, and it may exceed the previous reasons for the development of protective technologies (i.e., related to military-sponsored research and development efforts on fortifications and hardened facilities). Unlike the global politically and ideologically motivated conflicts of the past dominated by well organized military forces, most of the armed conflicts in the last two decades have been localized and dominated by social, religious, economic, and/or ethnic causes. We no longer face a traditional conflict with a well defined adversary and must now consider an amorphous evolving adversary. Furthermore, societies must learn to cope with a different type of warfare that is termed *low intensity conflict*. Well understood and reasonably predictable military operations have been replaced by much less understood and less predictable terrorist activities carried out by determined individuals or small groups that have a wide range of backgrounds and capabilities.

Such activities are directed against well selected targets and are aimed at inflicting considerable economic damage and loss of lives. Obviously, as demonstrated by recent tragic incidents in the United States, low intensity conflict is a misnomer. Such activities, despite involving a few individuals or small groups, can have devastating consequences. They can adversely affect national and international stability and cause worldwide serious economic, social, and political damage. Hence, such activities are termed asymmetric in recognition of the devastating results achieved from modest inputs.

Defending society against this form of rapidly evolving type of warfare will remain a challenge, at least through the first half of the 21st century, and probably longer. Any successful response will require a well planned multilayered approach that strikes a fine balance between ensuring a nation's security and maintaining the freedoms that a modern society enjoys. The causes for terrorism are related to a broad range of important areas (culture, history, sociology, geopolitics, economics, religion, life sciences and medicine, psychology, etc.). Therefore, besides the serious need for innovative developments in these areas, society must invest in the development of effective capabilities in intelligence, law enforcement, and military application to counter such threats.

Technology can and will play a major role in these efforts, and society must develop innovative and comprehensive protective technologies. Furthermore, we must not employ only empirical approaches (e.g., using tests to observe consequences) to address these issues. The free world must develop innovative

Introduction

theoretical, numerical, and experimental approaches to protect from traditional weapons and WMDs and must conduct these activities in a well coordinated collaboration involving government, academic, and private organizations. Such technologies represent the last layers of defense between society and the threats after all other layers of defense have failed. They are vital for ensuring the safety of people and the preservation of valuable national assets.

An assessment of historical terrorism activity data (U.S. Department of State, 2004) indicates that about 85% of recorded incidents involved explosive devices, about 5% involved ballistic attacks, and the rest were related to other activities such as arson and kidnapping. A similar picture emerges from similar reports on international terrorism (U.S. Department of State, 2003–2006). Such reports could be invaluable for those performing threat analyses because they contain lists of terrorist organizations, their histories, areas of activity, sources of support, and operational capabilities. These lists are watched closely by terrorism experts, foreign governments, and lobbyists who try to influence the lists. These reports also serve as important policy tools for the U.S. Department of State that can add or remove terrorist groups or their state sponsors as a result of the activities attributed to them. Unfortunately, recent editions of such reports raised criticism related to inaccurate statistics. We hope that a new Worldwide Incidents Tracking System maintained by the National Counterterrorism Center will correct such issues.

Although we must not overlook the possibility that terrorists may use unconventional (nuclear, radiological, chemical, and biological) WMDs instead of large explosive devices, the immediate and most expedient threats are the conventional ones. Therefore, required policy and investments in R&D must be shaped accordingly. Clearly, we must know how to accurately and rationally assess various threats, hazards, and accidental incidents. Engineers and emergency personnel must know how to design facilities to protect lives and property and conduct effective rescue operations and forensic investigations. They should address the known threats associated with explosive devices while preparing for possible unconventional WMDs. The study of heavily fortified military facilities may no longer be the main area of concern (although this technology must be kept relevant). Careful attention must be devoted to typical civilian facilities (office and commercial buildings, schools, hospitals, communication centers, power stations, industrial buildings, transportation infrastructures, etc.), whose failure could severely disrupt the social and economic infrastructures of nations. We lack essential knowledge on how such facilities behave under terrorist threats and how much protection they can provide against a broad range of WMD threats (blast, shock, impact, fire, chemical, biological, nuclear, and radiological effects). Many materials and components typically used in such facilities were never studied for these applications.

Preliminary recommendations for measures to minimize casualties and damage from such attacks in the future were made in the U.S. (Downing, 1996). The National Research Council (NRC) recommended areas for future research and action to implement technology transfer (National Research Council, 1995). More far-reaching recommendations were issued recently to address an anticipated enhanced threat from terrorist acts within the U.S. (USCNS/21, 1999–2001). These recommendations must be expanded to address current and anticipated needs for targeting facilities used for hostile activities.

The U.S. Department of Defense (DOD) has the responsibility for developing, maintaining, and applying effective technologies to protect its civilian and military personnel worldwide, and to launch effective operations against hostile entities to reduce such threats to acceptable levels. Outside of the DOD and the Department of Energy (DOE), experience in WMD effects exists in a limited number of R&D organizations and consulting firms that do government work. Since the mid-1980s, a gradual decline has been noted in protective technology related academic R&D activities and very few eminent academicians in this field are still available. This, combined with the lack of formal engineering training in protective science and technology at U.S. universities, has resulted in a shortage of experienced technical personnel in this field in many government and private organizations that must handle such activities.

This brief summary highlights the fact that the Free World has always been reacting to terrorism. The previously observed levels of anti-terrorist preparedness and proactive prevention measures have been unimpressive. Unless these conditions change dramatically, the consequences to society could be very grave. Combating terrorism must include the consideration of many unconventional aspects of warfare. These issues must be integrated with scientific and technical capabilities to provide a comprehensive approach that can be used against current and future low intensity conflicts. Most nations do not have the required resources to approach this problem independently. Moreover, since this evolving threat affects many countries and it endangers the stability of the entire world, the required R&D should be conducted in a collaborative multinational framework. Nations must adopt a considerably more proactive and collaborative approach to address this serious problem. We must employ all reasonable means to eliminate or at least reduce considerably the causes for terrorism. Such activities must address cultural, social, educational, political, religious, and economic aspects of the problem. Only then will law enforcement, security, and technical capabilities succeed in mitigating the effects of terrorist activities.

The rest of this chapter is aimed at developing the foundation for enabling effective mitigation of abnormal loads, such as those associated with either deliberate or accidental explosive incidents. The planning of such activities should take into consideration not only the physical environments associated with the detonation of explosive devices. It should address an overall hostile environment that may include a large number of parameters, one of which is a particular explosive device. Furthermore, because the definition of failure is related closely to performance requirements, (access and perimeter protection, available site, security measures, assets to be protected, functional and operational concepts, mission requirements, serviceability of a facility, service disruptions, and other factors), all related parameters should be considered. As noted previously, other important factors that can influence the effectiveness of such incidents and remedies could be related to nontechnical areas (social sciences, psychology and human behavior, political sciences, foreign relations, etc.) and they should be considered in the overall process of threat and risk assessment. The thrust of the present discussion is to develop realistic guidelines for analysis, design, assessment, retrofit, and research in the field of protected facilities that will be treated in subsequent chapters.

1.2 PROTECTIVE PLANNING AND DESIGN PHILOSOPHY

The fundamental goal of protective construction is to improve the probability of survival of people and other contents in a given facility for a given threat. It is important to realize that the protective building is the last layer of defense against a threat and that all other protective measure (intelligence, law enforcement, surveillance, barriers, etc.) have failed if the threat can be projected onto a facility. This implies that a designer must "know" the threat before conceptualizing the design and this may not be possible in many cases. Attackers can use various weapon systems in different combinations and such events cannot be predicted. However, using reliable information and objective threat and risk assessment can produce effective estimates of such incidents.

Usually, a facility design is based on a standard threat (for example, a specific bomb at a given stand-off distance). In other cases, a statistical approach, requiring that a specific percentage of facilities and contents will survive if a site is attacked, may be employed. Guidelines on how to perform threat assessments have been developed (ASCE, 1999; FEMA, 2003). In the present approach, only the deterministic method (designing a structure to survive a given threat) will be presented.

Physical security can be achieved by a variety of means and devices with a wide range of capabilities. These capabilities can be used to enable detection, deterrence, delay, and prevention of hostile activities. Structural hardening, as briefly discussed above, is a passive defense capability; it is only one aspect of these considerations and should be addressed in the broader context of physical security. As with any other fortification technology, passive defense alone cannot be used to protect against mobile and constantly varying threats. Although the emphasis of the present discussion is on fortification science and technology that can be integrated into protective architecture considerations, users should always perform comprehensive assessments of their specific physical security needs, as described in pertinent references (U.S. Army, TM 5-853, December 1988 and Department of the Army and Air Force, May 1994; FEMA, 2003; AIA, 2004).

A structure must be designed to prevent catastrophic failure and to protect its contents (personnel and equipment) from the effects of an explosion. Such effects may include nuclear and thermal radiation, electromagnetic pulse (EMP), air blast, ground shock, debris, fragments, and dust (protection from chemical and biological (CB) threats should be considered, as appropriate). In order for a military facility to survive, the continuation of its operational mission must be ensured. For civilian facilities, however, the main concern is protecting people and/or critical assets. Therefore, survivability requirements (criteria) vary from one type of facility to another.

Specific guidelines for achieving these goals are available from ASCE (1997 and 1999), DOD (June 2002, July 2002, and October 2003), and FEMA (2003). References are also made to the prevention of progressive collapse (Department of Defense, 2005; GSA, 2003). Generally, the following issues should be addressed to protect valuable assets.

The first is maximizing standoff distance to reduce the blast and fragmentation loads on the structure under consideration. If sufficient space is available, this may be the most effective mitigation approach, but it requires one to ensure that the protecting perimeter is secured to ensure the specified distances. The next most important issue to consider is the prevention of building collapse, and this requires careful attention to the structural layout and design details. Once these items are addressed, one must minimize hazardous flying debris such as glass, dislodged structural parts, and nonstructural components such as furniture and equipment. Providing an effective building layout also can contribute significantly to protecting valuable assets. This may be achieved by placing less valuable assets closer to hazards and more valuable assets farther from hazards. Providing protected spaces to enable people to take quick refuge until further instructed from security and/or rescue personnel is known to be very effective. Limiting airborne contamination should be considered, as appropriate. Additionally, one should address requirements for fire hazard mitigation and effective evacuation, rescue, and recovery operations. Providing mass notification is essential to prevent panic and assist in post-incident activities. Finally, one should consider options to facilitate future facility upgrades, as might be required by periodic risk assessments.

A protected facility consists of several components: a protected perimeter, a protective structure, essential subsystems, and nonessential support subsystems. After all non-structural considerations are addressed, another specific issue that must be considered to protect a facility is blast- and shock-resistant design of the structure to protect contents from the effects of blast, shock, radiation, fragments, debris, dust, etc. It should be clear that the survival of the structure alone may not be sufficient if damage to the contents or personnel exceeds the survivability criteria. In order to consider all these factors, a designer should know very accurately what is expected of a facility.

The terms *hardness* and *survivability* define the capability of a protective system (or facility) to resist the anticipated effects and meet the protection criteria. Survivability can be increased by enhancing a facility's hardness or other protective features. Other means for achieving the same goal can be employed as well (for example, redundancy, ensuring a larger standoff distance, location of other sites, etc.). Placing another protective layer between the facility or its contents and the weapon also can be very helpful because that layer will absorb some of the undesirable effects. Burying a facility in rock or soil will provide major

Introduction

benefits by reducing weapon effects on the system and/or by making it harder to locate the target.

A designer should employ protection criteria to ensure that the facility and its contents will not be subjected to environments (motions, stresses, etc.) beyond a certain limit. An engineer should incorporate fragility data for equipment and people into the design. The effects of soil–structure and equipment–structure interactions must be considered in estimating the various responses. The following components are associated with a protected facility:

- Perimeter: structure and shock isolation systems
- HVAC system: blast closures and air ducts
- Mounts, fasteners, anchors, and connectors
- Lifelines and penetrations (openings, communication lines, doors, water, sewer, fuel lines, etc.)
- Devices for EMP protection
- Radiation shielding
- Fire barriers, thermal shields, and fire suppression systems
- Hydraulic surge protective devices to provide a survivable environment for personnel, prime mission equipment, power supplies, HVAC (including purification equipment to deal with CB attacks), water supply, sewage disposal, and communication equipment.

The design process for protective facilities addressed in this chapter is similar to that described in the resources cited earlier, but includes additional steps that address the non-accidental aspects of design:

1. Define facility operational performance requirements.
2. Establish quality assurance (QA) criteria for analysis, design, and construction work, and assign responsibilities for various activities throughout the entire project.
3. Perform threat, hazard, and risk assessments and determine future risk assessment reviews.
4. Determine explosive sources and their locations and magnitudes.
5. Estimate corresponding loading conditions.
6. Establish general siting, facility layout, and design criteria.
7. Proportion members for equivalent static loads.
8. Compute blast loads on facility more accurately.
9. Compute loading from fragments.
10. Compute loading from crater ejecta.
11. Compute loading from ground shock.
12. Combine all dynamic loads and perform preliminary dynamic analyses.
13. Redesign facility to meet protective criteria under these dynamic effects.
14. Consider nuclear radiation, EMP, thermal effects, CB, etc., if appropriate.

15. Verify design by acceptable methods.
16. Prepare design documentation for shop and field work.
17. Embark on contracting and construction activities; activate appropriate QA.
18. End-of-construction inspection and review.
19. Facility begins its service life.

The design of a facility usually requires interaction among specialists in several disciplines, such as security, architectural, structural, mechanical, electrical, electronics, and hardening. This effort utilizes a team approach to ensure that all possible aspects of the problem have been considered and that the proposed combined plan optimizes the available solutions in the various areas. The analysis methods are divided into two main categories: (1) simplified methods usually performed by linear or nonlinear single-degree-of-freedom (SDOF) or multi-degree-of freedom (MDOF) methods and (2) advanced methods using finite difference (FD) or finite element (FE) techniques. For most design procedures, accurate and reliable simplified methods can be employed. However, it is recommended that advanced analysis be performed after a facility has been designed in order to verify that it will function as required. Advanced analyses (fully nonlinear dynamic) can be very expensive and should be employed with caution. These issues are addressed in greater detail in other sections and chapters.

1.3 PROTECTION METHODOLOGY, THREAT, AND RISK ASSESSMENT

A rational approach for selecting appropriate protective measures for an asset is based on comparing the cost of the mitigation with the cost of the consequence if no improvements are made in protecting the asset. Initially, small investments in protecting an asset can provide significant benefits. However, additional protection enhancements will become increasingly more expensive, as shown schematically in Figure 1.1.

We can identify the following four regions of the relationship. Region A describes the conditions at a very early stage in a project, when careful assessments and studies of protection options and their implementation must be performed, before any meaningful protection benefits can be gained. In region B, the protection enhancement is much larger than the required investment to implement it. Region C requires judgment to decide whether further investment in protection improvements is feasible or warranted. Finally, it would be uneconomical to consider situations that fall into Region D where the cost of very small improvement is very large. Clearly, one must compare the cost of protection enhancement with the value of the achieved benefit from such action (i.e., the gain from reducing the risk associated with the combination of threat and damage to the asset under consideration). A cost/benefit ratio can be defined by a

Introduction

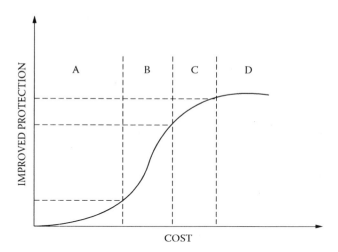

FIGURE 1.1 Cost versus improved protection.

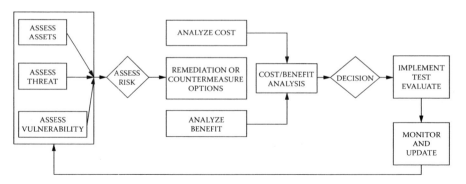

FIGURE 1.2 Risk management approach.

comprehensive risk assessment and management approach for each asset or inventory of assets as described next.

These activities must be performed within a framework of a comprehensive risk management approach that should start with a complete assessment of the problem, as described in Figure 1.2. This is a prerequisite for reaching any objective decision on future actions, and it underlines the rationale for future considerations.

A risk management approach is a structured risk analysis that considers both current and future assets, threats, vulnerabilities, countermeasures, and consequences before and/or after remediation. The approach enables one to develop a risk communication capability that is essential for project development. The benefits of a comprehensive risk management approach are direct outcomes of the analysis features aimed at ensuring that the greatest risks have been identified and addressed. Such an analysis determines and quantifies risks, leads to logical security recommendations, and contributes to the understanding of risks and costs considered for implementation in a decision support tool. The assessment phase consists of three activities (Figure 1.2): asset assessment, threat assessment, and vulnerability assessment. These activities are described briefly.

Risk analysis is applying the results from three separate assessments (asset, threat, and vulnerability) in a risk equation that will be discussed later. An asset is anything of value to an owner (people, information, equipment, facilities, activities, operations, etc.). Each asset can be classified with several levels of details to enable an appropriate degree of accuracy in its assessment. For example, at the lowest level of resolution, an asset might be a facility. At the second level, it might be described as a secure compound or an industrial building. The third level would be more specific and define the specific site such as an embassy compound or corporate campus. A fourth level would define a specific building in that compound, while higher levels of resolution could gradually zoom in to a single specific item. The approach is used to identify assets that may require protection and determine how critical they are. It also allows us to identify undesirable events that may involve such assets and the expected impacts of their loss on the owner. Such assessments can be used to value and prioritize assets based on consequences of loss or damage.

1.3.1 THREAT, HAZARD, AND VULNERABILITY ASSESSMENTS

Understanding and defining possible threats or hazards is essential. Such information is the first step for developing credible loading definitions that must be available to ensure the selection of acceptable mitigation approaches. Furthermore, threat assessments must be performed periodically to evaluate the suitability of existing mitigation measures based on previous threat assessments.

After the Khobar Towers attack in Saudi Arabia, the Secretary of Defense assigned retired General Wayne A. Downing to head a task force to examine the facts and circumstances surrounding the attack and recommend measures to minimize casualties and damage from such attacks in the future (Downing, 1996). The National Research Council (1995, 2000, and 2001) and FEMA (2003) conducted studies to explore the ability and feasibility of protecting against emerging threats and transferring blast-mitigating technologies and design concepts from military to civilian applications. These reports included specific recommendations that should be implemented for land-based facilities. More far reaching recommendations have been issued recently to address an anticipated enhanced threat to the United States from terrorism (USCNS/21, 1999–2001). However, in light of the terrorist attacks against the World Trade Center and the Pentagon in 2001 followed by incidents in North Africa and Europe, these recommendations must not only be taken seriously, but modified to protect against serious current and evolving threats. All these recommendations clearly call for a comprehensive approach in developing protective technologies and the establishment of protective design standards for both new and existing facilities. Furthermore, for the

Introduction 11

approach to be fully comprehensive, it is critical that an effective government–academic–industry partnership be developed to provide an institutional network to foster R&D, training, and technology transfer.

Conventional, nuclear, industrial, and terrorism hazards environments associated with explosive incidents can be defined by using appropriate analytical procedures to extract definitions of the required parameters, for example, load–time histories, radiation levels, temperature–time histories, CB conditions, etc. Threats can be posed by industrial accidents, military operations, terrorist activities, or other criminal undertakings. Although such threats may not be well defined, a design team must provide specific guidelines that address the related uncertainties. In assessing a threat (or hazard), one must employ a significant measure of probabilistic analysis and data available from various sources, and a good understanding of adversaries to obtain a reasonable estimate of the anticipated hostile environment (e.g., loading conditions) to be used in the design activities.

The threat assessment should identify and characterize potential threat categories and adversaries, assess the intent, motivation, and capability of each adversary, determine frequency of past incidents, and estimate threats relative to each critical asset. Analyzing intent requires understanding of the adversary's perspective, motivation, and willingness to accept risk and commit resources (people, money, facilities, etc.). For example, adversaries may be terrorist organizations, foreign intelligence services, criminal elements, vandals, or disgruntled employees, etc. Undesired events may be classified in the context of their sources or the adversaries. Foreign intelligence services may cause facility penetration, non-access attacks, recruiting staff, etc. Terrorist threats include kidnapping, bombing, sabotage, etc. Natural hazard threats are fires, floods, wind (storms, tornadoes, hurricanes), and earthquakes. Criminal threats may include fraud, theft, robbery, arson, vandalism, computer hacking, identity theft, etc. Insider threats could be results of espionage, misuse of equipment, or malicious acts by disgruntled staff. Finally, one might consider military threats including war, insurrection, or various types of operations. Information about threats can be obtained from law enforcement agencies and/or security consulting organizations. One must understand the adversary, its intent, capabilities, and operational history to determine the seriousness of a threat. Vulnerabilities can result from building characteristics, equipment properties, personal behavior, locations of assets, or operational and personnel practices.

1.3.2 Risk Assessment

After these assessments are performed, the next step is assigning numerical values and letter ranks that will be used for the risk assessment definition. The risk assessment is obtained based on the outcome from the following risk equation:

$$\text{Risk} = \text{Threat} \times \text{Impact} \times \text{Vulnerability} \tag{1.1}$$

Impact represents the criticality of the asset to the owner, and is defined by a number between 1 and 100. Threat is an outcome of the threat assessment, and is defined by a number between 0 and 1. Vulnerability describes the vulnerability of the asset and is also defined by a number between 0 and 1. One can also assign letter ranks to each of these parameters, for example, H for high, M for medium, L for low, and C for critical. This approach is illustrated in Table 1.1 for several hypothetical cases.

Once the risks have been quantified, we can estimate (1) the degree of impact relative to each critical asset, (2) the likelihood of attack by a potential adversary or threat, and (3) the likelihood that a specific vulnerability will be exploited, to determine relative degree of risk and prioritize risks based on integrated assessment. These activities can be accomplished as parametric studies to identify existing countermeasures and their level of effectiveness in reducing vulnerabilities and estimate the degree of vulnerability of each asset from the threat. Each countermeasure, enhancement, or retrofit may reduce the threat or vulnerability. Changes in the owner's use of the asset could change the impact factor.

If such changes are implemented, we can compute a new risk value and compare the cost of the modification with the cost associated with the original risk. This would enable us to obtain cost-to-benefit ratios for different options and prioritize such modifications accordingly. A countermeasure represents an action taken or a physical entity used to reduce or eliminate vulnerability, threat, or impact. A cost–benefit analysis is a process in which cost-to-benefit ratios of countermeasures are compared and the most appropriate combination is selected.

Each countermeasure has a tangible cost. The benefit represents the amount of risk reduction based on the overall effectiveness of countermeasures. This approach enables one to make objective and optimized investment decisions to

TABLE 1.1
Risk Assessment Cases

Asset	Incident; Consequence	Asset Rating		Threat Rating		Vulnerability Rating		Risk Rating
People	Attack; assassinate staff	H/C	95	H/C	0.90	M/H	0.75	M (64)
	Attack; kidnap staff	L/C	50	L/H	0.50	L/M	0.25	L (6)
Data	Loss; activity failure	H/C	95	H/H	0.85	H/M	0.95	H (77)
	Press leak; plan disclosure	H/M	15	M/M	0.35	M/M	0.4	L (2)
Equipment	Theft; lost function	L/M	95	L/M	0.8	H/M	0.85	M (65)
	Access; signal diversion	H/H	50	H/H	0.7	L/L	0.1	L (<4)
Facility	BC attack; denial of use	H/H	95	H/M	0.95	H/H	0.95	H (86)
	Mail bomb; damage	L/L	1	H/H	0.75	L/M	0.25	L (<1)
Activity	Sabotage; schedule delay	M/M	5	M/M	0.35	L\L	0.25	l (<1)
	Monitor; security breach	L/H	15	L/L	0.1	H/M	0.50	L (<1)

H = high. M = medium. L = low. C = critical.

Introduction

enhance the physical security of specific assets and/or on an entire inventory of assets. The selected solutions can then be implemented but the entire situation must be continuously monitored to ensure that all the requirements are met. Periodically, new risk assessments should be performed as described above to reevaluate the situation and changes should be recommended as appropriate. For terrorist threat assessments, one must employ an approach similar to that described in recent publications (Chapter 1 and Appendix A of ASCE, 1999; U.S. Department of State, 2005 and 2006). The outcome of such an assessment may be presented in a tabulated format (Table 1.2).

That assessment approach is more qualitative than the previous procedure. One can use the guidelines provided in that publication to assess the threats associated with a broad range of terrorist and/or criminal activities. The magnitude of the terrorism problem might be illustrated based on published data (U.S. Department of State, 2006), as shown in Table 1.3. To illustrate the proposed concept also for military applications, one could use the approach presented in Section 1.4.1 in which the military capabilities of the former Warsaw Pact nations (the former Soviet Union and its allies) are used to apply the approach. Although that military organization no longer exists, the information could still be useful for certain parts of the world and will be reasonably realistic as a case history.

Nevertheless, terrorism is not the only sources for trouble; many problems are directly related to domestic criminal activities, as illustrated by Table 1.4. All these data highlight a disturbing trend: a dramatic increase in the number and cost of threats and incidents related to criminal and terrorist activities in the U.S. or internationally against the interests of the U.S. and its allies. Interestingly, accidental incidents are decreasing, probably because of enhanced preventive procedures.

In order to develop an understanding of a threat, one needs data. For example, assessing a military threat requires data on military capabilities, equipment, and concepts of operations. Analyses of such data can yield a comprehensive threat definition that can then be translated into general requirements for the defending side: the required type of design and/or research needed to meet the stated objectives for a specific facility (or system of facilities). While such analyses are continually conducted by military analysts and intelligence services, in most cases they are not available to the academic and/or technical communities that are developing engineering and scientific solutions for some of these problems. However, as shown later, the data are obtainable from books and articles available to the general public. One can reach practical conclusions based on the analysis of such information.

After a threat has been assessed, assuming a focus on explosive and ballistic attacks, the next step is translating the findings to physical properties. This must be done after a better understanding of explosive and ballistic devices that could be used against a given target, as discussed in Chapter 2.

TABLE 1.2
Assessment of Typical Threats

Aggressor Tactic	Design Basis Threat Severity	Weapons	Tools
Moving vehicle bomb	Very high	2,000 lbm TNT[1]	12,000 lbm truck
	High	500 lbm TNT	5,000 lbm truck
	Medium	100 lbm TNT	4,000 lbm car
	Low	50 lbm TNT	4,000 lbm car
Stationary vehicle bomb	Very high	2,000 lbm TNT	12,000 lbm truck
	High	500 lbm TNT	50,000 lbm truck
	Medium	100 lbm TNT	4,000 lbm car
	Low	50 lbm TNT	4,000 lbm car
Exterior	High	IID, IED (100 lb TNT), and grenades	None
	Medium	IID, IED (2 lbm TNT), and grenades	
	Low	IID, Rocks and clubs	
Standoff weapons	High	Mortars (to 50 lbm TNT)	None
	Low	Antitank weapons	
Ballistics	Very high	30.06 AP	None
	High	7.62 M80 ball	
	Medium	44 Magnum handgun	
	Low	38 Super handgun	
Forced entry	Very high	Handguns and submachine guns (up to UL-SPSA)	Unlimited hand, power, thermal tools, and explosives[2]
	High		Unlimited hand, power, tools, and limited thermal tools/explosives[3]
	Medium		Unlimited hand tools, limited power/thermal tools and hand-held hydraulic jacks
	Low	None	Unlimited hand tools
	Very low		Limited hand tools

[1] To convert pounds (mass) to kilograms, multiply by 0.454.
[2] Up to 20 lbm TNT untamped breaching charge; tamped breaching charge equivalent to 20 lbm untamped charge, and 10,5000 grain/linear foot linear shape charge.
[3] Up to 10 lbm TNT untamped breaching charge, tamped breaching charge equivalent to 10 lbm untamped charged, and 600 grain/linear foot linear shape charge.

Source: ASCE, 1999.

TABLE 1.3
Worldwide Terrorism Incidents in 2005

Incident	Number
Worldwide incidents	11,111
Incidents resulting in death, injury, or kidnaping of at least one individual	8,016
Incidents resulting in death of at least one individual	5,131
Incidents resulting in no deaths	5,980
Incidents resulting in death of only one individual	2,884
Incidents resulting in deaths of at least 10 individuals	226
Incidents resulting in injury of at least one individual	3,827
Incidents resulting in kidnaping of at least one individual	1,145
Individuals killed, injured, or kidnapped worldwide as result of terrorism incidents	74,087
Individuals killed worldwide as result of terrorism incidents	14,602
Individuals injured worldwide as result of terrorism incidents	24,705
Individuals kidnapped worldwide as result of terrorism incidents	34,780

Source: U.S. State Department, 2006.

TABLE 1.4
Types of Explosive Attack Incidents in U.S.

	Number of Incidents		
Type of Incident	1989–1993	1991–1995	1993–1997*
Bombings	7,716	8,567	8,027
Attempted bombings	1,705	2,078	2,291
Incendiary bombings	2,242	2,468	2,294
Attempted incendiary bombings	557	782	898
Threats to treasury facilities	24	31	45
Hoax devices	2,011	2,245	2,833
Accidental–noncriminal	211	192	150
Total number of incidents	14,466	16,363	16,538
Reported killed	258	456	101
Reported injured	3,419	3,859	513
Reported property damage	US $641.3 M	US $1,257.3 M	US $896.9 M

Data from Bureau of Alcohol Tobacco and Firearms (ATF).

* The bureau has not published such complete data since 1997; some information is provided at: www.atf.treas.gov/aexis2/statistics.htm.

1.4 FROM THREAT AND HAZARD ENVIRONMENT TO LOAD DEFINITION

The threats described above may provide reasonable estimates of the environments that would be considered by facility designers. Understanding tactics and weapon systems capabilities is essential for developing realistic planning and design goals. Furthermore, developing research programs that enable us to reduce the magnitude of a specific threat would be cost effective. It is evident that terrorist attack (i.e., explosive) environments can be unpredictable and may arise from the detonation of multiple explosive devices. Nuclear, chemical, and biological (NCB) environments may be superimposed on conventional explosive effects, and we might have to consider a combined effects environment rather than a single detonation of a particular weapon. Nevertheless, this approach would be supported by separate studies that consider single weapon effects and the possibility of developing theoretical methods for combining such effects to create realistic environments in which specific systems are required to perform and survive. Furthermore, we must address the differences in time scales for the different loading environments (from microseconds to hours or longer), which might enable us to separate the loads and effects accordingly.

Although this chapter does not address all aspects of WMD threats, we must consider other non-conventional (radiological, chemical, and biological) threats when appropriate. Countermeasures and structural designs must be developed for a given type of threat-induced loading environment or a combination of threats. More background on explosions and their effects can be found in other sources (Baker, 1983; Baker et al., 1983; Henrych, 1979; Johansson and Persson, 1970; Persson et al., 1994; Bulson, 1997; Zukas and Walters, 1998), and this area is addressed in more detail in Chapter 2. It should be noted that design manuals and many related computer codes are based on empirical data. Care should be taken when using these tools. Also, the possibility of fire following an explosion must be considered in the design process, although this subject is not addressed herein.

Blast effects, especially from high explosive (HE) devices, are often accompanied by fragments, either from the explosion casement or debris propelled and engulfed by the blast wave. Blast parameters from spherical HE charges in open air are well known and can be found in various handbooks and technical manuals, as noted previously. When an air blast is reflected by a nonresponding structure, a significant pressure enhancement that acts as the load on a structure is achieved. For detonations in complex and/or nonresponding structural geometries, accurate determination of such parameters can be achieved with available computer programs. Currently, there are very few computer programs for the prediction of load parameters for responding structural models and their accuracy may not be always well defined.

Compared to nuclear detonations, typical HE devices must detonate relatively close-in to their targets in order to achieve severe effects; this would depend on the type of target. This proximity requirement is a shortcoming of HE devices

Introduction 17

and usually attackers will try to compensate by using multiple detonations and/or large amounts of HEs to achieve their objective.

Due to the lack of physical data, certain types of explosive loads cannot be determined accurately, and the information available in various design manuals is based on estimates. This is particularly true for cases where explosive charges are placed in contact with or very near a target and for nonstandard explosive devices. Explosive charges of shapes other than spherical or cylindrical will not give pressure distributions with rotational symmetry (especially in close-in conditions), and the information provided in the various design manuals must be used cautiously for such cases. This difficulty is particularly true for close-in explosions, but its importance diminishes significantly for more distant explosions.

In addition, the orientation of the weapon with respect to the target is important, since it would be combined with the geometry of form of the shock front that would interact with the target. Another observation in recent experiments with close-in detonations has been that the effects of shock wave focusing due to reflections from other surfaces could be important in the load definition. Furthermore, the effect of an explosive device's casing on the resulting detonation is clearly not as predicted by data in various manuals. The result is that a much larger amount of energy would be available to produce the blast and propel the fragments, thus causing a more severe loading environment than anticipated. The combination of pressure and fragment impulse as a function of the detonation distance from the target is another important issue that does not have reliable models at this time, and this topic must be studied.

Understanding tactics and weapon systems capabilities is essential for developing realistic goals. Furthermore, developing research programs that provide information that can be employed to reduce the magnitude of a specific threat would prove to be cost effective. It is evident that an anticipated conventional weapons environment in which a facility was supposed to function is rather unpredictable and might lead to the detonation of multiple weapon systems. NCB environments might be superimposed on conventional weapon effects, and one might have to consider a combined-effects environment rather than a single detonation of a particular weapon. Nevertheless, this approach would be supported by separate studies that consider single-weapon effects and the possibility of developing theoretical methods for combining such effects to create realistic environments in which specific systems were required to perform and survive.

In order to derive loading ranges that would be employed for structural analysis, it is necessary to assess the given target and to decide on the types of weapons that would be employed to defeat it. For example, an important target such as a major base or airfield could be subjected to massive air attacks, possibly combined with CB and/or tactical nuclear effects, whereas less important targets probably would be attacked by artillery, air support, and possibly CB weapons. For each of these cases, one could develop an equivalent pressure–time history that would be employed for the analysis. These types of analyses typically are performed by the attacking side in order to derive the firing requirements for damaging a target. One could employ the same approach for defensive purposes. Similarly, one can employ the approach for combining the effects of several weapon systems (artillery, tactical HE rockets, aerial bombs, etc.) or improvised HE devices for deriving the total load–time history. From such assessments, it becomes very clear that the structural analyst should consider the effects of multiple hits in the vicinity of the target rather than the effect of a single detonation. The consideration of a single-load history for structural analysis is an acceptable approach only when a nuclear environment is the threat; this may not be the case for most targets. The only other case where a single detonation could be justified is a demolition charge delivered by terrorists, but if the target is important it is quite probable that more than a single charge would be deployed.

As for structural analysis, it has been shown experimentally and numerically (Krauthammer, 1986; Krauthammer et al., 1986 and 1994) that for close-in HE detonations and for certain nuclear loads, the mode of structural failure is controlled by material failure or by direct shear. At present there is some understanding of these phenomena, but they are clearly not well understood. Nor is it understood whether and how the direct shear can be coupled with the flexural modes of response, as was achieved for the coupling between bending shear, flexure, and thrust. It is strongly recommended that such studies be performed, with the ultimate objective of a better understanding of structural behavior and, thus, improved pressure–impulse (PI) diagrams and design methods, as discussed in Chapter 8.

The area of equipment response under the effects of multiple detonations is very important. Usually tests are performed using a single explosion, wherein the time histories do not include the effects of closely timed explosions that could introduce additional frequencies into the loading environments. In addition, the increase in the deployment of "smart weapons" (including those placed by humans at a desired location) raises the issue of having to protect against very close-in or internal detonations that may not be randomly distributed over the general target area. Under these conditions, it would be possible to have several such weapons detonate in a well coordinated manner, thus subjecting the target to a most severe load–time history at critical locations. Current analysis and design methods do exclude such considerations, thereby introducing serious risks into the field.

Available information (ASCE, 1999; DOD, June 2002 and July 2002; Department of the Army, May 1994; FEMA, 2003; AIA, 2004) can be used for developing reasonable assessments of security needs and for defining potential loads. Such guidelines assist in determining design criteria for the physical protection of assets within either a new or a renovated facility (they can also be used for exposed assets). Detailed procedures for load determination are delineated for each type of attack. These sources include terminology definitions, as well as comprehensive discussions of aggressor threats and tactics, assets definition and description, threat determination, levels of protection, and design constraints. The loads include ballistic attack, explosive attack, and forced-entry attack. Detailed procedures for load determination are delineated for each type of attack. There is a basic difference between many of the tests conducted by DOD organizations on typical weapon systems and the tests on terrorist threats, as briefly discussed above. Data from typical weapons systems or very large ANFO devices may not provide much useful information for protective architecture considerations. On the other hand, the tests on "terrorist type" explosive devices can provide very useful information. Such tests have become more relevant and frequent in recent years and have been conducted in the U.S., Ukraine, Israel, and other countries.

Concerning airblast, it was concluded that environments associated with spherical charges within simple, no-responding geometries can be handled effectively. However, development, improvements, and validation of computer codes for nonspherical charges and complex and/or responding geometries are needed. Questions remain in the area of close-in environments from nonspherical charges. The combined effect of fragments and blasts from cased charges also requires further research. Although buried structures may not be of much interest for civilian protective architecture projects, it is unclear whether the issue of medium-structure interaction with back packing can be handled correctly. Furthermore, the effect of rock fill and medium–structure interaction due to an explosion in such a medium cannot be addressed with current computer codes, and this issue should be investigated further.

1.4.1 MILITARY THREAT ASSESSMENT

Performing an assessment of a military threat requires access to many classified resources, and using such data would prevent the information from being published in unrestricted forms. Nevertheless, one can illustrate the approach by employing information available to the general public, as shown next for data available on the former Soviet military.

1.4.1.1 Tactics and Strategy

The Soviet doctrine for attack of a defending enemy was based on several important strategic concepts discussed at length in the literature (Grechko, 1975; U.S. Army, 1977 and 1978; Haselkorn, 1978; Critchely, 1978; Defense Intelligence Agency, 1980; Ulsamer, 1982; Hines and Peterson, 1983; Mossberg, 1982; DOD, 1985). Based on these sources, it is understood that the following operations requirements would adequately define the military principles behind such a threat:

1. Achieve a desired force ratio by carefully choosing the attack zones (these ratios were estimated to be 3–5 to 1 for armor, 6–8 to 1 for artillery, and 4–5 to 1 for infantry).
2. Heavy artillery preparatory fires ("softening") for at least 30 min before attack and continued support as required during the attack (this phase may be combined with nuclear fires or CB weapons). This paramount concept was reflected by the "standard" distribution of artillery at the

front (about 100 launching tubes per 1 km for breakthrough of well prepared defenses). Such concentrations were typical of military operations in the Middle East during the 1973 Yom Kippur War, where more than 2,000 guns were employed to soften the Israeli fortification system. Artillery support would provide a pattern of fire about 200 to 400 m ahead of advancing troops and 100 m ahead of advancing tanks.

3. Tactical air power was to be fully integrated into the task of supporting the overall mission: to provide air cover for ground forces and achieve air superiority. Air strikes were considered an extension of artillery fires and would concentrate on preplanned targets including C^3I facilities, tactical nuclear delivery systems, and the neutralization of artillery (including tactical air) support and reserve elements within the operational zones. Air cover effort would be concentrated for maximum effect at a given time, and continuity of operations would be maintained, including repeat attacks to prevent repair of damaged targets. It is well known that Soviet tactical aircraft were generally less sophisticated and therefore simpler to maintain than Western aircraft. As a result, the Soviets would sustain a theoretical sortie rate of four or five missions per day under ideal conditions. One would expect two or three sorties per day for the initial three days of war, and a sustained rate of one or two sorties per day. In the event that air superiority was achieved by the Soviets, these rates might have increased because of an increased deployment of aircraft from forward (close-in) airfields. Long-range aviation (LRA) reinforcement by medium bombers would strike targets of strategic importance that were beyond the range of tactical aircraft. Those bombers would be the primary force available for attacks on various targets throughout the theater of operations. During the first days of combat, they would strike airfields, nuclear weapon storage sites, and C^3I facilities. In later stages of the war, the bombers would be employed for other targets as well and would also support frontal aviation. The capabilities of the Soviet air force were described in several publications cited earlier; see, for example, Defense Intelligence Agency (1980), from which it was possible to compute an anticipated range of weapon delivery to a given target.

4. The Soviets considered the use of toxic agents to be legitimate and effective for destroying the enemy under modern combat conditions. Chemical munitions were available to field units; the first-use decision is political while follow-up use could be made at the division level. This doctrine was supported by the fact that Soviet military forces were the best equipped in the world for the use of such weapons and it seemed safe to assume that such weapons could be delivered by artillery, rockets, and aerial bombs (discussed later). This conclusion was supported by findings in the Middle East, where those types of equipment were captured. Tactical nuclear strikes were categorized as weapons of mass destruction, together with chemical and bacteriological

armaments (defined herein as "special weapons"), and Soviet forces had known capabilities to employ such weapons. The Soviet literature was quite misleading discussing the use of nuclear weapons while stating that the use of CB weapons was forbidden by international law. Nevertheless, incidents in Vietnam, Cambodia (Kampuchea), Laos, Afghanistan, and in the war between Iran and Iraq suggest that chemical weapons were used; this supports previous statements on the matter. According to Soviet tactical doctrine, the use of special weapons did not preclude the simultaneous use of conventional weapons; hence the term "complementary warfare." Under such conditions, one would anticipate different requirements for forces, higher rates of advance, and enhanced depth of objectives. One also might have to reconsider tactical principles based on conventional weapons alone and that expected an increased emphasis on the use of Special Forces.

5. There is sufficient evidence that airborne assault was an integral part of the Soviet combined arms doctrine. Such operations were to be deployed for deep penetration and rapid exploitation and to secure objectives forward of advancing ground forces. Tactical airborne landing would be employed to capture and destroy enemy means of nuclear attack, airfields, depots, and other objectives. The Soviets had eight airborne divisions and additional air assault bridges for special operations behind enemy lines. In addition, it was estimated that at least one third of the Soviet combat forces were trained for commando operations. An important component of this category was the Special Purpose Forces (SPETSNAZ) trained to perform sensitive missions abroad under the former GRU and/or KGB direction. In wartime, those forces would operate deep behind enemy lines and attack various military, economic, and political targets, similar to special operation forces used by other countries.

After these operational concepts and capabilities were introduced, it became necessary to perform a similar analysis for obtaining information on the weapon systems that would be incorporated into the anticipated military actions. Such an analysis would lead to estimates of environments in which the structural systems have to function.

1.4.1.2 Weapons Systems

The Soviet arsenal included a variety of weapon systems that had to be considered for the definition of hostile environments in which a given facility would be required to function. Here, the weapon systems are classified and discussed according to the methods by which they are used and controlled, as presented in the literature (U.S. Army, 1977; Jane's, 1982–1983; DOD, 1985). Additional information on the subject is presented in Section 1.3.

Artillery — The integration of artillery support with missile systems and strikes of tactical aviation was the keystone of Soviet operational planning. Offensive operations would follow an effective suppression of enemy defenses and Soviet artillery would be employed under centralized control and in large numbers. Firepower capabilities were increased by providing large numbers of high-quality weapons systems; nevertheless, the 122 mm Howitzer was the basic cannon. However, self-propelled (SP) and improved versions of the 122 mm and the 152 mm weapons were introduced in substantial numbers, and it was expected that they would replace a large number of the towed versions. Under the general category of field artillery, one would consider the following weapon systems: Howitzers and other types of cannons, mortars (120 mm and above) multiple-rocket launchers (MRLs), tactical rockets, and operational missiles. This section addresses cannons and mortars; rocket and missile systems are discussed later.

Other artillery weapons that would be considered when Warsaw Pact forces were analyzed were the 203 mm Howitzer, 240 mm mortar, 180 mm gun, 100 mm field gun, 130 mm field gun, and a variety of anti-tank and anti-craft guns that could be employed against ground targets. Specifications for these weapons were mostly unclassified and could be obtained from publications available to the general public, as mentioned earlier. From that information, it was concluded that the HE ammunition weights varied from 88 kg for the 180 mm gun to 12.2 kg for the 100 mm field gun. Firing rates also varied from 1 round per min for the 180 mm gun to as fast as 10 rounds per min for the 100 mm field gun. Here, it is reasonable to concentrate on the specifications of what would have been the main weapon system used, the 122 mm Howitzer: ammunition weight, 21.8 to 27.3 kg; firing rate 6 to 8 rounds per minute. As far as the number of artillery, on the basis of available references it was noticed that the Soviet forces increased the number of SP field artillery faster than it decreased the number of towed systems.

In order to illustrate the environment that would have been typical of a Soviet attack, one should understand the "firing norms" that correspond to Soviet tactics and the target damage criteria:

1. **Harassment:** Attempt to achieve a 10% damage rate in order to reduce enemy fire effectiveness.
2. **Neutralization:** Inflict 20 to 30% damage for a temporary reduction of enemy capabilities.
3. **Annihilation:** Try to inflict 50 to 60% damage in order to destroy fighting capabilities completely, until it is reconstituted.

It should be noted that these criteria were more severe than the corresponding Western definitions, further illustrating the Soviet concept of artillery support: massive in scope and aimed at inflicting terminal damage on the defending side. Fire coverage is computed as rounds per hectare per minute (1 hectare = 10,000 m^2 = 2.47 acres). A typical fire plan for a 122 mm Howitzer battalion supporting the attack of a motorized rifle battalion included preparatory fires of about 1,600

Introduction

rounds over a period of 40 min and 1,200 rounds as supporting fires to follow "on call" during a shorter period. Therefore, one would consider the possible effect of about 2,800 detonations with a projectile weight of 21.8 to 27.3 kg, in combination with tactical air support and/or missile fires; these are discussed later.

Missiles and rockets — The Soviets were using MRLs before WWII and continued to develop such systems for various applications. All Warsaw Pact members deployed such systems, which were either supplied by the Soviet Union or developed and modified domestically. A brief summary of MRL systems was provided in Jane's (1982–1983), and from that information it is clear that there were ten such systems having 20 (BMD-20) to 40 (BM-21, M-1972) barrels. The rocket calibers were from 122 to 250 mm and rocket weights were 39.6 to 455 kg. Between 1980 and 1984, an estimated 4,400 additional MRLs were produced, an indication of the number of such systems that would be present during hostilities.

In addition, one would consider battlefield support rockets that could deliver relatively large warheads (conventional, CB, or nuclear). The Soviets deployed the FROG series (FROG 1 through FROG 7) that could deliver 450 kg warheads (or even heavier versions) over ranges of 40 to 70 km, and the SCUD A, B, or C that could deliver large warheads up to 450 km. (Such systems have been further improved by other countries and are currently supplied to various customers.) The accuracy of such systems was improved gradually over time. One would also consider the SS-21 and SS-23, for which information can be found in various publications.

Aerial bombs — A typical sample of bombs (free fall) in the Soviet arsenal was discussed in the above literature. The number of weapons to be dropped on a target depended on operational requirements, but on the basis of flying constraints it was safe to assume that a typical fighter-bomber would deliver two weapons per target in a single pass and that repeated attacks would be expected to ensure target annihilation. Regarding the type of weapon systems that had to be considered, a concentration on the FAB designation series (100, 250, 500, and 1,000, corresponding to the weight of the bomb in kg) and on the boosted concrete demolition bombs under the designation BETAB (250 and 500 kg) was recommended. Another bomb was the 500 kg M62 designed for target penetration and demolition bombs in the 100 to 1,000 kg range. Again, neither the CB weapons nor the tactical nuclear systems were listed, and they would be considered for specific conditions.

Demolition charges and special weapons — Under this general title one would include special weapon systems designed to accomplish specific tasks, systems that are more sophisticated than the weapons previously described. Many such systems were operated by Special Forces (airborne and/or commando units) to achieve a desired outcome. Typically, one would expect the deployment of such weapons against primary targets, as discussed earlier. In order to protect against these weapons, one must consider the deployment of active defense together with conventional hardening techniques, and the designer should be aware of specific requirements based on sound operational considerations.

1.4.1.3 Summary

Accordingly, considering an artillery softening, 1,600 rounds of 122 mm shells would be fired over a period of about 40 min. From this it is clear that about 40 rounds per minute would hit the target, and since the target area is known it would be possible to employ a statistical approach to compute the probability for a hit in a given area. One could then compute the number of such hits per unit time per unit area, and from the charge weights the associated loading functions. Similarly, one can estimate the loading environment from a typical aerial attack, based on the number of planes, sorties, and bombs used for such missions.

1.4.2 Terrorism and Insurgency Threat Assessment

This activity is much more complicated because applicable operational manuals are not available. Both terrorists and insurgency groups will usually improvise their operations and tools. These are limited by imagination, boldness and motivation, resources, and constraints imposed by the society against which such operations are launched. Such groups can change operational parameters more quickly than expected of a typical military adversary. They will operate with total disregard of accepted norms of engagement (e.g., the Geneva Convention), and will attack any target that can serve their purpose. Therefore, one must perform continuous threat assessments, and work closely with all other organizations that could have similar interests (e.g., law enforcement, security, intelligence, etc.). The resulting comprehensive security plan must be reviewed after each threat assessment, and appropriate measures must be implemented to remedy any possible uncovered deficiency.

1.5 TECHNICAL RESOURCES AND BLAST MITIGATION CAPABILITIES

The information presented next was combined with details from briefings organized by the sponsors of the present study to derive an assessment of current blast mitigation technologies and their applicability to the protection of civilian facilities. Additional information from the following two activities assisted in deriving a better understanding of the state of knowledge in blast mitigation: a three-day workshop sponsored by the Norwegian Defence Construction Service (Krauthammer, 1993) and an ASCE-sponsored study to collect information and provide guidelines on how to design and/or upgrade civilian facilities to resist terrorist activities (ASCE, 1999); Smith (2003) presented guidelines on how to respond to bomb threats.

The findings of these activities are classified according to several technical areas: threat and load definition, computational analysis and assessment, and experimental analysis and assessment. Since the indications are that explosive devices will continue to be primary hazards, the emphasis in this section is on blast, shock, and impact. Structural design for safety and physical security

Introduction

requires a sound background in fortification science and technology. Loading environments associated with many relevant threats (impact, explosion, penetration, etc.) are extremely energetic, and their duration is measured in milliseconds (about one thousand times shorter than typical earthquakes).

Structural response under short-duration dynamic effects could be significantly different from the much slower loading cases, requiring a designer to provide suitable structural details. Therefore, one must explicitly address the effects related to such severe loading environments in addition to considering the general principles used for structural design to resist conventional loads. One must be familiar with the background material on structural consideration and design and the experience gained from recent terrorist bombing incidents. A brief summary of a few frequently used references is provided next.

As noted above, R&D activities in the fortification area followed the perceived threats that were directly related to international geopolitical activities. Soon after WWII and until the early 1980s, most of these R&D efforts concentrated on nuclear threats. Gradually, since the early 1970s, more attention has been given to threats from conventional weapons. Since the late 1980s, most of the fortification activities seem to have shifted to the conventional weapons area, and since the early 1990s it seems to be the only topic of concern. Accordingly, the attention shifted from the global behavior of structures (either buried or above ground) to the local response. Most design manuals, however, still emphasize responses to uniformly distributed loads that use classical resistance models for the structural elements (elastic–plastic material models, yield line strength theories, etc.).

Most of the experimental work in this area has focused on the behavior of structures for clearly military objectives, although some civilian structures had been tested under nuclear effects (either actual or simulated). In addition, because of budget and scheduling considerations, many experimental efforts in this general area did not look at several scales of structural modeling. With respect to fortification technology that could be useful to the civilian protective architecture community, one would need to extract data on the behavior of above-ground structures subjected to HE detonations. Unfortunately, most of the studies sponsored by DOD were directed at the behavior of underground structures; many of these studies were sponsored by Defense Nuclear Agency (DNA) (renamed Defense Special Weapons Agency [DSWA] and later merged into Defense Threat Reduction Agency [DTRA]), and were performed by Waterways Experiment Station (WES) (later merged into ERDC) and Airforce Weapons Laboratory (AFWL) (AFWL is no longer in existence). Nevertheless, useful information has been generated from a limited number of tests on above-ground structures, and structural components (Coltharp, 1985, among others). Such studies addressed the behavior and design of openings, doors, blast valves, glazing, protective perimeter walls, features to defeat Rocket Propelled Grenades (RPGs), etc. Studies for AFWL on the design of specific structural detailing for hardened facilities provided insights into the localized nature of structural response (Krauthammer and DeSutter, October 1989). Although these studies were discontinued when that organization disbanded, some of the work addressing concerns in explosive safety was carried on later by the U.S. Navy (Naval Facilities Command [NAV-FAC]) and Department of Defense Explosive Safety Board (DDESB) (Otani and Krauthammer, 1997; Ku and Krauthammer, 1999; Krauthammer, 1999). Given the number of presentations in recent conferences, it is evident that there is great interest in the effects of penetrating weapons on buried facilities. Although some useful information could be extracted from these studies, the bulk of the data may not be very useful to the civilian protective architecture community. As a result, various other studies were initiated during the last few years addressing the behavior of civilian buildings under simulated terrorist attack.

The predominant method of testing has been by loading structures (in many cases only one size per study) with either weapons or HE devices. It should be noted, however, that even identical HE devices may produce different loading environments, and such experiments are difficult, risky, and very expensive. These experimental approaches could be justified for the assessment of structures subjected to the specific threat, but under certain conditions they may not be very useful for obtaining precise information on loads and structural response mechanisms. Because of the high uncertainty in the load data, measurements from such tests may not be very helpful for the verification of computational tools. Therefore, design manuals have to employ simple yet safe assumptions with respect to loads and structural behavior.

1.5.1 DESIGN MANUALS

Of all the manuals previously cited, TM 5-1300 (Department of the Army, November 1990; this is a tri-service manual with corresponding Air Force and Navy designations) is the most widely used by both military and civilian organizations. It has been used for the design and assessment of various facilities both in the U.S. and abroad and includes step-by-step analysis and design procedures. Various users have adhered to its recommendations as closely as possible. The *Security Engineering Manual* (U.S. Army, TM 5-853, December 1988 and May 1994) is more specific for physical security measures; it also contains information from TM 5-1300, but its use is limited to authorized personnel.

Tri-Service Manual TM 5-1300 (Department of the Army, 1990) — This manual is intended primarily for explosives safety applications. It is the most widely used manual for structural design to resist blast effects. One reason for its widespread use by industry is that it is approved for public release with unlimited distribution. For predicting the mode of structural response, the manual differentiates between a "close-in" design range and a "far" design range. On the basis of the purpose of the structure and the design range, the allowable design response limits for the structural elements (primarily roof and wall slabs) are given as support rotations. This publication is most widely used by both military and civilian organizations. Several concerns have been raised about some issues addressed in this manual, as will be discussed later, and the manual is currently being revised.

Introduction

Army Technical Manual 5-855-1 (Department of the Army, 1986) — This manual is intended for use by engineers involved in designing hardened facilities to resist the effects of conventional weapons. It includes design criteria for protection against the effects of penetrating weapons, contact detonations, or the blast and fragmentation from standoff detonations. The recommended response limits are given only as ductility ratios, not support rotations. Nonetheless, a more recent supplement to TM 5-855-1 (Department of the Army, ETL 110-9-7, 1990) provides response limit criteria based on support rotations. A more updated manual should be used whenever possible (DOD, June 2002).

ASCE Manual 42 (1985) — The manual was prepared to provide guidance in the design of facilities intended to resist nuclear weapon effects. It presents conservative design ductility ratios for flexural response. Although an excellent source for general blast-resistant design concepts, it lacks specific guidelines on various issues (e.g., structural details).

ASCE Guidelines for Blast-Resistant Buildings in Petrochemical Facilities (1997) — This reference contains detailed information and guidelines on the design of industrial blast-resistant facilities, with emphasis on petrochemical facilities. It includes considerations of safety, siting, types of construction, material properties, analysis and design issues, and several detailed examples. The information is not limited to blast-resistant industrial facilities; it is very useful for all aspects of blast-resistant design, including physical security.

ASCE Structural Design for Physical Security (1999) — This is a state-of-the-practice report addressing a broad range of topics in this field. It starts with a detailed procedure for threat assessment, followed by an overview of how to define loads associated with typical attacks. The behavior of structural systems under the anticipated loads, and the corresponding design of structural elements are treated next. Separate sections address the behavior and design of windows and doors for the same loading environments. Also, a dedicated section addresses the retrofit of existing structures. This report is about to be revised and published as a guideline in coordination with the revision of TM 5-1300.

DoD and GSA Criteria — United Facilities Criteria (4-010-01 and 4-023-03, Department of Defense, October 2003 and January 2005, respectively) and GSA criteria (June 2003) are aimed at meeting minimum antiterrorism standards for buildings and the prevention of progressive collapse. Both DoD and GSA publications recognize that progressive collapse could be the primary cause for casualties in facilities attacked with explosive devices, and they include guidelines to mitigate them.

DOE Manual 33 — This is similar to TM 5-1300 described above, but it contains updated material, based on more recent data.

FEMA Guidelines (December 2003) — This publication is not a technical or design manual, but it contains clear and comprehensive guidelines on issues that need to be addressed and simple explanations of such steps. This may be the most effective starting point for people who wish to learn how to handle protective construction projects.

1.5.2 Design and Construction Considerations

Existing publications (U.S. Army, TM 5-853, December 1998 and May 1994; ASCE, June 1999; FEMA, 2003) provide guidelines on the structural engineering topics for physical security design. Discussions of structural systems behavior under the effects of common threats are used as bases for comparisons of different types of loading effects on structures; general comparisons of different types of structures and their expected response characteristics; and guidelines on effectively selecting the structural type, materials, and structural components for enhancing safety and physical security.

Specific information is also given on the behavior, design, and analysis of components typically used as part of a facility's structural system. The discussions include both global and localized responses. The principal objectives of protective design are to protect personnel and assets and to minimize the operational disruption of a facility. The approaches include tables and charts from the various design manuals described above.

The combined effects of material properties, loads, support conditions, and structural detailing are understood, at least empirically, and this state of knowledge is reflected in the current design codes. The current quasi-static design approaches are reasonable for implementation. However, the application of pressure–impulse diagrams should be re-evaluated and the transition between different behavioral modes should be better defined.

User-friendly and physics-based, single-degree-of-freedom codes that include various structural response capabilities should be developed and incorporated into the design process. Design activities should be supported by review of existing data, analysis, and testing, and design methods should be re-evaluated to include more precise criteria. Unlike many current procedures, all designs should be based on acceptable design criteria that include construction ability, performance, maintenance, and repair requirements for the facility under consideration. Guidelines on construction aspects and cost control should be provided.

Robustness and response levels should be related to a facility's contents and its mission requirements (for civilian facilities, the mission requirements parameter would address instead considerations of safety). It is also desirable to introduce cost/benefit criteria for various design options. Designers should be guided with respect to design tradeoffs, but the design process should be well defined.

Although current design procedures give guidelines on how to enhance the breaching resistance, it could be impractical to protect against breaching and direct shear effects by conventional means. Alternative reinforcement details should be permitted for cases in which lacing would be required. The use of various materials and combinations of materials (e.g., high-strength concrete, possibly in combination with conventional and fiber reinforcement and damage absorption devices) should be studied, and future design guidelines should address such options. Guidelines and recommendations should be provided on how to evaluate future capacity of previously loaded structures.

Introduction

Design procedures for special details are also included in existing publications (U.S. Army, TM 5-853, December 1998 and May 1994; U.S. Army, TM 5-1300, November 1990; ASCE, 1997 and 1999; FEMA, 2003), for example, the selection and design of security windows, doors, utility openings, and other components that must resist blast and forced-entry effects. General design approaches for hardening typical commercial construction to resist physical security threats are also included. Threats considered relate specifically to external or internal explosions. Concepts for changing the essential quantities for dynamic resistance include mass and strength increase, support conditions modification, span decrease, replacement of inadequate components, and loaded area reduction. Additionally, retrofit effects on forced-entry resistance are discussed. The discussions include analysis techniques for predicting retrofit requirements, retrofit materials and techniques, forced-entry resistance retrofit, and cost.

It was noted that TM 5-1300 (U.S. Army, November 1990) is the publication most widely used by both military and civilian organizations. The security engineering manual (U.S. Army, TM 5-853, December 1988 and May 1994) is more specific for physical security measures, but its use is limited to authorized personnel. Despite the wide acceptance of these publications, recent studies (Krauthammer et al., 1994; Woodson, 1994) show that current design procedures in TM 5-1300 may not be adequate for connections or plastic hinge regions, and raise questions about recommendations for both flexural and shear resistance models in slabs.

Considerable attention has been given to the behavior of subsystems typically found in hardened facilities (generators; air, water, and fuel supply equipment; communication and computer equipment; etc.). Here too, the emphasis shifted from nuclear threats (57 reports from the SAFEGUARD series from 1972 to April 1977) to conventional threats, as mentioned above. Crawford et al. (August 1973) provided guidelines on the installation of such equipment based on the information obtained from studies related to the nuclear threat. There is no comparable source of information related to HE effects, but one may use data from individual studies for such purpose.

Marquis and Berglund (July 1990) and Marquis et al. (August 1991) explored the issues of predicting equipment response and providing fragility spectra for typical equipment in military facilities based on data from both seismically induced responses of equipment in nuclear power plants and shock responses of ship-mounted equipment. Dove (March 1992), comparing different in-structure shock prediction methods, showed that many exhibited serious limitations and that the ISSV3 code could be used to obtain good predictions. That information, combined with findings by Harris and Tucker (August 1986), Ball et al. (March 1991 and April 1991), and Mett (April 1991), could be more relevant to the protection of civilian facilities against terrorist threats than was the information obtained from earlier (nuclear weapon effects) studies.

The important findings indicate that most mechanical or electromechanical types of equipment are sufficiently rugged to survive anticipated in-structure shock environments. Problems were encountered primarily with faulty wire installations or inadequate attachment procedures for the structure. Although shock isolation is quite feasible, it was noticed that certain shock isolation devices may not provide the expected protection. These issues should be addressed by the facility designers.

1.6 ANALYSIS REQUIREMENTS AND CAPABILITIES

1.6.1 COMPUTATIONAL ANALYSIS

A detailed discussion of computational techniques in this field is presented later in this chapter; the following brief discussion is provided only as a general summary. Closed-form solutions are limited to simple geometries, simple loading and support conditions, and linear materials. When approximate, simplified methods are used, one must assume a response mode and the corresponding parameters. It is recommended to use such methods together with data from computer codes based on current design manuals. Current medium-structure interaction models are too simplistic and do not include nonlinear effects.

To accommodate a practical range of numerical capabilities, simple, intermediate, and advanced computer codes are needed. Advanced numerical methods require significant resources, and they should be used in the final stages of detailed structural analyses for obtaining design guidelines and/or in the detailed evaluation of the anticipated structural response. However, such advanced codes must be validated against precision test data to ensure their reliability (Krauthammer et al., May 1996 and May 1999). These publications note that the best results are obtained when a structure is analyzed gradually, employing a range of numerical approaches from simple to advanced. Structural response calculations may not be valid for scaled ranges less than 1 (ft/lb$^{1/3}$) and structural breaching calculations should be performed separately from structural response analyses.

1.6.2 EXPERIMENTAL ANALYSIS

There are basic differences among many of the tests conducted by defense organizations with typical weapon systems, and tests with terrorist-type explosive systems that are usually IEDs. Data from typical weapons systems or very large ammonium nitrate fuel oil (ANFO) charges may not provide much useful information for protective architecture considerations. On the other hand, tests with terrorist-type explosive devices can provide very useful information.

Such tests have become both more relevant and frequent in recent years, and have been conducted in several countries. Furthermore, it is anticipated that numerical simulations could be used more frequently instead of some experiments. However, data from precision tests are needed for the calibration, validation, and verification of the various computer codes. The combination of experiments with continuum mechanics theories to clarify behavior, damage, and transitions between response modes are also anticipated. Moreover, there is a need to obtain constitutive relations for various materials up to very high pressure levels, and a need to define and better explain strain rate effects. So far, there is confidence in scaled tests of structural concrete systems, as long as real construction materials can be used; however, it is not clear whether smaller scaled tests on typical construction materials can provide useful information. Furthermore, it is very difficult to develop small scale steel members that enable one to preserve the geometric characteristics of commercial steel shapes.

When scaled tests are to be performed, more than one scale should be used in order to verify proper behavior and account for size effects. Furthermore, there are serious questions about using scaling laws to study breaching and other severe structural responses, where the behavior might be more related to material properties than to structural behavior. Also, recent studies showed that size effects are coupled with loading rate effects to significantly influence material behavior (Krauthammer et al., October 2003; Elfahal et al., 2005; Elfahal and Krauthammer, March–April 2005), and these findings need to be incorporated into advanced computational tools and design recommendations.

1.7 PROTECTIVE TECHNOLOGY: CURRENT STATE AND FUTURE NEEDS

The current fortification and physical security technologies may be evaluated in light of the information presented above. First, it should be noted that most of the current design procedures are highly empirical. This is not intended as a negative comment. The problem of structural behavior under severely short-duration dynamic loads is a difficult one for the following reasons: it cannot be treated with closed-form solution procedures; experimental approaches are complicated and not as precise as one would wish; and numerical solutions, although increasingly more widely used, still lack reliable load simulation and well defined material properties, and require large computational resources.

Nevertheless, fortification engineers must provide solutions to such complicated, and in many cases not well defined, problems. Thus, many analysis and design recommendations are simplified, helping to produce usually safe structures whose safety margins are not well defined. This could be acceptable for military facilities, where risks and certain casualty rates are acceptable. However, the requirements for military facilities may be significantly different from those for civilian structures and design recommendations must reflect such differences. Therefore, the direct adoption of military design recommendations for civilian applications should be carefully examined. Basically, current protective design procedures are assumed to be "reasonably safe" for military applications. However, their safety margins are not well defined, and various issues still require attention, as discussed below.

There are questions concerning load definition for various events (here one must also consider the difficulties in medium-structures interaction, combined blast and fragment effects, and the uncertainties associated with such events). Design calculations should be supported by advanced simplified computer codes, and the traditional concept of pressure–impulse diagrams should be re-evaluated. Better material models should be made available for advanced computations (which should be performed at final design stages), and structural response calculations should be performed separately from breaching evaluations.

Precision testing should be promoted and scaled tests should be conducted within the ranges that allow the use of real construction materials. In design, the general concept of safety factors should be re-evaluated to ensure structural robustness and compatibility with different structural resistance mechanisms. Design guidelines should enable the use of material combinations, advanced reinforcements, efficient reinforcement details, considerations of life cycle costs, and should provide the capability to evaluate alternative designs. Guidelines and recommendations should be delineated on how to evaluate future capacity of previously loaded structures.

Recent events and developments highlight three other areas in which there is a serious lack of knowledge about and/or need for protection technology. First is the issue of secondary threats to a building and its contents. It is typical to have fires, smoke, progressive collapse, etc., following an explosive attack. Although some work has been done in this area, the related information is not readily available in design manuals (especially those mentioned above). Second, there are no clear recommendations on how to perform post-attack recovery operations in a safe manner. This ties directly into the issue of post-attack damage assessment and the secondary threats, but it goes much further. During recovery operations, it is vital to enter a damaged facility quickly and rescue individuals and/or assets trapped inside. Such activities may require the rapid, safe removal of debris and temporary shoring up of specific elements.

Civil defense personnel must have decision support tools and guidelines on how to perform such work without causing further collapse. This is a complicated problem, especially during a period of crisis and confusion, and it is much more difficult than the traditional bomb damage assessment (BDA) activities performed by military personnel. It also should be noted that, based on recent experiences, military BDA is a critical area that requires significant improvements (Krauthammer, January 1999; Miller-Hooks and Krauthammer, August 2002 and May 2003). As noted previously, one recommendation from the NDCS workshop (Krauthammer, 1993) was to develop guidelines on damage assessment in design manuals. It seems reasonable that such development could be extended to include "rescue-related design guidelines" as an integral part of future design manuals. The third area in which there is a serious lack of knowledge about or need for protection technology is arson. Arson (including very hot fires) has emerged as an easy weapon, and this method of attack is gaining increasing acceptance among antisocial groups (both criminal and subversive elements). Because this threat could become a very serious problem in the future, new fire protection technologies for integration into various civilian facilities are needed.

The evolution in fortification technology, as briefly described here, reflects the gradual change in design capabilities as additional information becomes available. Now that the Cold War is over (researchers in this area do not have to

Introduction 33

rush from one crisis to another and the existing fortification infrastructure may need a new area of activity) it is time to solidify the state of knowledge in this field. Careful re-evaluation of data from previous research will help significantly to achieve this goal. Additional studies should be conducted in areas for which such data are either insufficient of unavailable. Finally, the generated knowledge base should be incorporated into a spectrum of computational tools in support of all design, construction, operation, and recovery activities.

Current activities aimed at the development of an updated design manual could be extremely useful if "civilian" versions of such a manual would be made available for non-DOD design activities. Such design manuals should include the features that are most desirable to the non-DOD design community, as briefly described above. Furthermore, current developments in virtual reality (VR)-based, or simulation-based (SB) design technologies dictate that such a manual should be developed in such a way as to enable its incorporation into VR-based or SB design tools. Linking the manual with advanced computational artificial intelligence and modern visualization capabilities will ensure the ultimate utilization of military fortification technology for civilian applications. Such a capability must be combined with educational and training activities to guarantee effective integration of protective architecture technologies into the civilian marketplace.

The discussion presented in the previous sections raised several issues regarding current methods for hardened systems analysis and design. Clearly, significant knowledge in this area has been gained since WWII, and the history of the last 50 years teaches that the threat from WMDs should be taken very seriously. The re-emergence of nuclear/radiological, chemical, and biological threats must not be overlooked. Obviously, additional research in many areas is badly needed, and such needs must be adjusted to reflect changes in threats, weapon systems, strategic and tactical concepts, and the capabilities of experimental and analytical tools.

Unfortunately, the technological changes in the development of conventional and unconventional weapons are much more rapid than the progress in the development of adequate protective technologies. Furthermore, society has been typically reacting to such changes after specific incidents, rather than developing effective measures in preparation for possible attacks. It would be wise to address such issues urgently and continuously, and to become much more proactive in the development of protective technology that can help avert very serious consequences.

1.7.1 Relationships to R&D

Most studies on the responses of protective structures to detonation effects are still in the domain of the single detonation event. Unfortunately, because many of the problems related to this kind of event are not yet fully understood, it is logical to address them in a disconnected manner. Furthermore, there is a clear separation between nuclear effects and conventional effects research — a separation that may no longer be justified in light of the growing evidence that the two environments may be combined. A combined approach requires the study of the responses of both damaged and undamaged systems. Several studies have been conducted, but a serious effort must be made to plan and conduct more.

One of the important issues to be faced by facility designers is load definition. It has been shown before [ASCE, 1985; Department of the Army, 1986 and 1990 (rev. 1998)] that a rational estimate of an equivalent pressure–time history can be determined by combining the effects of blast, fragments, etc. Nevertheless, significant questions remain regarding the resulting loads on the structures. The empirical methods used for the derivation of the present estimates and the purely analytical methods used in the literature (Lakhov and Polyakova, 1967; Rakhmatulin and Demy'anov, 1966; Henrych, 1979) do not consistently provide accurate estimates. One of the primary reasons for this inconsistency is the lack of reliable methods for performing accurate computations of detonation–soil interaction and/or soil–structure interactions under highly nonlinear conditions wherein the stresses vary with time and location. The development of reliable analytical/numerical tools for such computations and their validation against experimental data is significant.

Engineers and researchers who have worked on the development of protective structures against nuclear detonations tend to forget that typical conventional weapons must detonate relatively close-in to their targets in order to achieve the planned effects. This proximity requirement is also the shortcoming of these weapons, and usually the attacking force will try to compensate by using a large number of detonations and hoping that a few will achieve their objectives.

The detonation itself is not a simple spherical or cylindrical charge. It is important to be able to define load as a function of shape. Furthermore, the effect of the case on the resulting detonation is clearly not as was predicted and reported by Crawford et al. (1971), who recommended that the weight of the explosive be reduced by about 50% to reflect the energy loss due to fragmentation of the case. That recommendation was based on the assumption that the case would be uniformly disintegrated into very small particles. However, those who are familiar with the actual behavior know that it is possible to have much larger and fewer fragments than anticipated. The result is that a much larger amount of energy would be available to produce the blast and propel the fragments and thus cause a more severe environment than anticipated. In addition, the orientation of the weapon with respect to the target is important, since it would be combined with the geometry to form the shock front that would interact with the target.

Another observation in recent experiments with close-in detonations is that the effects of shock-wave focusing due to pressure rays between fragments and reflections from the ground surface could be important in the load definition. All of these issues need to be addressed in order to provide rational models that would be employed for structural analysis and design. The combination of pressure and fragment impulse as a function of the detonation distance from the target is another important issue that does not have reliable models at this time, and this topic must be studied.

Introduction

As for structural analysis, several areas of research are essential. It was shown experimentally and numerically (Krauthammer, 1986) that for close-in HE detonations and certain nuclear loads, the mode of structural failure is controlled by material failure or by direct shear. At present we have some understanding of these phenomena, but they are clearly not well understood. We also do not understand whether and how direct shear can be coupled with the flexural modes of response, as was achieved for the coupling of bending shear, flexure, and thrust (Krauthammer, 1987). It is strongly recommended that such studies be performed, with the ultimate objective of a better understanding of structural behavior and thus improved design methods.

The area of equipment response under the effects of multiple detonations is very important. Usually tests are performed with the single-detonation approach, wherein the time histories do not include the effects of closely timed explosions that could introduce additional frequencies into the loading environments. In addition, the increase in the deployment of smart weapons raises the issue of having to protect against very close-in or internal detonations that may not be randomly distributed over a general target area. Under these conditions, it would be possible to have several such weapons detonate in a well coordinated manner, thus subjecting a structure to a most severe load–time history at critical locations. Current analysis and design methods do exclude such considerations, thereby introducing serious risk into the field.

1.7.2 Policy and Technology Needs

According to the Downing Assessment Task Force recommendations (Downing, 1996), DOD was to develop a comprehensive approach to force protection that included the designation of an existing DOD organization to be responsible for force protection. It further stated that consideration should be given to a national laboratory to provide this expertise, and that this entity should be provided funds and authority for research, development, test, and evaluation (RDT&E) efforts to enhance force protection. From the NRC committee's findings (1995), and the USCNS/21 report (March 2001), it is clear that the potential hazards for both civilian and military systems and facilities (land-, air-, and sea-based) are considerable, and the construction technologies and design principles developed for hardened military structures are not directly applicable to these facilities. These conclusions and recommendations were confirmed by recent terrorist attacks.

Although several countries made significant investments in research related to terrorism effects (including blast effects on structures), the focus through the early 1990s was primarily on weapons performance and the design of hardened military facilities. Consequently, we lack knowledge and experience in how many typical civilian facilities and systems behave under blast-induced loads. Furthermore, we are gradually gaining the required knowledge and experience in how to cost effectively retrofit existing systems and facilities to make them blast-resistant.

New and advanced materials, novel retrofit and design concepts, and expedient design concepts, and certification and validation processes must be developed in response to these challenges. The NRC reports (1995, 2000, and 2001) included specific recommendations that should be implemented for land-based facilities. More far reaching recommendations have been issued by the USCNS/21 (2001) to address anticipated enhanced threats to the U.S. from terrorism. In light of the 2001 terrorist attacks against the World Trade Center and the Pentagon, these recommendations must be modified to address both the protection requirements faced by society, including land-, sea-, and air-based systems and facilities, and the protection of civilian populations, as follows:

- Expand current defense programs of both short- and long-term research on terrorist threat protection to include theoretical, experimental, and numerical studies of the resistance of military and civilian land- and sea-based facilities (buildings, ships, etc.) to anticipated threats and include considerations of both structural and nonstructural subsystems.
- Adapt existing technology developed for military use and disseminate it to civilian design professionals (emphasizing land-, sea-, and air-based facilities and systems) through professional organizations and academic curricula.
- Establish both national and multinational government–academic–industry partnerships whose purpose is to enhance and facilitate the development and implementation of such technologies. Furthermore, expose design professionals to the range of measures that can be taken to protect facilities and systems (buildings and compounds, aircraft, ships, etc.) from terrorist attack. The partnership should provide a network to foster cooperation and effective protective technology transfer among government, academia, and private industry.

The various departments of defense and/or homeland security have the responsibility for developing, maintaining, and applying effective technologies to protect civilian and military personnel worldwide. Currently in the U.S., the bulk of the research related to this problem is addressed through special groups [e.g., the National Security Council's Technical Support Working Group (TSWG) that focuses on rapid solutions to interagency problems and the Army's Technology Based Program in Force Protection against Terrorist Threat that is focusing on military and government installations]. Also, the U.S. Department of State supports a limited program addressing threats to its embassies.

The contributions to current protective measures from these research programs have been limited because of a lack of funding. While such organizations were assigned the responsibility for conducting the technology-based research mission for the services for survivability and protective structures, the technology-based funding has remained practically at the same level as before. Outside of the defense establishments, experience in terrorism effects (e.g., blast effects on facilities) exists in a limited number of consulting firms involved in such work.

Introduction

Most of these firms, like their government sponsors, suffer from lacks of experienced technical personnel and their levels of interest and involvement depend directly on the availability of funding.

The recommendations of the Downing Task Force, the NRC studies, and the USCNS/21 reports clearly call for a comprehensive approach in developing protective technologies and the establishment of DOD design standards for new construction, hardening of existing facilities, and ship survivability. Furthermore, for the approach to be fully comprehensive, it is critical that an effective government–academic–industry partnership is developed to provide an institutional network to foster R&D, training, and technology transfer. Consistent with these recommendations, an integrated and multinational systems approach to the terrorist threat problem should be explored seriously.

1.7.3 Development and Implementation of Protective Design Methodology

The following summarize the recommended actions required to address the issues raised previously:

- Mobilize the international scientific community (including government, private, and academic) for this effort.
- Establish comprehensive and complementary long- and short-term R&D activities in protective technology to ensure the safety of international government, military, and civilian personnel, systems, and facilities under evolving terrorist threats.
- Develop innovative and effective mitigation technologies for the protection of government, military, and civilian personnel, systems, and facilities from terrorist attack.
- Launch effective technology transfer and training vehicles that will ensure that the required knowledge and technologies for protecting government, military, and civilian personnel, systems, and facilities from terrorist attack will be fully and adequately implemented.
- Establish parallel and complementary programs that address the nontechnical aspects of this general problem (i.e., culture, religion, philosophy, history, ethnicity, politics, economics, social sciences, life science and medicine, etc.) to form effective interfaces between technical and nontechnical developments and implement a comprehensive approach for combating international terrorism.

The following sequence of activities that address the critical scientific and technical needs, as described above, is fully compatible with ongoing activities focused on land-, air-, and sea-based systems and facilities. This compatibility is essential for ensuring that the required protective technologies will meet the broad range of critical national needs. The process is expected to involve a sequence of complementary activities, from basic research through implementation, as described below.

1.7.3.1 Basic and Preliminary Applied Research

- Define and characterize design threats and corresponding loads, and perform hazard and/or risk analysis, and determine corresponding consequences on assets, missions, and people.
- Predict and/or measure facility and/or system response to design threat loads, including considerations of characteristic parameters (material properties, geometry, structure and/or system type, assets, mission requirements, etc.).
- Develop theoretical/numerical retrofit and/or design concepts to bring a facility and/or system response and consequences within acceptable levels.

1.7.3.2 Applied Research, Advanced Technology Development, and Demonstration and Validation Tests

- Develop retrofit and/or design concepts to bring the facility/system response and consequences within acceptable levels.
- Verify retrofit and/or design concept through laboratory and/or field tests and develop final design specifications.

1.7.3.3 Demonstration, Validation, and Implementation

- Conduct certification and/or validation tests.
- Implement retrofit and/or design technology transfer, guidance, and/or training.

These activities should be conducted internationally through national centers for protective technology research and development (NCPTR&D). These centers will direct, coordinate, and be supported by collaborative government, academic, and industry consortia that will perform various parts of the activities. National academic support consortia (NASC) should be established to engage in this critical effort through both research and education activities. These NASCs will identify and mobilize faculty members from universities with appropriate scientific and technical capabilities and lead some of the required R&D.

The NCPTR&Ds and their consortia members will be staffed by a unique team of internationally known experts in all scientific and technological areas relevant to protective technology and will have access to advanced research facilities at all sites. Team members should have documented extensive experience in protective technology, developing and managing major research initiatives, developing and implementing innovative protective technologies, and training of military and civilian personnel in the application. Each collaborative effort should be conducted under the general guidance and oversight of an advisory committee

Introduction 39

consisting of internationally recognized technical experts and senior public, government, and military leaders in relevant fields. Specific technical guidance will be provided through combined government–academic–industry management teams with subgroups formed to focus on key science and technology areas.

1.8 SUMMARY

The information presented above raised several practical issues regarding current methods for hardened systems analysis and design in light of our understanding of anticipated hostile environments. Clearly, significant knowledge has been gained since WWII, and the events since then teach us that threats from conventional and unconventional weapons should be taken very seriously. Nevertheless, additional research in many areas is badly needed and such needs must be adjusted to reflect changes in weapon systems, local and international conflicts, strategic and tactical concepts, and the capabilities of experimental and analytical tools. Unfortunately, both geopolitical and weapons technology changes are much more rapid than the progress in the development of protective measures. It would be wise to address such issues actively and continuously in order to avoid a very tragic outcomes.

The following chapters focus on specific technical issues and on the various analysis and design activities required for blast, shock, and impact mitigation.

2 Explosive Devices and Explosions

2.1 INTRODUCTION

The comprehensive risk management approach as presented in Chapter 1 may indicate that a threat to be addressed involves the use of explosive devices. Therefore one must understand explosive devices and their capabilities and have a good understanding of explosive effects and the processes that generate them. This chapter provides a general background on these topics.

2.2 CHARACTERISTICS OF EXPLOSIVE PROCESSES, DEVICES, AND ENVIRONMENTS

2.2.1 NUCLEAR WEAPONS

The principles of modern nuclear weapons are similar to those of the first nuclear devices. Once a critical mass of a nuclear material (such as uranium or plutonium) is achieved, the chain reaction will result in the release of thermal, radioactive, and other energies that will create a nuclear environment. The weapons are complicated, but should be viewed as sources for vast amounts of energy that, when released, will create the effects that must be resisted by protective systems. More information on nuclear effects can be found in Glasstone and Dolan (1977).

2.2.2 CONVENTIONAL WEAPONS

This section is a general summary of conventional weapons (CW) properties based on TM 5-855-1 (Department of the Army, 1986); for additional information, see other publications, such as *Jane's*, etc. For information on accidental explosions, consult other sources (Baker et al., 1983; Department of the Army, 1990; Mays and Smith, 1995; Bulson, 1997). The designer or analyst must have reasonable assessments on the type of weapon systems that could be applied to the structure under consideration. Addressing military ordnance is important not only for the design of military-type protective structures, but also because many improvised explosive devices (IEDs) are produced with either stolen or discarded weapons. Typical military devices are introduced first, and explosion processes and phenomena are treated, as the foundation for defining physical threat environments.

2.2.2.1 Small Arms and Aircraft Cannon Projectiles

These projectile types may include ball, tracer, armor-piercing (AP), and armor-piercing incendiary (API), as defined in Table 2.1; ball and AP types are illustrated in Figure 2.1.

2.2.2.2 Direct and Indirect Fire Weapon Projectiles

These projectile types are used with various artillery weapons. They contain various amounts of high explosives (HEs) to cause severe blast and fragmentation effects. These effects, when associated with close-in detonations, can cause severe damage to structures. The designer must consider the effects of multiple hits, despite the fact that the literature contains no provisions for such considerations. Special attention should be given to HE antitank (HEAT) munitions that are very effective against exposed structures. Some of these weapons are defined and illustrated, respectively, in Table 2.2 and Figure 2.2.

The following is some general background on the use and effectiveness of such munitions. AP projectiles are effective against armor plate and reinforced

**TABLE 2.1
Characteristics of Small Arms and Aircraft Cannon Projectiles**

	Projectiles		
Caliber	Type	Weight (grains)	Muzzle Velocity (fps)
	Pistols		
7.62 mm	Ball	74–60	950–1,500
9 mm	Ball	125–160	990–1,275
0.45 in.	Ball	208–234	820–850
	Rifles		
5.56 mm	Ball, tracer	43–56	3,100–3,300
7.62 mm	Ball, tracer, AP	74–187	1,380–2,880
	Machine Guns		
7.62 mm	Ball, AP, tracer	149–182	2,680–2,850
12.7 mm (0.50 cal)	Ball, AP, tracer, incendiary	620–710	2,750–3,500
14.5 mm	Ball, AP, tracer, incendiary	919–980	3,200
	Aircraft		
23 mm	HE, AP, incendiary	3,080	2,260
30 mm	HE, AP, incendiary	6,320	2,560
37 mm	HE, AP, incendiary	11,350	2,260
40 mm	HE, AP, incendiary	15,000	3,280

Source: TM 5-885-1, 1986.

Explosive Devices and Explosions 43

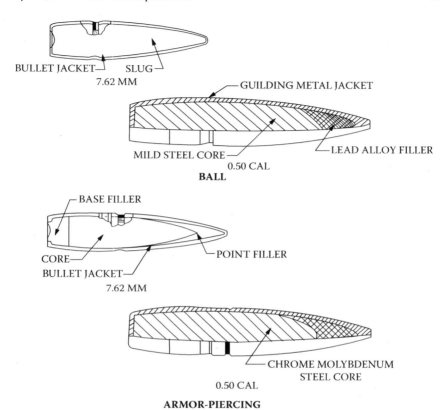

FIGURE 2.1 Typical small arms projectiles (TM 5-885-1, 1986).

concrete. AP solid shot is less effective against face-hardened and homogeneous plate at more than 10° obliquity unless coupled with a very high striking velocity. AP capped projectiles are more effective against concrete targets especially if they contain HE charges. These are fused to detonate at maximum penetration.

Oblique impact decreases penetration and causes a projectile to ricochet; therefore, exposed surfaces of protective structures should be designed to present the maximum angle relative to the direction and trajectory of fire.

HE shells contain about 15% explosives and 85% shell. They can be fitted with delayed fuses for penetration of soil and/or concrete. Mortar shells are similar to HE shells, but contain more explosive (20 to 40% HE). Some mortar shells such as the Soviet 240 mm device are designed to penetrate structures with special 630 lb shells.

The development of rocket-assisted projectiles should be of special interest for those concerned with penetrating weapons.

TABLE 2.2
Characteristics of Typical U.S. and Soviet Mortar, Artillery, and Tank Rounds

Caliber	Rate of Fire (rounds per minute)	Muzzle Velocity (fps)	Maximum Range (m)	Type	Total Weight (lb)	Explosive Type	Explosive Weight (lb)
				U.S. Mortars			
60 mm	18–30	552	1,814	HE	3.2	TNT	0.34
81 m	18–30	875	4,595	HE	9.4	COMP B	2.1
4.2 in.	5–20	960	4,650	HE	27	TNT	7.1
				Soviet Mortars			
82 mm	15–25	693	3,000	HE	6.8	TNT/AMATOL	0.91
120 mm	12–15	893	5,700	HE	35.2	AMATOL	3.48
160 mm	2–3	1,126	8,040	HE	90.7	AMATOL	17.03
240 mm	1	1,189	9,700	HE	288.2	TNT	70.34
240 mm	1	—	—	Concrete penetrating	632.0	TNT	—
				U.S. Artillery			
105 mm	1–3	1,621	11,500	HE	33	COMP B	5.08
155 mm	1–2	1,852	14,600	HE	94.6	TNT	15.4
175 mm	1	3,000	32,700	HE	147	COMP B	31.0
8 in.	0.5	1,950	16,800	HE	200	TNT	36.75
				Soviet Artillery			
122 mm	6–7	2,956	21,900	HE	47.8	AMATOL	10.2
130 mm	6–7	3,054	31,000	HE	73.6	TNT	12.7
152 mm	4	2,150	17,300	HE	95.8	TNT	—
180 mm	1	2,600	30,000	HE	225.0	—	33.9
203 mm	0.5	1,990	18,000	Concrete penetrating	220.5	TNT	
				U.S. Tank			
90 mm	8–9	2,400	17,900	HEAT	23.4	TNT	2.15
				Soviet Tank			
115 mm	5	—	—	HEAT	39.08	TNT	6.0

Source: TM 5-885-1, 1986.

2.2.2.3 Grenades

Grenades are hand-delivered or launched by other means (rifle or RPG). They are popular with special forces and infantry, as described in Table 2.3.

Explosive Devices and Explosions

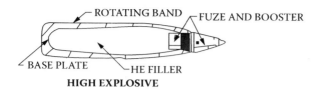
HIGH EXPLOSIVE

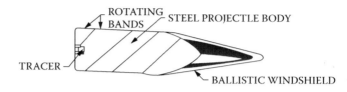
ARMOR-PIERCING SOLID SHOT

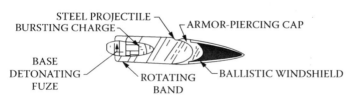
ARMOR-PIERCING CAP FED

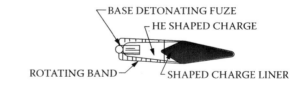
HIGH-EXPLOSIVE ANTITANK

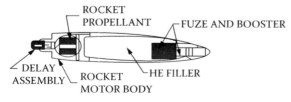
HIGH-EXPLOSIVE ROCKET ASSISTED

FIGURE 2.2 Typical cannon projectiles (TM 5-885-1, 1986).

TABLE 2.3
Characteristics of Typical Grenades

Type	Projectile Weight (lb)	Range (m)	Filler Type	Weight (lb)	Remarks
Fragmentation	0.86	40	COMP B	0.40	Hand-delivered
Concussion	0.97	30–40	TNT	0.50	Hand-delivered
Rifle-HEAT	1.48	115–195	COMP B	0.61	Penetrates approximately 10-in. armor and 20-in. concrete
40 mm launcher	0.61	350–400	COMP B	0.08	Shoulder-fired

Source: TM 5-885-1, 1986.

2.2.2.4 Bombs

Bombs are divided into four principal groups: HE, fire and incendiary, dispenser and cluster, and special purpose applications.

2.2.2.4.1 HE Bombs

The characteristics of some types of HE bombs are presented in Table 2.4 and Figure 2.3. There are several types, all containing HE material primarily for demolition:

- General purpose (GP) bombs are designed to perforate light reinforced concrete (RC) and thin armor. They then detonate to cause general destruction by blast and fragments.
- Light case (LC) bombs are usually used for contact explosions; the effect is primarily blast.
- Fragmentation (FRAG) bombs are designed for heavy concentration of fragments around the point of the explosion. These are effective against personnel and light equipment, but with lower blast effects.
- Armor piercing (AP) bombs have heavy cases and are designed to be used against heavily protected targets (by armor or RC).
- Semi-armor piercing (SAP) bombs are similar to AP types, but for lighter armor.
- Fuel–air–explosive (FAE) bombs release a flammable gas (or liquid) that, when ignited, will cause shock waves in the air. The effects are similar to the detonation of a HE device at a certain altitude.

2.2.2.4.2 Fire and Incendiary Bombs

There are several types of such weapons based on a thin case filled with a chemical or petroleum mixture. They are effective against fire-sensitive targets. Gasoline

TABLE 2.4
Characteristics of Typical Generic High-Explosive Bombs

Designation and Classification	Total Weight W_T (lb)	Diameter d (in.)	Length L (in.)	Charge-Weight Ratio (%)	Slenderness Ratio L/d (in.)	Sectional Pressure $\frac{4W}{d^2}$ (pp psi)
GP	100	110	8	29	51	3.6	2.2
GP	250	260	11	36	48	3.3	2.7
GP*	250	280	9	75	35	8.3	4.4
GP	500	520	14	45	51	3.2	3.4
GP*	500	550	11	90	35	8.2	5.8
GP*	750	830	16	85	44	5.3	4.1
GP	1,000	1,020	19	53	54	2.8	3.6
GP*	1,000	1,000	14	120	42	8.6	6.5
GP	2,000	2,090	23	70	53	3.0	5.0
GP*	2,000	2,000	18	150	48	8.3	7.9
GP*	3,000	3,000	24	180	63	7.5	6.6
SAP	500	510	12	49	30	3.9	4.5
SAP	1,000	1,000	15	57	31	3.8	5.6
SAP	2,000	2,040	19	66	27	3.5	7.2
AP	1,000	1,080	12	58	5	4.8	9.5
AP	1,600	1,590	14	67	15	4.8	10.3

* High slenderness ratio bombs.

Source: TM 5-855-1, 1986.

gel firebombs are designed to be detonated at the openings of protective structures for deoxidation of the interior.

2.2.2.4.3 Dispenser and Cluster Bombs

These are designed to release many bomblets that may be HE, HEAT, chemical, FAE, etc. They are used primarily against personnel and softer surface targets.

2.2.2.4.4 Special Purpose Bombs

These include a wide variety of weapons (leaflet, smoke, flare, etc.) that may not harm structures. The anti-runway (rocket-assisted) bombs should also be considered for hardened systems design.

2.2.2.5 Rockets and Missiles

2.2.2.5.1 Tactical Rockets and Missiles

These can be used against armor and RC, and use various types of launching and guidance systems, as shown in Table 2.5.

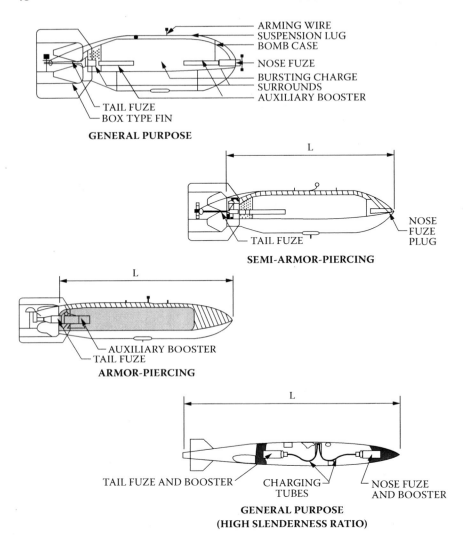

FIGURE 2.3 Typical high-explosive bombs (TM 5-855-1, 1986).

2.2.2.5.2 Battlefield Support Missiles

These include the LANCE, SCUD, FROG, etc. They have HE, CB, and nuclear warheads for long-range delivery.

2.2.2.6 Special-Purpose Weapons

2.2.2.6.1 Fuel–Air Munitions

The fuel (such as propylene oxide) is dispersed by a central buster charge to form a fuel–air cloud that can be detonated with a second charge. The formed vapor

TABLE 2.5
Characteristics of Typical U.S. and Soviet Surface-Launched Rockets and Missiles

Diameter (mm)	Warhead			Muzzle Velocity (fps)	Maximum Effective Range (m)	Weapon System
	Type	Explosive Filler				
		Weight (lb)	Type			
66	HEAT	0.66	OCTOL	476	200	U.S. LAW
85	HEAT	1.25	RDX	985	300	Soviet RPG-7
102	HEAT	3.5	OCTOL	250	1,000	U.S. DRAGON
120	HEAT	—	—	394	3,000	Soviet SAGGER
140	HE	8.10	TNT	1,320	9,810	Soviet MRL
221	HEAT	5.2	OCTOL	657	3,750	U.S. TOW
240	HE	59.81	TNT	969	10,300	Soviet MRL

Source: TM 5-855-1, 1986.

cloud must have a fuel–air concentration within well defined limits (known as the upper and lower detonation limits) to ensure that the mixture can detonate. As will be discussed below, the correct mixture concentration is necessary to ensure complete combustion of the fuel and the release of sufficient thermal energy that can create a shock wave that drives the detonation. The fuel–air mixture may either burn or deflagrate if the mixture concentration is below the lower detonation limit (however, very low mixture concentrations may not even combust). Combustible vapor clouds can be formed also by the release of liquified natural gas (LNG), coal dust, grain dust, etc. The peak pressure in the cloud is up to about 300 psi, and the pressure–distance curves are similar to those associated with HE. Fuel–air munitions create large area loadings that are different from anticipated localized HE detonations.

2.2.2.6.2 Thermobaric Devices

Thermobaric explosive devices are designed to provide longer duration pressure pulses. This can be achieved by using conventional HEs in combination with additives that continue to burn after the initial detonation is completed. The longer burning process elevates the temperature and causes an increased pressure pulse, as will be discussed below.

2.2.3 CONVENTIONAL WEAPONS SUMMARY

The designer should use updated material when obtaining data on weapons. Most libraries have publications that contain such information.

2.3 EXPLOSIVES, EXPLOSIONS, EFFECTS AND THEIR MITIGATION

2.3.1 BLAST EFFECTS AND MITIGATION

An explosion should be defined in the broadest way as a function of explosive sources. Such a definition includes the overall spectrum of explosive effects that include both conventional (chemicals, various vapors, and high explosives) and nuclear devices, and both military and nonmilitary systems. For purposes of the present discussion, the treatment of nuclear threats may not be relevant (although it may become relevant as a function of nuclear weapon proliferation); rather, the emphasis here is on nonnuclear explosive devices that are typical of terrorist activities. The following discussion briefly addresses the nuclear issue, then focuses on nonnuclear devices.

Although explosive devices have been in use for several hundreds of years, comprehensive treatments of blast effects and their mitigation appeared only after WWII. Between 1945 and the early 1980s, the main thrust was the nuclear threat. The reasons for this are clear. WWII was the first international conflict that resulted in massive destruction of vast areas, including numerous population centers of various sizes. Furthermore, the tremendous destruction potential of newly introduced nuclear weapons had been demonstrated, and the nuclear arms race was accelerating. These circumstances led to the introduction of several sources that contained information on the nuclear threat and its mitigation (Glasstone and Dolan, 1977; U.S. Army, 1957; Newmark et al., 1961 and 1962; Allgood, 1970; Crawford et al., 1973 and 1974; Gut, 1976; Defense Civil Preparedness Agency, 1977; ASCE, 1985; Schuster et al., 1987). The two most recent publications in this group represent the state-of-the-art design of protective structures to resist nuclear blast effects.

Prior to the early 1970s, the nonnuclear explosion threat was addressed in a less rigorous manner. The first effort to document the effects of conventional weapons and provide some guidance on mitigation appeared immediately after WWII (NDRC 1946). While various other early accounts of HE devices and their effects are known, they all refer to specific events and are limited in scope. As noted above, limited attention was given to conventional explosion effects in the U.S. before the early 1970s, largely because of the preoccupation with the nuclear threat (until that time, other countries had more advanced capabilities in this area). The American experience in Vietnam and the real needs for explosive safety gradually shifted more attention to this issue and a variety of related publications appeared [Crawford et al., 1971; U.S. Army, 1977; Baker, 1983; Baker et al., 1983; Department of the Army, 1986 (rev. 1998) and 1990; Drake et al., 1989; U.S. Department of Energy, 1992). Increases in terrorist activities with special emphasis on U.S. targets created the need to address the issue of physical security and the requirements for protection against these additional threats. Pertinent information and recommended procedures were developed and incorporated into two additional manuals (U.S. Navy, 1988; U.S. Army, 1988). As in the publications on the nuclear issue, most of these sources contain considerable amounts of empirical information and practical guidelines. While the main focus of many publications in this group is the mitigation of conventional explosion effects, another group has the chief purpose of presenting the scientific foundation in this general area.

Several publications treat specific areas of scientific foundation in considerable depth. Biggs (1964) wrote a book on structural dynamics, with special attention to blast effects and their mitigation. Although the primary emphasis was on the nuclear blast problem, the thorough treatment of the related dynamics problems also enabled consideration of other cases. This is an excellent resource for those who wish to be introduced to the field. Johansson and Persson (1970) provided an extensive treatment of HEs, detonation processes, and the corresponding physical effects. Their book gives significant insight into the nature of explosions, energy release, and the generation of blast loading sources. Some aspects of detonation are treated rigorously by Fickett (1985), who presented mathematical models for both physical and chemical phenomena; he did not, however, include much information on experimental observations and their relationships to the theory.

Henrych (1979) wrote a comprehensive treatise on both nuclear and nonnuclear explosion phenomena and their effects on various media and structural systems. The book contains many problems, their governing differential equations, and theoretical solutions. Although this book may not be a design aid, engineers with advanced training can use it effectively for a wide range of related issues.

Baker et al. (1983) gave special attention to accidental explosions of various chemical compounds, and their book contains both scientific and practical treatments of related blast mitigation. The concept of pressure–impulse diagrams is explained and is used to explain structural response to explosive loads, and the issue of blast-resistant design is also included. Batsanov (1994) provided a comprehensive discussion on explosion effects on both physical and chemical properties of various materials and processes, delineating how this knowledge can be employed by industry.

Although ground shock (the propagation of shock waves through geologic media) may not be a serious issue in physical security (the main threat is still an explosive device detonated near an above-ground structure), it could be an important issue in general structural protection considerations. Blast effects on structures are directly related to stress wave propagation, especially to shock wave effects. In all close-in explosions wherein shock waves must travel through the surrounding medium to cause damage to a facility (including shock propagation through gases, liquids, and solids; it appears also with impact and penetration), a precise description of the wave propagation phenomena is essential. The literature on this general subject can be divided into two principal groups. The first group addresses the classical issue of wave propagation, with emphasis on linear or linearizable problems (Kolsky, 1963; Achenbach, 1973; Davis, 1988). The second group is more focused on nonlinear problems, with special emphasis on shock waves (Rakhmatulin and Dem'yanov, 1966; Whitham, 1974; Rinehart, 1975; Ben Dor, 1992; Han and Yin, 1993).

It should be noted that the treatment of shock waves in various materials became more pronounced beginning in the mid 1960s, and this is reflected in the publications. Henrych (1979) and Lakhov and Polyakova (1967) also wrote extensive theoretical treatments of this important topic, although their emphasis is more on the phenomena and effects related to explosions. These publications mainly contain detailed discussions of theoretical, closed form, solution approaches. It is interesting to note that the works by Rakhmatulin and Dem'yanov (1966), Lakhov and Polyakova (1967), and Henrych (1979) are based primarily on research in the former Soviet Union, and they represent the general approaches employed there in the area of fortification. However, given the nature of these publications, such approaches are limited to a small domain of real problems, and one must employ numerical solution procedures for addressing the more complicated (and more realistic) cases. As far as shock propagation in gases is concerned, the books by Henrych (1979), Baker (1983), Baker et al. (1983), Ben Dor (1992), and Han and Yin (1993), among others, provide substantial information on this topic.

Although the emphasis of this report is on blast effects, two related issues may be important under certain conditions: the impact associated with objects propelled by the blast environment and the problems of penetration by such objects. These issues are addressed in the manuals cited above, but the reader may need additional information to supplement the limited data provided therein. Jones (1988) presented a comprehensive treatment of the problem of mechanical impact, although his emphasis is more on crash worthiness than blast-induced effects. Zukas (1990) edited a collection of chapters by well known researchers in the areas of high velocity impacts and penetration mechanics, and this source contains much valuable information.

This summary provides a brief historical background to the current state of knowledge of blast, mitigation, and design technologies. It describes and assesses the state of the art and highlights capabilities that could be useful for defensive architecture needs in the civilian sector. Recommendations are presented on required future R&D activities that would benefit the civilian defensive architecture community.

Explosive materials are designed to release large amounts of energy in a very short time (the characteristic duration of an explosion in measured in microseconds). This tremendous energy release causes the generation of shock waves in the surrounding media that propagate away from the center of the explosion. A careful examination of the literature provides the required background for understanding the nature of explosions and the ways in which the resulting shock waves interact with structures and affect the integrity of materials and structures. The combined effects of pressure pulses and the corresponding particle velocities determine the characteristic local response, typical to impact or close-in conventional explosions, and separate it from global structural behavior.

2.3.2 EXPLOSIVES AND EXPLOSIONS

Excellent information on the chemical processes of detonation and deflagration is found in various books (Johansson and Persson, 1970; Dobratz, 1974 and 1981; Baker et al., 1983; Smith and Hetherington, 1994; Persson et al., 1994; Cooper and Kurowski, 1996; Zukas and Walters, 1998). A general fuel may contain a combination of carbon, hydrogen, oxygen, nitrogen, and sulfur. To understand the burning process of such a material, one can assume first that the fuel molecule is separated into its chemical components, as follows:

$$C_cH_hN_nS_sO_o \Rightarrow cC + hH + nN + sS + oO$$

Next, one needs to address the oxidation processes for each of these components by the oxygen available in the fuel. Nitrogen atoms will combine to form nitrogen molecules, N_2. However, some traces of nitrogen oxides (NO_X) will be formed, and this knowledge is used by law enforcement in forensic investigations to determine the type of explosive event that caused the observed damage. Hydrogen combines with oxygen to form water (H_2O). Carbon will combine with oxygen to form carbon dioxide, CO_2. However, if the fuel molecule does not have sufficient oxygen (is under-oxidized), the burning will produce only carbon monoxide, CO. If sulfur is present in the fuel, it will combine with available oxygen to form sulfur dioxide, SO_2. Any remaining oxygen atoms will combine into O_2 molecules. One may write the chemical equation for the complete (stoichiometric) oxidation of such a fuel as follows:

$$C_cH_hN_nS_sO_o + \{c + h/4 + o/4 + s\}O_2 \Rightarrow$$
$$c\ CO_2 + (h/2)\ H_2O + (n/2)\ N_2 + s\ SO_2 + Q \quad (2.1)$$

The indexes c, h, n, s, and o indicate how many atoms of each compound are used. Q or Q_d is the *heat of reaction, of O_2 detonation*, or *the heat of combustion*. This is the enthalpy that must be added to the system to enable the reactants to produce the final products at the same pressure and temperature. In the case of fuel burning, the reaction is *exothermic*. This means that the chemical reaction produces more heat than it needs to sustain itself, and the change in enthalpy is negative (measured in kJ/mol). Since the number of moles (one mole is defined as 6.023×10^{23} molecules) per unit weight are known for a given material, one can specify its energy release in kJ/kg. In the case where the enthalpy change is accompanied by a large entropy change and in the case of gaseous product formation, it is positive. Clearly, adding or removing heat from a chemical reaction can affect the process.

A decomposition (or oxidation) process progresses from the point of initiation at a rate that depends on various parameters, e.g., pressure, temperature, specific properties of the fuel material, mixture properties (lean or rich), etc. Generally, if the reaction zone (flame) progresses at velocities lower than the speed of sound in the medium, the process is defined as combustion. The speed of combustion can be increased with confining pressure. When an explosive material burns at speeds below the speed of sound, the process is defined as deflagration. At velocities higher than the speed of sound for the medium, the chemical reaction is defined as detonation.

One may calculate the oxygen balance for an explosive by computing the number of carbon, nitrogen, sulfur, and hydrogen atoms and comparing these with the number of oxygen atoms that would be required to form the fully oxidized molecules on the right side of Equation (2.1). One can see the requirement to provide two oxygen atoms for each carbon atom, half an oxygen atom for each hydrogen atom, and two oxygen atoms for each sulfur atom. Accordingly, one must determine whether the amount of available oxygen in the fuel is sufficient for a complete chemical reaction. This procedure can be accomplished by subtracting the quantity $(2c + h/2)$ from o, the number of oxygen atoms available in the fuel. Accordingly, the oxygen balance (OB) is derived by multiplying that result by the ratio of the atomic weight of oxygen (AW_o) and the molecular weight of the fuel (M). A further multiplication by 100 will produce the OB in percent, as shown below (Smith and Hetherington, 1994; Cooper and Kurowski, 1996):

$$OB\ (\%) = 100(o - h/2 - 2c)(AW_o/M) \qquad (2.2)$$

The nominal atomic weights of oxygen, carbon, hydrogen, nitrogen, and sulfur are 16, 12, 1, 14, and 32, respectively. One can illustrate this approach for several typical explosive materials, as follows. In the case of nitroglycerine, $C_3H_5N_3O_9$, one can write the following chemical reaction:

$$C_3H_5N_3O_9 \Rightarrow 3\ CO_2 + 2.5\ H_2O + 1.5\ N_2 + 0.25\ O_2 + Q \qquad (2.3)$$

Note that the complete combustion of each molecule of nitroglycerine will leave free oxygen, indicating that this explosive should have a positive OB as shown below:

$$OB = 100(9 - 5/2 - 2*3)$$
$$[16/(3*12 + 5*1 + 3*14 + 9*16)] =$$
$$100(5.833)[16/227] = 3.52\%$$

According to Johansson and Persson (1970), this reaction is accompanied by the heat release of 6,700 kJ/kg and 740 liters of gas. This reaction takes place over a very short time (the changes in pressure and temperature occur within 10^{-12} s for a 0.2 mm reaction zone). The release of heat and gas will result in the following consequences: At the end of the reaction zone, the temperature will be about 3,000°K, the pressure will be about 220 kbar, and the density will be 30% higher than before. Computing the OB for nitroglycerine will show that nitroglycerine has about 3.5% more oxygen than is needed for complete combustion. Other explosives have different values of OB. For example, we can repeat this calculation for TNT:

$$C_7H_5N_3O_6 \Rightarrow 7\ CO_2 + 2.5\ H_2O + 1.5\ N_2 - 3.25\ O_2 + Q \tag{2.4}$$

It is known that ΔH for TNT is 4.520×10^3 kJ/kg. The minus sign in front of the amount of oxygen indicates that a full combustion of TNT will produce a deficit of oxygen, which means that this chemical does not have sufficient oxygen for full combustion. The OB calculations will show the following:

$$OB = 100(6 - 5/2 - 2 * 7)$$
$$[16/(7 * 12 + 5 * 1 + 3 * 14 + 6 * 16)] =$$
$$100(-10.5)[16/227] = -74\%$$

Clearly, TNT has an oxygen deficiency of 74%, which will prevent it from being an efficient explosive material. To remedy this situation, one will have to add to it another explosive that is rich in oxygen to ensure a complete combustion that can release all the internal energy. Other explosives can be analyzed similarly to define their OB values, as shown elsewhere (Smith and Hetherington, 1994; Cooper and Kurowski, 1996). A similar analysis can be done for conventional fuels such as gasoline and propane as shown below.

The complete chemical combustion of propane in air is defined below (Hardwick and Bouillon, 1993):

$$C_3H_8 + 5\ O_2 \rightarrow 3\ CO_2 + 4\ H_2O \tag{2.5}$$

The thermodynamic expressions for computing the energy for a chemical explosion are the following:

$$\Delta A = (\Delta A°_f)_P - (\Delta A°_f)_R$$

$$\Delta A = \Delta U - T\Delta S$$

$$\Delta U = (\Delta u°_f)_P - (\Delta u°_f)_R \tag{2.6}$$

$$\Delta S = S_P - S_R$$

$$S = S° - R_g \Sigma x_i \ln x_i$$

P represents products, R represents reactants, $\Delta A°_f$ is the Helmholtz free energy at standard state, $\Delta u°_f$ is the internal energy at standard state, x_i is the mole fraction of species I, $S°$ is the entropy at standard state, and R_g is the ideal gas constant. The explosion energy is computed at 25°C and 1 atm pressure. Also, it is possible to assume that the following relationship can be used (Crowl and Louvar, 1990):

$$\Delta U \approx \Delta H \quad (2.7)$$

ΔU is the internal energy change or $(H_{Products}) - (H_{Reactants})$, and ΔH is the enthalpy change = $(\Delta u°_f)_P - (\Delta u°_f)_R$. Therefore,

$$\Delta H = [4 \text{ mol } (-241.8 \text{ kJ/mol}) + 3 \text{ mol } (-393.5 \text{ kJ/mol})] -$$
$$[5 \text{ mol } (0 \text{ kJ/mol}) + 1 \text{ mol } (-103.85 \text{ kJ/mol})]$$

$$\Delta H = -2.04385 \times 10^3 \text{ kJ/mol of } C_3H_8$$

$$C_3H_8 \text{ (molecular weight)} = 44.1 \text{ g/mol @} 25°C$$

$$C_3H_8 \text{ (density)} = 0.493 \text{ g/cm}^3 \text{ @} 25°C$$

If one considers a commercial 5-gallon propane container, it is usually filled by 20 lb (weight) of propane gas. The number of moles in a 5-gallon propane container is derived as follows:

$$20 \text{ lb } (453.6 \text{ g/lb}) (1 \text{ mol}/44.1 \text{ g}) = 205.7 \text{ moles}$$

$$\text{Total internal energy} = 205.7 \text{ moles} \times -2.04385 \times 10^3 \text{ kJ/mol} =$$
$$-420.5 \times 10^3 \text{ kJ}$$

The minus sign (–) indicates energy release from the combustion process. The TNT equivalency can be computed from:

$$W_{TNT} \text{ (kg)} = [E \text{ (kJ)}/4.520 \times 10^3 \text{ kJ/kg}] =$$
$$[-420.5 \times 10^3 \text{ kJ}/4.520 \times 10^3 \text{ kJ/kg}] = 93 \text{ kg}$$

The release of 20 lb of propane, given sufficient oxygen, can produce an explosion equivalent to that of 93 kg TNT. Therefore 1 kg of propane is equivalent to 10.25 kg of TNT. Similarly, one can compute the TNT equivalency of gasoline as follows. The complete chemical combustion of gasoline in air is shown below (Hardwick and Bouillon, 1993):

$$2 C_8H_{18} + 25 O_2 \rightarrow 16 CO_2 + 18 H_2O \quad (2.8)$$

The TNT equivalency can be obtained in two ways. The first is the approach used for propane:

$$\Delta H = [18 \text{ mol } (-241.8 \text{ kJ/mol}) + 16 \text{ mol } (-393.5 \text{ kJ/mol})] -$$
$$[25 \text{ mol } (0 \text{ kJ/mol}) + 2 \text{ mol } (-250.1 \text{ kJ/mol})]$$

$\Delta H = -1.01482 \times 10^4$ kJ for 2 moles of C_8H_{18} or

-5.0741×10^3 kJ/mol of C_8H_{18}

C_8H_{18} (molecular weight) = 114.23 g/mol @ 25°C

C_8H_{18} (density) = 0.676 g/cm^3 @ 25°C

Assuming that the gas tank contains 5 gallons of gasoline,

5 gallons = 18,925 mL

18,925 mL * 0.676 g/mL = 12,793 gr = 12.793 kg

The number of moles in 5 gallons of gasoline is derived below:

12.793 kg [1000 g/kg] [1 mol/114.23 g] = 111.99 moles

Total internal energy = 111.99 moles × −5.0741 × 10^3 kJ/mol = −568.53 × 10^3 kJ

Accordingly, one can find the following TNT equivalency:

W_{TNT} (kg) = [E (kJ)/4.520 × 10^3 kJ/kg] =

−5686.53 × 10^3 kJ/4.520 × 10^3 kJ/kg = 126 kg

The release of 5 gallons of C_8H_{18}, given sufficient oxygen, can produce an explosion equivalent to that of 126 kg of TNT. Therefore, 1 kg C_8H_{18} is equivalent to 9.85 kg of TNT.

The second method was illustrated by Pounder (1998). The gross calorific value for gas oil is 10,900 kcal/kg. Therefore, one can compute the internal energy for 5 gallons of gasoline as follows:

1 gallon = 3.785 L, so 5 gallons of gasoline = 18.925 L, or 18,925 mL

18,925 mL [0.676 g/cm^3] = 12,793 g = 12.793 kg gasoline

The amount of heat given off from the full combustion of 5 gallons of gasoline is computed as follows:

12.793 kg [10,900 kcal/kg] = 138,209 kcal = 583.823 × 10^3 kJ

The TNT equivalency is computed as before:

W_{TNT} (kg) = E (kJ)/4.520 × 10^3 kJ/kg =

583.823 × 10^3 kJ/4.520 × 10^3 kJ/kg = 129 kg

The release of 5 gallons of C_8H_{18}, given sufficient oxygen, can produce an explosion equivalent to that of 129 kg of TNT. Therefore, 1 kg C_8H_{18} is equivalent to 10.1 kg TNT; this is very similar to the result obtained earlier.

2.3.3 EXPLOSIVE TYPES AND PROPERTIES

One may classify single molecule explosive materials into three general groups: primary, secondary, and tertiary. The classification is based on the difficulty of setting off an explosive reaction. Primary explosives or initiators are very sensitive to both heat and shock. Such explosives will ignite easily, and most likely detonate. They are used as initiators for explosives that are harder to initiate. For example, mercury fulminate, lead azide, and silver azide are primary explosives. Secondary explosives, some of which are still very volatile (e.g., nitroglycerine) are less sensitive than those in the previous group and their relative stability allows their use for various industrial or military applications. Nitroglycerine, nitromethane, nitrocellulose, PETN, RDX, and TNT belong to this group.

Tertiary explosives are very stable and require significant initiation to produce an explosion. Although they provide for enhanced safety in both their production and application, the right conditions will cause their detonation and one should follow safety guidelines to avoid accidental detonation. Ammonium nitrate is one such explosive, and when mixed with fuel oil it forms ANFO. Characteristic properties for several explosives are provided in Table 2.6 and information for many others can be found in the literature (Johansson and Persson, 1970; Dobratz, 1974; Persson et al., 1994; Cooper and Kurowski, 1996; Zukas and Walters, 1998).

TABLE 2.6
Characteristic Properties of Typical Explosives

Explosive	ρ_{max} (g/cm³)	D_i at ρ_{max} (km/sec)	Q_d (kJ/g)
AN (Amon./Nit.)	1.73	8.51	1.59
Composition C-4	1.59	8.04	5.86
Nitroglycerin	1.60	7.58	6.30
Nitromethane	1.13	6.29	6.40
Nitrocellulose	1.66	7.30	10.60
Pentolite (50/50)	1.70	7.53	5.86
PETN	1.77	7.98–8.26	6.12–6.32
RDX	1.76–1.80	8.7–8.75	5.13–6.19
TNT	1.64	6.95	4.10–4.55

ρ_{max} = explosive density. D_i = detonation velocity. Q_d = heat of detonation (Q in Equation 2.1).

2.3.4 COMBUSTION PHENOMENA AND PROCESSES

In the previous sections, we addressed the exothermic chemical reaction that converts a fuel into oxidized products and heat. Such a process may be defined as *burning* when generic combustible materials are considered. For energetic materials, one could distinguish between two burning processes: *deflagration* occurs when the burning process rate is slower than the speed of sound in that material and *detonation* is a burning process at rates exceeding the speed of sound in the material. If one defines the rate of combustion by the reaction front speed of propagation along the material, the supersonic propagation results in the formation of a shock wave. For subsonic burning (for a fuel or propellant) one can relate burning rate to pressure as follows (Cooper and Kurowski, 1996):

$$R = a\, P^n \tag{2.9}$$

R is the burning rate (in length/time units), P is the pressure, a is the burning rate coefficient (in units of length/time/pressure^n), and n is the dimensionless burning rate exponent. Typical propellants burn at rates between 0.2 and 0.66 in./s (5 and 16 mm/s) at normal atmospheric pressure, but at much higher rates under higher pressures. The burning rate is also temperature-dependent, as follows:

$$R = R_0\, e^{S(T - T_0)} \tag{2.10}$$

R_0 is the burning rate at normal temperature T_0 (70°F), e is 2.718, S is a material constant (about 0.002), and R and T, respectively, are the rate and temperature at which the new rate is to be computed.

In deflagration and detonation, a fuel is transformed from either a solid or liquid to a gas, and one needs to address this phenomenon by thermodynamics. A rapid reaction in which the system does not have sufficient time to lose heat before that process completes is termed *adiabatic*. An adiabatic reaction is *isochoric* if the volume of the system in which the adiabatic reaction takes place does not change before the reaction is over. The temperature associated with such a condition is defined by T_v; the index v indicates constant volume. However, if the gases produced by the adiabatic reaction can expand to maintain a constant pressure, that reaction is termed *isobaric*. The temperature associated with such a condition is defined by T_p and the index p indicates constant pressure. If one assumes ideal gas conditions for the reaction products, one may employ the following relationship:

$$PV = n\mathfrak{R}T \tag{2.11}$$

P is pressure in atmospheres, V is volume in liters, n is the number of moles of gas (one can use gas tables that provide the number of moles per gram of gas), $\mathfrak{R}$ is the universal gas constant defined as R_g in Equation (2.6). Its value is 0.08205 liter * atmosphere/gmole * degrees K. T is temperature in °K. Equation (2.11) can be rewritten as:

$$PV/T = n\mathfrak{R} = \text{constant} \qquad (2.12)$$

Equation (2.8) indicates that the value of PV/T is constant for a closed system (i.e., the system prevents losses of mass and/or heat). One may employ Equation (2.11) or (2.12) for simplified calculations to illustrate the effects of combustion processes. If one considers an ideal gas in a closed system with defined values of $(PV/T)_1$ that is forced to change conditions to $(PV/T)_2$, Equation (2.12) indicates that $(PV/T)_1 = (PV/T)_2$. For example, consider an ideal gas in a closed container with a volume of 1 liter at room temperature of 24°C (297°K) and a pressure of 1 atm. If the temperature of the gas is changed quickly to 3000°C (3273°K) without a change in the volume, the pressure will have to change as follows.

Since the value of PV/T is constant, $(1 * 1/297) = (P_2 * 1/3273)$, from which one derives the value of $P_2 = (3273/297) = 11$ atm (since 1 atm is 14.7 psi, the pressure will be 162 psi). As one can see, a sudden large temperature increase caused a large increase in the pressure. If the temperature increase in the closed system is not applied uniformly to the entire volume, it may cause large pressure differences between the higher and lower temperature regions. Such pressure differences will induce gas flow from the higher to the lower pressure regions. If the pressure differences are sufficiently large, they can produce shock waves in the gas.

Equation (2.12) can be used to explain the basic approach of a thermobaric explosive device. If one mixes a conventional HE with materials that continue to burn after the detonation is over (e.g., aluminum powder), the burning will keep the temperature elevated for a longer period. Obviously, for a constant volume, the pressure must increase proportionally to the elevated temperature so that the value of PV/T remains constant. Conversely, if one can reduce the temperature during an explosion, the pressure will have to become smaller.

Although wave propagation will be addressed in more detail in Chapter 3, a brief discussion is provided here to explain the relationships between the explosion process and shock waves.

2.3.5 Detonation Process and Shock Waves

We showed in the previous section that the combustion of an explosive material releases an intense thermal pulse that causes sharp increases in pressure and density. One can treat the corresponding physical phenomena as a one-dimensional (1-D) stationary detonation as illustrated in Figure 2.4.

One may assume that the transition zone is very narrow, representing the detonation front or reaction zone. The chemical reaction is assumed to occur in the transition zone, where the explosive material burns, the resulting gas products are formed, and the thermal energy is released in the form of the heat of

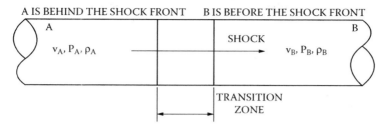

FIGURE 2.4 One-dimensional shock flow.

detonation. Furthermore, instead of considering the detonation front propagating into the unburned explosive, one can keep the front stationary and assume the material is flowing to the left. That means that, as the material flows through the transition zone, it transforms from the steady unexploded state to turbulent and unsteady gaseous products. The detonation front represents the boundary between steady and unsteady material flows. The shock front velocity in the high pressure region is much higher than the particle velocity in the unexploded region. It has been shown (Johansson and Persson, 1970; Henrych, 1979; Persson et al., 1994; Cooper and Kurowski, 1996; Zukas and Walters, 1998) that the steep increase in flow velocity across this front is termed the Chapman-Jouguet (C-J) plane. One can apply the laws of conservation across the shock front as follows:

$$\text{Conservation of mass: } \rho_A(U - u_A) = \rho_B(U - u_B) \quad (2.13)$$

$$\text{Conservation of momentum: } P_A - P_B = \rho_B U u_B \quad (2.14)$$

$$\text{Conservation of energy: } E_A - E_B = (1/2)(P_B + P_A)(u_B - u_A) + Q \quad (2.15)$$

P is pressure, U is the shock front velocity, u is the particle velocity, ρ is the density, and E is the internal energy. The indices A and B correspond to the material states before or after the shock front, respectively. Equations (2.13) through (2.15) are termed the *Rankine-Hugoniot* equations that define the flow discontinuity across the shock front. These equations can be simplified by assuming that $u_B = 0$ and that P_B is negligibly small. One may plot the pressure along the 1-D material as shown in Figure 2.5.

Note on Figure 2.5 that a sharp pressure jump exists at the front of the reaction zone, with a peak pressure termed the Von Neuman spike. The pressure decays behind the front and reaches the C-J pressure state at the end of the reaction zone, after which the pressure continues to decay exponentially to represent gas expansion. The shock front moves to the right at shock speed U. It is customary to plot thermodynamic processes in the pressure–volume space, and one may apply this concept schematically also to the detonation process (Persson et al., 1994; Smith and Hetherington, 1994; Cooper and Kurowski, 1996; Zukas and Walters, 1998). The Hugoniot equation (2.15) can be rewritten in the following form:

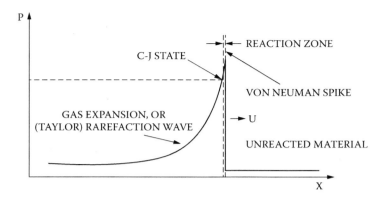

FIGURE 2.5 Pressure versus distance relationship.

$$E_1 - E_0 = (1/2)(P_1 + P_0)(V_0 - V_1) + Q \qquad (2.16)$$

The indices 1 and 0 represent the compressed and initial properties, respectively. Also, one may employ the thermodynamic expression $H = E + PV$ that relates enthalpy, energy, pressure, and volume to obtain:

$$H_1 - H_0 = (1/2)(P_1 + P_0)(V_0 - V_1) + Q \qquad (2.17)$$

One can adopt the previous assumptions in which a 1-D solid explosive in equilibrium is transformed to 1-D gas in equilibrium and the reaction zone has a zero width. Note that the detonation front moves to the right at a steady state velocity and that the detonation process at each point is completed instantaneously. Each material has its own Hugoniot curve, as shown in Figure 2.6. The solid material must transform instantly from a solid state to the gas state and one must switch from one Hugoniot curve to the other. Points 1 and 3 on the Hugoniot for the solid explosive represent the initial and fully shocked conditions without accounting for the change of state due to the detonation. The straight line between these points is termed the Rayleigh line and it is tangent to the Hugoniot for the gas state at point 2, which defines the C-J state for the explosive material. Applying the conservation Equations (2.13 through 2.15) to these conditions will define the sharp transition between the solid explosive state and the post-detonation gas state, as shown by Zukas and Walters (1998).

Two additional points of interest can be defined on the gas Hugoniot (Persson et al., 1994). Point 4 represents a constant volume detonation state, while point 5 represents a constant pressure detonation state. Locations on the gas Hugoniot to the right of point 5 represent only deflagration states, while locations to the left of point 4 represent detonation states. Moreover, locations between points 4 and 5 represent conditions that cannot support either deflagration or detonation.

However, these three conservation equations are insufficient, and one needs an equation of state for the gases that form the expanding detonation products to

Explosive Devices and Explosions

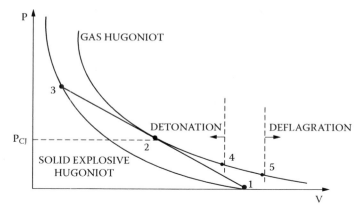

FIGURE 2.6 Detonation process in pressure–volume plane.

enable a complete definition of the adiabatic process in which the entropy is maintained constant, i.e., an *isentropic* process. Although one might consider using Equation (2.11) for an ideal gas, various empirical equations were proposed that fit the actual post-detonation gas state more accurately. One such formulation is the Jones-Wilkins-Lee (JWL) that has the following format:

$$P = A \exp(-R_1 V/V_0) + B \exp(-R_2 V/V_0) + C(V/V_0)^{-(1+\omega)} \quad (2.18)$$

P is the pressure, V and V_0 are the specific volumes at any state and the initial state, respectively, and A, B, C, R_1, R_2, and ω are empirical constants based on experimental data (Dobratz, 1981).

These idealized expressions can be employed with appropriate initial and boundary conditions to define the environment behind the propagating shock front. In many cases, however, only experimental approaches can be used for defining real conditions near an explosion. Such experimental data can then be combined with some aspects of the above-mentioned theoretical approach for deriving empirical solutions, which, as a result of inevitable and frequent deviations between data from different computational and experimental approaches, must rely on correction coefficients.

If the explosion is confined by another material (e.g., soil), one must repeat the above process for a soil element. Again, imposing the conservation of mass, momentum, and energy, one derives the idealized governing equations that describe the shock wave propagation in the surrounding material. Again, significant differences could appear between solutions based on such expressions and experimental data. One can compute the thermal energy losses corresponding to the mechanical changes in the surrounding medium induced by the passage of the shock front. The discussion becomes significantly more complicated when three-phase media (solid, liquid, and gas) are considered. This issue is not expanded upon here, but, the treatment of such problems in the nonlinear domain requires the application of advanced numerical algorithms and the extensive use of supercomputers.

Each explosion induces a series of different shock waves in the medium, such as longitudinal pressure (P-wave), longitudinal rarefaction (N-wave), shear (S-wave), and surface Rayleigh (R-wave). These waves represent different mechanisms of energy transfer and they travel at different velocities. Initially these waves are grouped closely together, but at some distance from the source they separate and arrive at the site of interest at different times. In the case of close-in detonations, it is reasonable to assume that such separation is insignificant and that all the waves reach the target at approximately the same time. If one considers several sequential explosions, it is reasonable to expect some interference between their effects. Further discussions of wave propagation will be included in Chapter 3.

3 Conventional and Nuclear Environments

3.1 INTRODUCTION

As discussed in Chapter 1, the environment created by an explosion involves several effects. The final five effects listed below result only from nuclear explosions. All seven effects are discussed in preparation for defining the corresponding loads on structures.

- Airblast
- Ground shock
- Ejecta
- Fragments
- Fire, thermal, and chemical
- Radiation
- Electromagnetic pulse (EMP)

An explosion generates a shock wave (a high-pressure front propagating outward from the point of detonation) that has a pressure that decays with distance. Conventional high explosives (HEs) tend to produce different magnitudes of peak pressure, heat production, etc. As a result, the environments they produce will be different from each other. In order to establish a base for comparison, various explosives are compared to "equivalent TNT" values. The term used is the *free air equivalent weight* of an explosive, that is, the weight of TNT required to produce a selected shock wave parameter of a magnitude equal to that produced by a unit weight of the explosive.

Since the values for such comparisons are different at different pressure levels, and since the comparison can be for either pressure or impulse, average equivalency factors must be used as shown in Table 3.1. Another approach to compare the effects of different explosives is based on their heat of reaction, Q, as discussed in Chapter 2. Dividing the heat of reaction of any explosive by the heat of reaction for TNT will provide an equivalent TNT weight for that explosive. For example, the equivalent TNT weight for nitroglycerine is obtained by dividing 6,700 kJ/kg by 4,520 kJ/kg for TNT for a value of 1.48. This means that, based on the heat of reaction ratio, each kilogram of nitroglycerine is equivalent to 1.48 kg TNT.

The pressure ranges given in the table are rather low (relatively large standoff distances); therefore, the comparisons may not be very accurate for close-in detonations. Because close-in detonations are of particular interest for HE weap-

TABLE 3.1
Averaged Free Air Equivalent Weights Based on Blast Pressure and Impulse

Explosive	Equivalent Weight Pressure	Equivalent Weight Impulse	Pressure Range (psi)
ANFO (9416)[a]	0.82		1–100
Composition A-3	1.09	1.07	5–50
Composition B	1.11	0.98	5–50
Composition C-4	1.37	1.19	10–100
Cyclotol (70/130)[b]	1.14	1.09	5–50
HBX–1	1.17	1.16	5–20
HBX–3	1.14	0.97	5–25
H–6	1.38	1.10	5–100
Minol II 70/30[c]	1.20	1.11	3–20
Octol 75/25	1.06		
PETN	1.27		5–100
Pentolite	1.42	1.00	5–100
	1.38	1.14	5–600
Tetryl 75/25[d]	1.07		3–20
Tetrytol 70/30 65/35	1.06		
TNETB	1.36	1.10	5–100
TNT	1.00	1.00	Standard for pressure ranges shown
Tritonal	1.07	0.96	5–100

[a] Ammonium nitrate plus fuel oil.
[b] RDX/TNT.
[c] HMX/TNT.
[d] TETRYL/TNT.

Source: Department of the Army, 1986.

ons, a designer must be very careful with the selection of equivalent TNT weights. Experiments have shown that pressures at full contact with explosives may be in the range of 200 to 500 kbar (2900 to 7250 ksi), as discussed in Chapter 2.

An important issue in the analysis and design of protective structures is the comparison of effects from various weapons detonated at various distances. Such a comparison can be made by employing acceptable scaling laws. One approach is cube root scaling used to relate explosive effects on the basis of the corresponding energy levels.

If R is the distance from a reference explosion of a charge having a weight of W, then parameters such as over-pressure, dynamic pressure, particle velocity,

Conventional and Nuclear Environments

etc., for the reference explosion can be related to those arising from another charge W_1 at a distance R_1 as follows:

$$\frac{R}{R_1} = \left(\frac{W}{W_1}\right)^{1/3} \tag{3.1}$$

or

$$\lambda \equiv \frac{R}{W^{1/3}} = \frac{R_1}{W_1^{1/3}} = const \tag{3.2}$$

The term λ is defined as the scaled distance. Similarly, impulse can be scaled as follows:

$$i_s = \frac{i}{W^{1/3}} \tag{3.3}$$

or time of arrival:

$$t_{a_s} = \frac{t_a}{W^{1/3}} \tag{3.4}$$

The concept of cube root scaling implies that all physical quantities with dimensions of pressure and velocity remain unchanged in the scaling. Scaling relationships apply when identical ambient conditions exist, the compared charges have identical shapes, and the charge-to-surface geometries are identical. Under other conditions, the same scaling laws can be used for obtaining approximate comparisons, but they must be used with caution. It has been shown experimentally that these scaling laws are correct for charge weights ranging from a few ounces to hundreds of tons.

The basic principle of cube root scaling is that the energy released from a point explosion will propagate with an expanding sphere of the shock wave. In other words, the various blast effects will be proportional to the energy per unit volume (specific energy). Since the volume of the sphere is proportional to R^3, the scaling will contain the cube root. If, however, the explosion is not similar to a point explosion (or spherical charge), the scaling laws may change. For example, for an explosive line or cylindrical charge, square root scaling should be used because the energy will propagate with the expanding cylindrical shock front. These differences are important in the case of close-in explosions, for which the charge shape is essential in defining the shock front. For explosions far from a target, the assumption of a point detonation is quite reasonable and cube root

3.2 AIRBLAST

3.2.1 HE Devices and Conventional Weapons

3.2.1.1 External Explosions

The general shape of an airblast shockwave pressure–time history for an open air explosion is illustrated in Figure 3.1. The shock front is essentially vertical, reflecting the sudden rise in pressure due to the explosion. The peak incident pressure P_{so} is at the end of this initial phase (*rise time*). The incident pressure is the pressure on a surface parallel to the direction of propagation. The propagation velocity v decreases with time (and distance), but it is typically greater than the speed of sound in the medium.

Gas particles (molecules) move at the particle velocity, u, which is lower than v. The particle velocity is associated with the dynamic pressure, which is caused by the "wind" generated from the blast shock fronts. Since the shape of the shock wave depends on the energy released into the volume that is defined by the front location, as the shock front propagates away from the explosion center, the peak (incident pressure) decreases and the duration decreases.

The following observations can be made about the pressure–time history: The shock front arrives at the target at time t_a and reaches the peak incident pressure P_{so} at t_r, after t_a. Since t_r (rise time) is very short, an instantaneous rise to the peak pressure can be assumed. The peak pressure decays to the ambient value in time

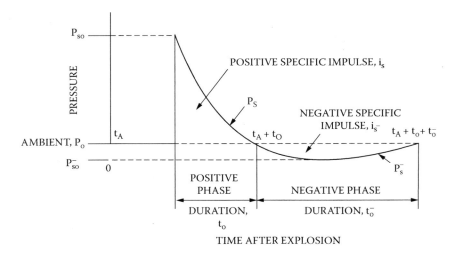

FIGURE 3.1 Free-field pressure–time variation (TM 5-855-1, 1986).

t_o, and this is defined as the positive phase of the pressure pulse. Following is the negative phase for a duration of t^-_o, characterized by a pressure lower than the ambient pressure and a reversal of the wind (particle flow). The negative phase is not important for the design and is usually ignored. The impulse delivered to the target (i.e., associated with the blast wave) is the area under the positive phase of the pressure–time curve, and is defined as i_s.

The shock wave propagates in the manner described above as long as no obstacles are encountered. However, if the wave reaches a surface that is not parallel to the direction of propagation (such as a wall or a structure), a reflected pressure is generated. The reflected pressure has the same general shape as the incident pressure, but the peak is higher than that of the incident wave, as shown in Figure 3.2. The reflected pressure depends on the incident wave and on the angle of the inclined surface.

The duration of the reflected pressure depends on the size of the surface, which determines the rate of flow around it (i.e., the flow will try to go around the wall in order to continue behind the obstacle). This secondary flow from the high pressure to the lower pressure regions reduces the reflected pressure to the

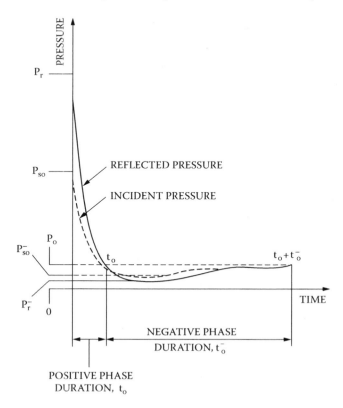

FIGURE 3.2 Typical reflected pressure–time history (TM 5-855-1, 1986).

stagnation pressure, a value that is in equilibrium with the pressure in the high velocity associated with the incident wave. If the reflected pressure cannot be relieved by the secondary flow (such as in the case of an infinite plane wave impinging on an infinitely long wall), the incident wave is reflected at every point on the surface, and the duration of the reflected pressure (positive phase) is the same as that of the incident wave.

The unit impulse for a completely reflected wave is defined as i_n, for the positive phase. The positive phase wave length L_w is the distance from the detonation center experiencing positive pressure, at a given time. P_r is calculated from Figure 3.3 as:

$$P_r = C_{ra} \cdot P_{so} \qquad (3.5)$$

As mentioned earlier, different explosions can be compared by employing the cube root scaling method and charge equivalency to TNT. Such comparisons for the blast effects of spherical TNT charges are presented in Figure 3.4.

One should remember that for close-in explosions, the charge shape could be important, because it defines the shock front geometry. Therefore, these calculations should be used with caution when real weapons are considered. Furthermore, the effects of the casing (the metal container in which the explosive is cast) have not been considered, and since this issue is still under extensive study, it is not clear how the effects should be treated. On one hand, part of the explosive energy will be used for breaking the case and propelling fragments. In the past (see Crawford et al., 1971), it was assumed that about 50% of the energy was

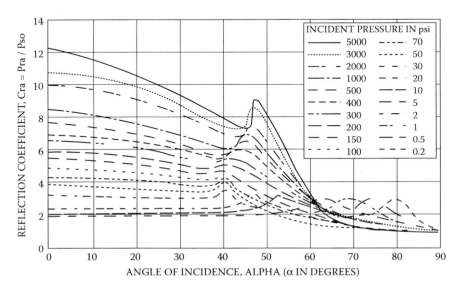

FIGURE 3.3 Reflected pressure coefficient versus angle of incident (TM 5-855-1, 1986).

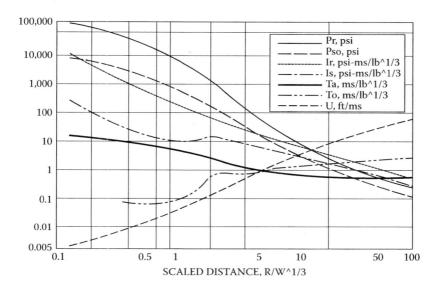

FIGURE 3.4 Shock wave parameters for spherical TNT explosions in free air (TM 5-855-1, 1986).

exhausted by these phenomena; however, a close examination of test data has revealed that such estimates could lead to unconservative solutions.

For an explosion that occurs in the air above a target, the relationships in Figure 3.4 can be used to obtain the blast parameters. Figure 3.3 can be used to derive the reflection coefficients for the reflected shock, and one can obtain a sequence of spheres to represent the wave propagation toward a target, as shown in Figure 3.5. The problem is more complicated when an explosion occurs a short distance from the target. The spherical shock wave that propagates away from the center of the explosion is also reflected from the ground before reaching the target. The reflection from the ground (an enhanced shock) interacts with the original shock wave to produce a resultant shock front known as the *Mach front*. For design purposes, it is assumed that the Mach front is a plane wave with a uniform pressure distribution, and that the pressure magnitude is about the same as that of the incident wave (actually, it is somewhat larger). The point at which the initial wave, the reflected wave, and the Mach front meet is the *triple point*. The triple point marks the top of the Mach front, and the path of the triple point defines the height of the front (which grows as the front moves away from the explosion center). To simplify the problem, it is assumed that the target is affected by the Mach front. Otherwise, one needs to use advanced numerical simulations to define more complicated blast–structure interaction conditions.

If the triple point is above the structure (i.e., the Mach front is taller than the structure), it can be assumed that the structure is loaded by a uniform pressure distribution. If, however, the triple point is below the structure's full height, the

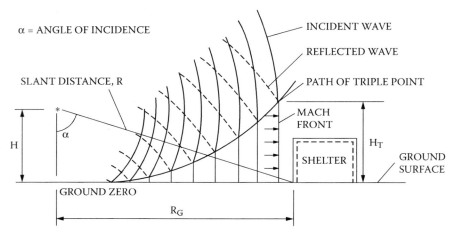

FIGURE 3.5 Blast environment from an airburst (TM 5-855-1, 1986).

pressure distribution must be adjusted accordingly (a uniform pressure up to the triple point and the incident pressure above it). The calculation of pressure acting on a structure is performed as follows (see Figure 3.5):

- Determine the slant distance R.
- Find the blast parameters from Figure 3.4.
- Compute P_r as a function of the incident angle α (Figure 3.3).

When an explosion occurs at the ground surface, simultaneous reflections from the ground are obtained. It is assumed that the shock waves are hemispherical, as shown in Figure 3.6. The parameters for a surface explosion are obtained from other relationships, as shown in Figure 3.7; they are similar to those shown in Figure 3.4, but in Figure 3.7 the parameter values are larger to account for the effects of an instantaneous reflection from the surface. As with air bursts, the parameters are calculated for the assumption of a planar wave (i.e., the structure is assumed to be far from the explosion). As noted, data from relevant experiments or reliable computations should be used for close-in detonations.

The dynamic pressure must be derived also, since it represents the effects of "wind" from the explosion. The peak dynamic pressure q_o is related to the peak incident pressure P_{so}, as shown in Figure 3.8. For approximate analyses, a triangular pulse rather than the actual pressure–time history may be employed. The duration of the approximated positive phase t_{of} is computed as shown in Equation (3.6a) where i_s is the total positive impulse and P_{so} is the peak pressure.

$$t_{of} = \frac{2i_s}{P_{so}} \qquad (3.6a)$$

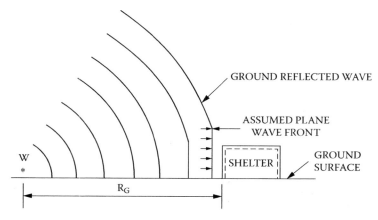

FIGURE 3.6 Surface burst blast environment (TM 5-855-1, 1986).

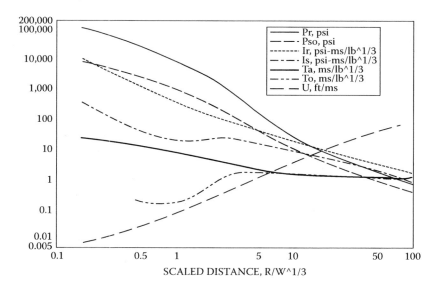

FIGURE 3.7 Shock wave parameters for hemispherical TNT surface bursts at sea level (TM 5-855-1, 1986).

For an explosion that occurs after a weapon has penetrated the ground and some of the gases are released (typical in shallow burial depths), Figure 3.9 can be used to estimate the peak airblast pressure as a function of distance, in which $\bar{x} = \lambda_x e^{p'\lambda_D}$ is the adjusted scaled ground range, $\lambda_x = R/W^{1/3}$ is the scaled ground range, $\lambda_D = R/W^{1/3}$ is the scaled depth of burst, R is the ground range (feet), D is the depth of explosion (feet), W is TNT weight (pounds), and ρ' is the specific

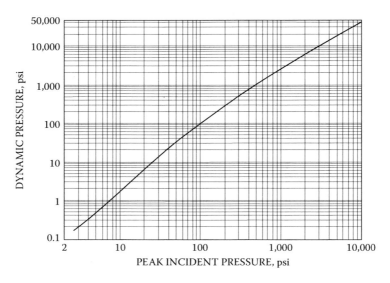

FIGURE 3.8 Peak incident pressure versus peak dynamic pressure (TM 5-855-1, 1986).

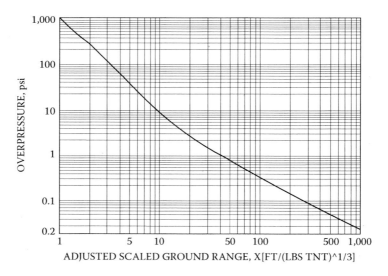

FIGURE 3.9 Peak incident air blast pressure from underground explosion (TM 5-855-1, 1986).

gravity of soil. Note that this curve is valid for values $\lambda_D < 2$ ft/lb$^{1/3}$. Penetrations, cratering, and ground shock are discussed in later sections.

Conventional and Nuclear Environments

3.2.1.2 Internal Explosions

A more complicated blast environment exists when an explosion occurs inside a confined space. The early time–blast phenomena would be very similar to either the spherical or hemispherical conditions characterized by a sharp high pressure spike and defined as the *shock pressure* phase of the event. The shock phase will load the various surfaces that define the confined space, and reflected shock waves will be formed from those surfaces, as shown in Figure 3.10. The duration of the shock pressure phase is very short, and can be estimated from the shock front velocities and the distances between the charge and the various surfaces.

After the shock phase, the blast environment becomes very complicated to define. The reflected shock waves will propagate and interact with various other surfaces. Each such interaction will generate new reflected shock waves and the process will continue for a considerable time. At the same time, the high pressure and high temperature gasses produced by the explosion expand into the confined space. This phase of the blast environment is defined as the *gas pressure* phase that will decay to ambient pressure as a result of gas leakage from the confined space and/or the cooling of the hot gasses. The duration of the gas pressure phase is considerably longer than the duration of the shock pressure phase. Because of its relatively long duration, this phase is also known as *quasistatic*, or *pseudostatic*.

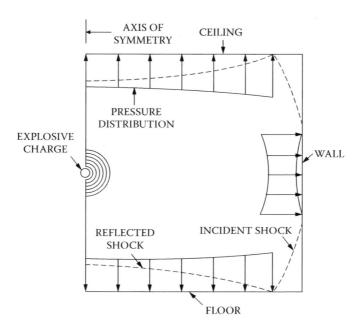

FIGURE 3.10 Shock reflections from walls during internal detonation (TM 5-855-1, 1986).

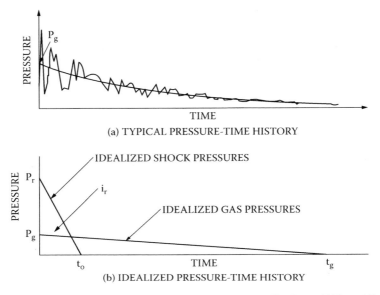

FIGURE 3.11 Shock and gas pressure phases (Department of Defense 1999, and Department of the Army, 1990).

The combined pressure versus time history for a typical internal explosion is illustrated in Figure 3.11(a). The initial short duration-reflected shock pressure spike is followed by a series of many lower pressure reflections, forming the gas pressure phase that gradually decays to ambient pressure. The gas pressure phase can be approximated by a smooth curve that passes through the many gas pressure reverberations. Furthermore, one may adopt an even simpler approximation of the two phases, as shown in Figure 3.11(b).

Both the shock and the gas pressure phases are approximated by triangular pressure pulses. The shock pressure is characterized by the peak reflected pressure P_r and duration t_r. The gas pressure phase has a peak value of P_g and a duration t_g. The two triangles overlap near the origin, indicating that this area is accounted for twice, once for the shock pressure phase and again for the gas pressure phase. Clearly, this implies that the total impulse (i.e., the area under the pressure versus time history) could be about 10% larger than measured experimentally. The following computation procedures are expected for the blast pulse definition; see Figure 3.11(b).

Shock pressure phase — One can use the equivalent TNT charge weight, distance to the target surface, and obliquity to determine the blast pulse parameters. The procedures were described earlier in this section for spherical or hemispherical detonations (Figure 3.3 through Figure 3.5). Knowing the magnitude of the peak reflected pressure P_r and impulse i_r, one can approximate the reflected pulse duration t_r as follows:

$$t_r = 2\, i_r/P_r \tag{3.6b}$$

Conventional and Nuclear Environments

This procedure, however, is very approximate, and it does not take into account the contribution of reflections from other surfaces to the pressure pulse on the target surface. A more accurate approach requires a more detailed wave propagation analysis and possibly the application of advanced computational fluid dynamics (CFD) computer programs.

Gas pressure phase — This phase of the pressure pulse is defined by the equivalent TNT charge weight W, the room volume V_I, vented area A_i, ambient pressure P_a, speed of sound C, and vent area ratio $\partial_c = A_v/A_w$, in which A_v is the total area of all openings in the room and A_w is the total internal area of the walls. Note that the dimensions of these parameters must be in consistent units (i.e., SI or standard). Furthermore, one can use empirical charts for estimating the required values, as shown in Figure 3.12 and Figure 3.13. One obtains the peak gas (quasistatic) pressure, P_{qs} (pounds per square inch), from Figure 3.12 with the appropriate charge weight (pounds) to internal volume ratio (cubic feet) and can then extract the corresponding duration t_0 (seconds) from the scaled duration $(t_0 C \partial_c A_i)/V_I$ obtained from Figure 3.13 (in which the parameter units are feet per second, cubic feet, square feet, and pounds per square inch). The maximum pressure is defined as $P = (P_{qs} + P_a)/P_a$. It is assumed that the duration of the gas pressure phase is t_0, the time required for the quasistatic pressure to decay to the ambient pressure.

3.2.1.3 Leakage Blast Pressure

The previous sections were devoted to conditions for explosion incidents where both the charge and the target are on the same side of the structural envelope. These included spherical or hemispherical blast pressure definitions for

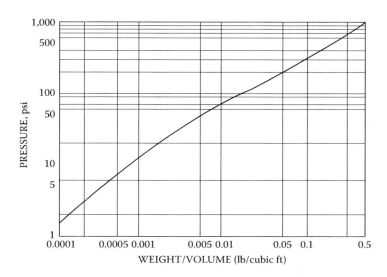

FIGURE 3.12 Peak gas (quasistatic) pressure (TM 5-855-1, 1986).

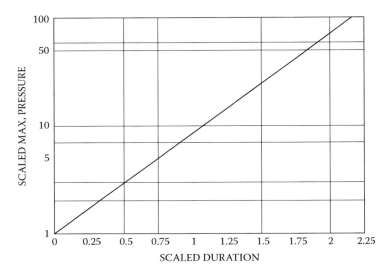

FIGURE 3.13 Scaled duration versus scaled maximum pressure (TM 5-855-1, 1986).

explosions outside a structure or internal blast pressures from an internal explosion. Other interesting blast pulse conditions include cases where the location of the charge and the target are on different sides of the structural envelope. For example, the blast pressure outside a building due to an internal explosion and the blast pressure inside a building due to an external explosion are such cases. They are defined by the phenomena associated with blast pressure leakage through an opening in the structural envelope and are briefly discussed next.

Leakage into protected space — The internal pressure increase can be represented by the following expression for external blast pressures below 150 psi (Department of the Army, 1986):

$$\Delta P_i = C_L (A_0/V_0) \Delta t \tag{3.7}$$

in which C_L is the internal volume, A_0 is the area through which the leakage occurs, V_0 is the internal volume, and Δt is the duration for the leakage. If one performs this calculation in standard units (cubic feet, square feet, and pounds per square inch), the leakage pressure coefficient is obtained from Figure 3.14.

This approach is applied over a series of time increments, from the time of arrival of the blast pressure on the building envelope until any desired time after that. Practically, one can use 10 to 20 time increments for the positive phase duration of the external pressure pulse (t_0). For the first time increment, the pressure differential $P - P_i$ is the magnitude of the external pressure, since P_i is still at the ambient value. One finds the corresponding value of C_L from Figure 3.14 and ΔP_i is derived from Equation (3.7). This procedure is repeated for the next time increment. P is the magnitude of the external pressure at that time and

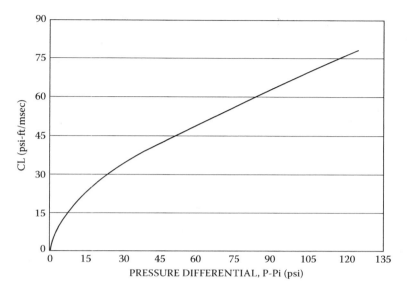

FIGURE 3.14 Leakage pressure coefficient versus pressure differential (TM 5-855-1, 1986).

P_i is the internal pressure computed for the previous step. The new internal pressure is enhanced by the new value of ΔP_i.

Leakage from internal explosion — Although the leakage process is similar to the previous case, this phenomenon is somewhat more complicated. It involves also the relative location of the external point of interest with respect to the opening. For example, one may need to calculate leakage blast pressure at a certain distance behind a building, while the opening is at the front of the building. One should employ the procedures in Chapter 2 of TM 5-1300 (Department of the Army, 1990) for such calculations.

3.2.2 Nuclear Devices

When a nuclear device is exploded at an altitude below 100,000 ft (30.5 km), approximately 50% of the released energy will result in blast and shock. The discussion here focuses on the effects from air burst, surface burst, and shallow burst. High altitude bursts (those above 100,000 ft) are of little interest to the structural engineer because they generate only strong EMP effects (that should be considered in facility design).

It is assumed that an air burst at an altitude under 100,000 ft will create a fireball that will not reach the ground, and that all other effects will be present as functions of the height of burst (HOB). Usually the cratering effects if any, are small. When the fireball from a nuclear explosion can reach the ground, the case is defined as a *near-surface burst*. It will be assumed that the fireball results from a contact explosion and that the center of explosion is about 2 ft (0.6 m) above the ground. In this case, all the induced effects must be considered for design. An underground explosion is defined as an explosion whose center is below the ground surface. Underground explosions can range from shallow to deep (fully contained).

With nuclear explosives and HE explosions, one can notice the incident wave, the reflected wave from the ground, the Mach stem, and the triple point, as shown in Figure 3.15. In a near-surface burst, the hemispherical shock is reflected from the ground and the reflected shock (enhanced by the reflection) travels faster than the incident wave. Eventually, the reflected pressure overtakes the incident wave to form the Mach stem that extends from the ground to the triple point. The air flow behind the Mach stem causes the formation of a second shock wave, resulting in double-shock loading on structures.

The characteristics of the blast pressure wave are similar to those of an HE explosion and are functions of the weapon yield, the HOB, and the distance from the burst. There is also the dynamic pressure that results from the mass flow behind the shock. The dynamic pressure is a function of the gas density and the flow velocity. The gas density is greater than air density because of compression from the shock and the inclusion of other particles such as dust and smoke. The time of arrival depends on the yield and the distance from the point of burst. As with HE explosions, one observes positive and negative phases for the over-pressure and for the dynamic pressure, as shown in Figure 3.16. The peak values of the negative phase rarely exceed 4 to 5 psi (28 to 34 kPa) below the ambient pressure. Peak over-pressure parameters can be obtained from Figure 3.17. The relationships in Figure 3.17 provide an approximation of ± 8% for peak over-pressures less than 100,000 psi (700 MPa). The shock front velocity can be evaluated by:

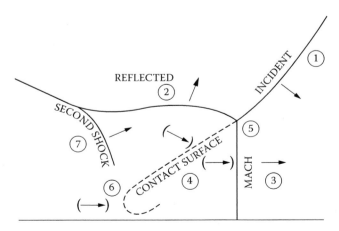

FIGURE 3.15 Second-shock phenomenon in strong Mach reflections (ASCE, 1985).

Conventional and Nuclear Environments

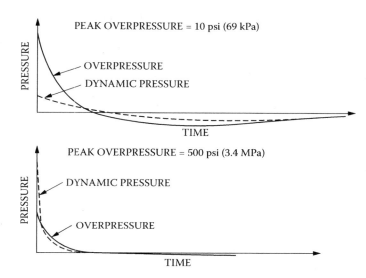

FIGURE 3.16 Qualitative variation of over-pressure and dynamic pressure with time at point (ASCE, 1985).

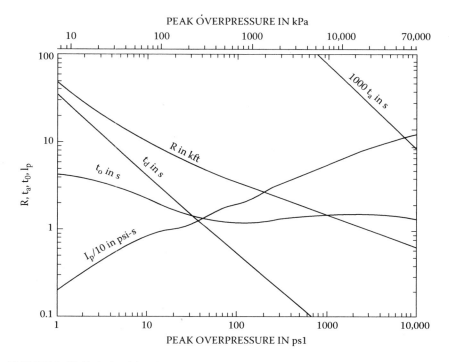

FIGURE 3.17 Relationship of peak over-pressure to range, time of arrival, positive phase duration, and impulse for 1 mt (ASCE, 1985).

$$U = C_0 \sqrt{1 + \frac{6}{7}\frac{P_s}{P_0}} \tag{3.8}$$

in which

$$C_0 = (\gamma g \mathfrak{R} T)^{\frac{1}{2}} \tag{3.9}$$

and U is the shock front velocity, C_o is the ambient sound speed, P_0 is the ambient pressure, P_s is the peak over-pressure, g is the gravity acceleration, $\mathfrak{R}$ is the gas constant, T is absolute temperature, and γ is the specific heat ratio (7/5 for an ideal gas).

As for HE charges, the application of cube root scaling is important for comparisons of explosion effects, and the following relationships are assumed:

$$\frac{R}{R_1} = \left(\frac{W}{W_1}\right)^{1/3} = \frac{t}{t_1} = \frac{I}{I_1} \tag{3.10}$$

Plots of ideal surface HOB peak over-pressure curves for three pressure ranges (1000 to 100 psi, 100 to 10 psi, and 100 to 1 psi) are shown in Figure 3.18 through Figure 3.20. The dotted line in Figure 3.18 indicates where the double peaks will be equal (in the Mach reflection region at high over-pressures) if they occur. To the right of the dotted line, the first peak is the maximum; between the dotted and dashed curves, the second peak is the maximum. These values are based on a confidence factor of 2. The predicted range at which a given over-pressure will occur is accurate within 20%.

For employing pressure–time histories in analysis and design, one may use analytic or approximate expressions, as shown in Figure 3.21. There are three types of approximations, each having the same peak value but varying in their durations:

- If the maximum structural response occurs after the pressure reaches the ambient pressure, use a duration of t_i in order to match the total impulse.
- If the maximum response occurs early after the arrival, use t_{00} to ensure that the tangent to the curve is matched at the peak (structures sensitive to high frequencies).
- t_{50} is chosen for intermediate responses and the triangular line passes through the value of 1/2 P_s on the curve.

These values can be obtained from Figure 3.22.

The peak dynamic pressure for an ideal gas ($\gamma = 7/5$) can be computed from various empirical equations or derived from Figure 3.23. Similarly, one can obtain

Conventional and Nuclear Environments

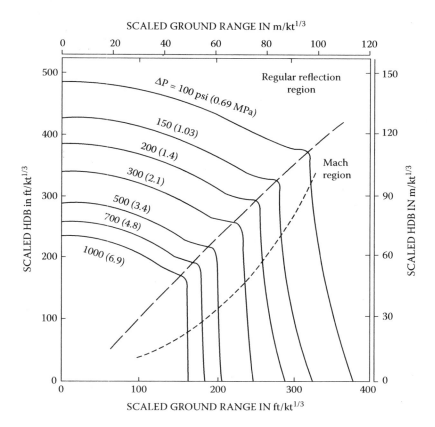

FIGURE 3.18 Ideal surface HOB peak over-pressure curves (ASCE, 1985).

empirical expressions for peak dynamic pressure as a function of scaled HOBs and scaled ranges, as shown in Figure 3.24 through Figure 3.27. By integrating the pressure–time histories, the corresponding relationships for the impulse may be obtained. Finally, Figure 3.28 shows the approximated durations for these time histories.

3.3 PENETRATION

The discussion that follows will be devoted entirely to the issue of penetration for conventional weapons. It is assumed that most nuclear weapons are not designed to produce such effects; therefore, it is reasonable to ignore the problem for nuclear devices.

A protected structure must be designed to prevent the detonation effect from hurting people and/or adversely affecting the mission of the facility. On the other hand, the weapon is designed to reach as close as possible to the site that will be damaged by the explosion; therefore, the detonation may occur at the weapon's

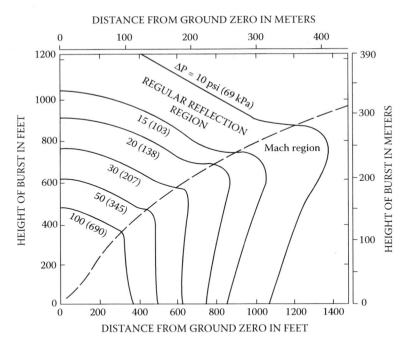

FIGURE 3.19 Height of burst and ground range for intermediate over-pressures for 1 kt (ASCE, 1985).

maximum penetration depth. The protective system must reduce the depth of penetration in order to maintain a safe standoff distance from the center of the explosion. It must also absorb the kinetic energy of the impacting weapon in order to stop it, and this is achieved by damage to the protective system (crushing of brittle materials, plastic deformations of ductile materials, etc.). The system must be designed for longer impulse durations that will allow the structure to absorb more energy (softer systems, thicker systems, etc.). If the weapon penetrates quickly, the structure did not function in an optimal manner. The penetration process is characterized by three typical phases:

- Impact
- Travel through protective materials
- Post-penetration conditions

The impact may cause damage due to the crushing or yielding of material at the point of contact and scabbing at the back side. The scabbing is produced by the reflection of stress waves at the rear face. An explosion at the front face may cause similar effects. It was shown by tests that scabbing becomes a problem for concrete when a 50% penetration of the protective layer thickness is achieved; for 63% or more penetration, one may expect full perforation (Pahl, 1989). However, there are many deviations from these empirical values.

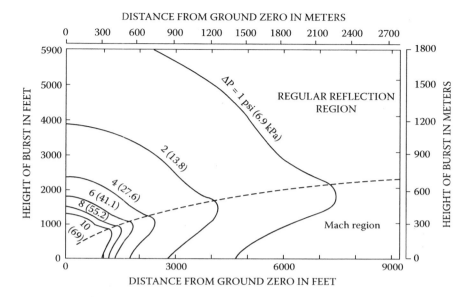

FIGURE 3.20 Height of burst and ground range for low over-pressures for 1 kt (ASCE, 1985).

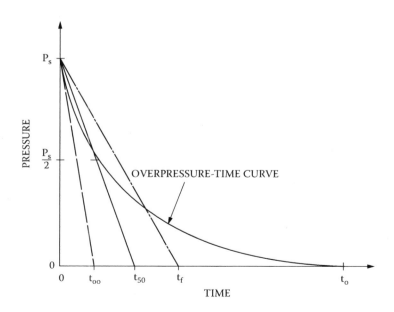

FIGURE 3.21 Triangular representations of over-pressure–time curves (ASCE, 1985).

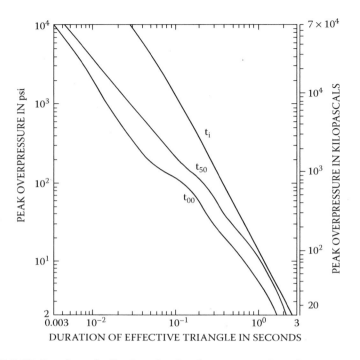

FIGURE 3.22 Duration of effective triangles for representation of over-pressure–time curves for 1 mt surface burst (ASCE, 1985).

TABLE 3.2
Penetration Parameter

Missile Characteristic	Striking Condition (Figure 3.29)	Property of Target
Weight (W_T)	Impact velocity	Strength or hardness
Caliber or diameter (d)	Angle of incidence	Density
Shape	Yaw	Ductility
Fuzing		Porosity
Structural resistance		

Structural response in the dynamic domain may allow the absorption of a significant amount of energy without severe damage. Although stress levels may be higher than the materials' ultimate strengths, their short durations could prevent more severe consequences. These issues are discussed and demonstrated in later sections. The parameters that affect penetration (TM 5-855-1, 1986) are shown in Table 3.2 and Figure 3.29.

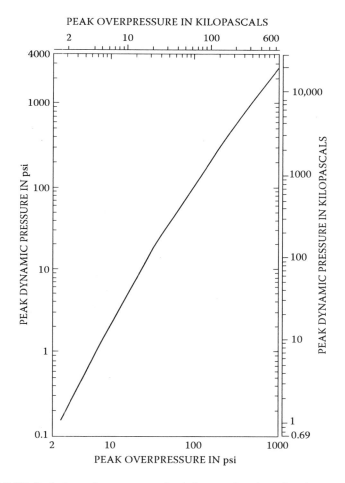

FIGURE 3.23 Peak dynamic pressure at shock front as function of peak over-pressure at sea level (ASCE, 1985).

The general effects of penetration on concrete and steel plates are illustrated in Figure 3.30. It should be noted that the projectile is arrested when its kinetic energy is completely removed. Thus, a deformable projectile is less effective than a rigid one because the deformation allows the removal of kinetic energy. In general, a "soft" projectile may be 50% less effective than a similar hard projectile.

3.3.1 Concrete Penetration

Concrete has compressive strengths that are about ten times higher than its tensile strengths. The impact of a projectile will cause severe cracking and crushing that must be supported by reinforcement in order to prevent failure. It was observed that the penetration into concrete is inversely proportional to $(f'_c)^{0.5}$, where f'_c is

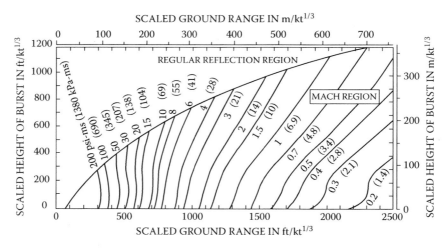

FIGURE 3.24 Peak dynamic pressure as function of scaled height of burst and scaled ground ranges from 0 to 25,000 ft/kt$^{1/3}$ (ASCE, 1985).

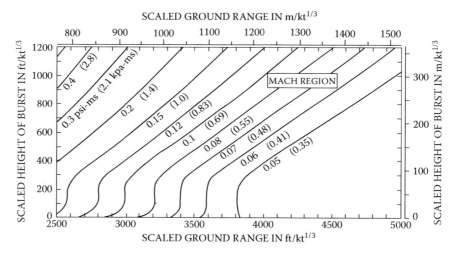

FIGURE 3.25 Peak dynamic pressure as function of scaled height of burst and scaled ground ranges from 2,500 to 5,000 ft/kt$^{1/3}$ (ASCE, 1985).

the uniaxial strength of concrete. An increase in aggregate size (especially if larger than the projectile diameter) also helps to reduce the penetration depth. Steel reinforcement improves concrete behavior by adding tensile resistance for controlling cracking, crushing, scabbing, and spalling. It is recommended that the following reinforcements be provided, as shown in Figure 3.31:

- Front face mats for reducing the effects of spalling near the front crater.
- Back face mats to provide flexural strength and control scabbing.

Conventional and Nuclear Environments

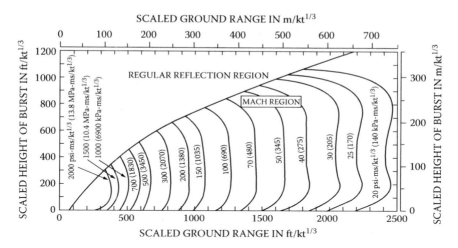

FIGURE 3.26 Dynamic pressure impulse as function of scaled height of burst and scaled ground ranges from 0 to 25,000 ft/kt$^{1/3}$ (ASCE, 1985).

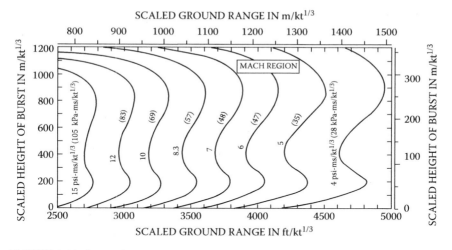

FIGURE 3.27 Dynamic pressure impulse as function of scaled height of burst and scaled ground ranges from 2,500 to 5,000 ft/kt$^{1/3}$ (ASCE, 1985).

- Shear reinforcement for reducing the effects of punching shear and flexural shear and for supporting the other reinforcing bars.
- Anti-scabbing plates placed at the back face to prevent scabbing effects from flying into the facility. They must be well enclosed into the structure since impact may fracture stud welds. A buried structure (under soil and burster slab) may not need such plates, but if modern penetration weapons are anticipated, this issue must be carefully evaluated.

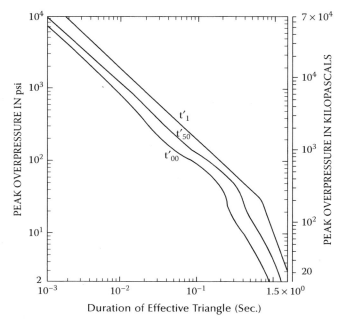

FIGURE 3.28 Duration of effective triangles for representation of dynamic pressure–time curves, for 1 mt surface burst (ASCE, 1985).

In some publications, layered structural elements are recommended rather than a single thick layer. This may not be practical for many structures, but could be effective for soil cover and burster slabs.

Several advanced computer codes may be used for the computation of penetration effects. Such computations are not recommended for design, however, since they are expensive and require extensive computational resources. For rapid assessments, one should try Equation (3.11) and the graphic approach shown in Figure 3.32 through Figure 3.34 (TM 5-855-1, 1986). For normal penetration of inert AP projectiles or AP or SAP bombs into reinforced concrete, the following empirical formula may be used:

$$X = \frac{222 \cdot P_p \cdot (d)^{0.215} \cdot V^{1.5}}{(f_c')^{0.5}} + 0.5d \tag{3.11}$$

in which X is penetration depth (inches), P_p is the sectional bomb or projectile pressure weight (pounds) divided by the maximum cross-sectional area (square inches), d is the penetrator diameter (inches), V is the striking velocity in units of 1000 fps, and f_c' is the uniaxial compressive strength of concrete (pounds per square inch). This formula is considered accurate within ±15% for single hits.

When considering modern, high slenderness ratio bombs at striking velocities under 1000 fps, the penetration depth should be increased by 30%. The nomograms in Figures 3.32 through 3.34 may be used for other nonnormal strike cases.

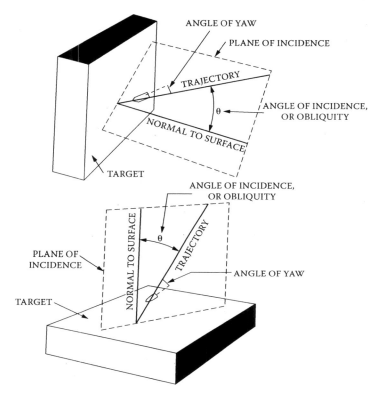

FIGURE 3.29 Geometry of impact (TM 5-855-1, 1986).

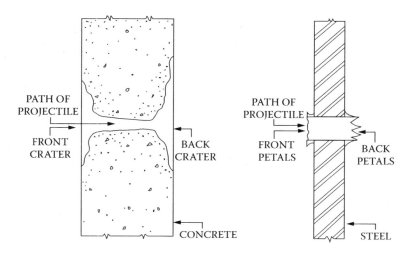

FIGURE 3.30 Perforation of concrete and steel (TM 5-855-1, 1986).

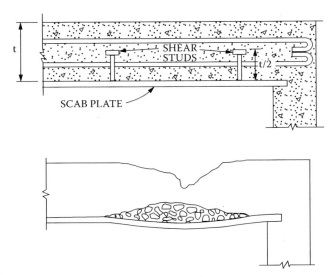

RETENTION OF SCAB MATERIAL

FIGURE 3.31 Scab plate (TM 5-855-1, 1986).

3.3.2 ROCK PENETRATION

Data on the penetration characteristics of rock are limited. However, a practical estimate can be obtained by employing the rock quality designation (RQD) in the following formula:

$$X = 6.45 \frac{W_T}{d^2} \frac{V_s}{(\rho Y)^{0.5}} \frac{100^{0.8}}{RQD} \tag{3.12}$$

in which X is the penetration depth (inches), W_T is the projectile weight (pounds), d is the projectile diameter (inches), V_s is the striking velocity (feet per second), ρ is the target bulk density (pounds per cubic foot), Y is the unconfined compressive strength of intact rock (pounds per square inch), and RQD is the percentage of core recovery (total length of all pieces that are more than 4 in. long) from a given core run. The following limitations should be noted:

1. The calculated depth x must be ≥3d.
2. For RQD >90 (nearly intact rock), $1 \leq d \leq 12$ in. and an accuracy of ± 20%.
3. For RQD <90, projectile range only $4 \leq d \leq 12$ in. and an accuracy of ± 50%.
4. Not valid for RQD <20.
5. Not recommended for blunt or near blunt projectiles.
6. Not valid for "mushroomed" (broken) projectiles.
7. Not valid for sharply curved penetration path or for tumbling projectiles.

Conventional and Nuclear Environments

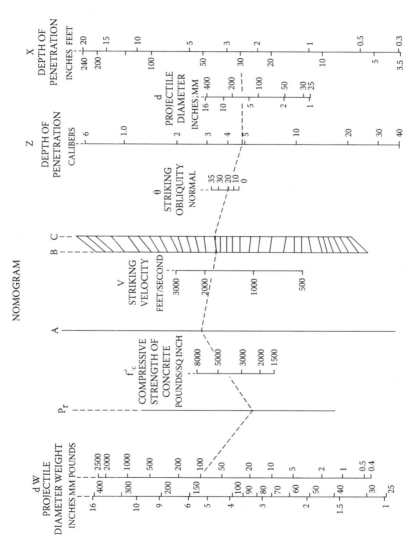

FIGURE 3.32 Penetration by inert AP or SAP projectiles or bombs in reinforced concrete (TM 5-855-1, 1986).

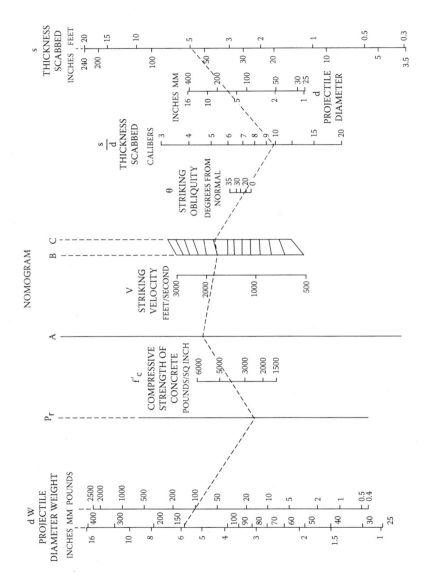

FIGURE 3.33 Scabbing by inert AP or SAP projectiles or bombs in reinforced concrete (TM 5-855-1, 1986).

Conventional and Nuclear Environments

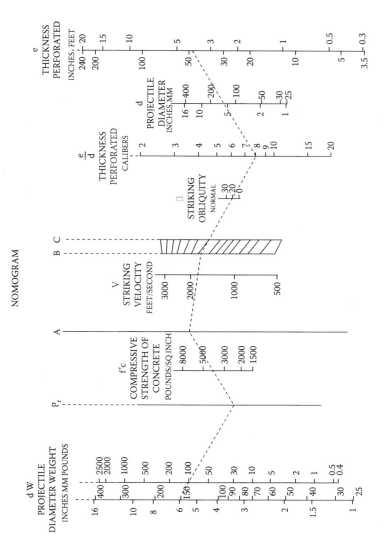

FIGURE 3.34 Perforation by inert AP or SAP projectiles or bombs in reinforced concrete (TM 5-855-1, 1986).

TABLE 3.3
Rock Quality Designations

RQD (%)	Rock Quality
0–25	Very poor
25–50	Poor
50–75	Fair
75–90	Good
90–100	Excellent

Source: TM 5-855-1, 1986.

TABLE 3.4
Engineering Classifications for Intact Rock

Class	Description	Compressive Strength (psi)
A	Very high strength	Over 32,000
B	High strength	16,000–32,000
C	Medium strength	8,000–16,000
D	Low strength	4,000–8,000
E	Very low strength	Less than 4,000

Source: TM 5-855-1, 1986.

If no accurate data are available for a site, one may employ information from Table 3.3 through Table 3.5.

3.3.3 SOIL AND OTHER GRANULAR MATERIAL PENETRATION

The penetration of granular material is not understood as well as that of other materials. However, in general, the following observations about granular penetration were made:

- Penetration decreases as the material density increases.
- Penetration increases as the water content increases.
- For the same density, the finer the grain, the greater the penetration.

Penetration paths of bombs into earth have J shapes, as shown in Figure 3.35. The straight part is approximately 2/3 of the total length, and the curvature radius is about 1/5 to 1/3 of the total path length. Figure 3.36 can be used to estimate penetration depths.

Most standard bombs have slenderness ratios (length to diameter) of 3 to 6; they are terra dynamically unstable and produce J-shaped penetration paths.

TABLE 3.5
Common Intact Rock Characteristics

Rock Type	Typical Density ρ (pcf)	Strength Range Y (psi)
Soft shale (clay shales, poorly cemented silty or sand shales)	143	200–2,000
Tuff (nonwilded)	118	200–3,000
Sandstone (large grain, poorly cemented)	125	1,000–3,000
Sandstone (fine to medium grain)	130	2,000–7,000
Sandstone (very fine to medium grain, massive, well cemented)	143	6,000–16,000
Shale (hard, tough)	143	2,000–12,000
Limestone (coarse, porous)	143	6,000–12,000
Limestone (fine grain, dense massive)	162	10,000–20,000
Basalt (vesicular, glassy)	162	8,000–14,000
Basalt (massive)	180	>20,000
Quartzite	162	>20,000
Granite (coarse grains, altered)	162	8,000–16,000
Granite (competent, fine to medium grain)	162	14,000–28,000
Dolomite	156	10,000–20,000

Source: TM 5-855-1, 1986.

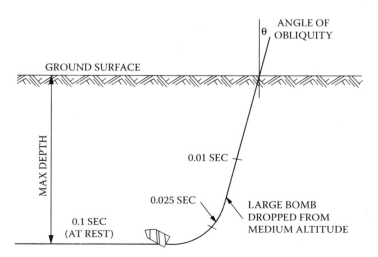

FIGURE 3.35 Courses of bombs that form J-shaped paths in earth (TM 5-855-1, 1986).

However, for slenderness ratios of 10 or larger, the path is almost straight and can be approximated by the following empirical equation:

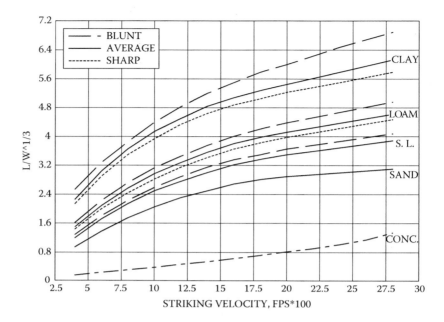

FIGURE 3.36 Penetration of bombs and projectiles into soil.

$$X_f = 0.0031\, S_i N_s\, (W_T / A_m)^{0.5}\, (V_s - 100) \tag{3.13}$$

where X_f is the final penetration depth (feet), S_i is the soil penetration index (from test or Table 3.6), N_s is the nose shape factor (Table 3.7), W_T is the projectile weight (pounds), A_m is the maximum cross-sectional area of the projectile (square inches), V_s is the striking velocity (feet per second). This equation has some limitations:

1. Valid for 200 fps $<V_s<$ 3,000 fps.
2. Valid for 60 lbs $\leq W_T \leq$ 5,700 lbs.
3. Slenderness ratios must be larger than 10.
4. Not valid for shallow penetration depths (of less than 3 bomb diameter + 1 nose length).
5. Accuracy at best is ±20%.
6. Not valid for sharp curvatures or broken bombs.

3.3.4 Armor Penetration

The most accurate penetration data available are for steel. Several types of steel could be considered for protection, as shown in Table 3.8. Steel is the most effective structural material for protection against various weapon effects, but it is not used for many types of stationary systems. The application of a class of

TABLE 3.6
Typical Soil Penetrability Index for Natural Earth Materials

Soil Index S_i	Material
2–3	Massive gypsite deposits. Well-cemented coarse sand and gravel. Caliche, dry. Frozen moist silt or clay.
4–6	Medium dense, medium or coarse sand, no cementation, wet or dry. Hard, dry, dense silt or clay. Desert alluvium.
8–12	Very loose fine sand, excluding topsoil. Moist stiff clay or silt, medium dense, less than about 50% sand.
10–15	Moist topsoil, loose, with some clay or silt. Moist medium-stiff clay, medium dense, with some sand.
20–30	Loose moist topsoil with humus material, mostly sand and silt. Moist to wet clay, soft, low shear strength.
40–50	Very loose, dry sandy topsoil. Saturated very soft clay and silts, with very low shear strengths and high plasticity. (Great Salt Lake Desert and by mud at Skaggs Island). Wet lateritic clays.

Source: TM 5-855-1, 1986.

TABLE 3.7
Nose Shape Factors

Nose Shape	Nose Caliber[a]	N_s
Flat	0	0.56
Hemisphere	0.5	0.65[b]
Cone	1	0.82[b]
Tangent ogive	1.4	0.82
Tangent ogive	2	0.92
Tangent ogive	2.4	1.0
Cone	2	1.08
Tangent ogive	3	1.11
Tangent ogive	3.5	1.19
Cone	3	1.33

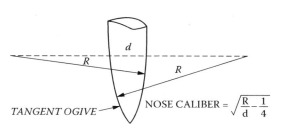

$$\text{NOSE CALIBER} = \sqrt{\frac{R}{d} - \frac{1}{4}}$$

[a] Nose length/diameter.
[b] Estimated.

Source: TM 5-855-1, 1986.

TABLE 3.8
Properties of Armor Plate and Mild Steel

Name	Abbreviation	BHN	Remarks
Face-hardened armor	Class A FHBP	Face 550–650	Unmachinable
		Back 250–440	Unmachinable
Homogeneous hard armor	BP	400–475	Unmachinable
Homogeneous soft armor	Class B STS	220–350	Machinable
Mild steel	MS	110–160	Machinable

* Unmachinable, except with special tools.

Source: TM 5-855-1, 1986.

TABLE 3.9
Plate Thicknesses of Homogeneous Soft Armor, Class B, STS BHN 250 to 300

4 Calibers	2 Calibers	1-1/2 Calibers	1 Calibers
Practical immunity	Favorable to defense	Favorable to attack at short range and small obliquity	Distinctly favorable to attack at medium range and large obliquity
Thickness required to resist attack at normal impact at striking velocity of 3500 fps or less; hypervelocity weapons perforate thicknesses up to 8 calibers based on caliber of core or at about the value above based on gun caliber	Thickness required to resist attack at normal impact at striking velocity of 2400 fps or less	Not immune against attack at obliquities below 20 degrees and striking velocities above 2200 fps	Not immune against attack at normal impact for striking velocities above 1700 fps and just resists attack at 40 degree obliquities at 2000 fps

Source: TM 5-855-1, 1986.

steels for protection is summarized in Table 3.9. Here, too, one can employ various computational methods for the analysis of penetration into steel. For design purposes, however, it is recommended that empirical formulas and nomograms be employed, as shown below.

For small-caliber projectiles (diameter ≤ 0.5 in.), the ballistic limit velocity (lowest velocity of a given projectile to defeat a given type and thickness of plate) can be calculated from:

Conventional and Nuclear Environments

$$V_1 = 19.72 \left\{ \frac{7800 d^3 \left[\frac{e_h}{d} \sec \theta\right]^{1.6}}{W_T} \right\}^{0.5} \quad (3.14)$$

in which V_1 is the ballistic limit velocity (feet per second), d is the caliber (inches), e_h is the thickness (inches) of homogeneous armor (BHN 360 to 440), θ is the angle of obliquity, and W_T is the projectile weight (pounds). For intermediate-caliber projectiles or bombs, one can use Figure 3.37 and Figure 3.38.

3.3.5 Penetration of Other Materials

Some of the other materials that could be used are plastic concrete (asphalt, ground limestone, and crushed granite, 10:30:60 by weight), concrete–earth (concrete covered with soil), and different types of fiber-reinforced concrete including SIFCON or FIBCON. Data on these materials can be found in various sources and should be evaluated before design. The relative penetration resistance of various materials, as compared to steel armor, is shown in Table 3.10.

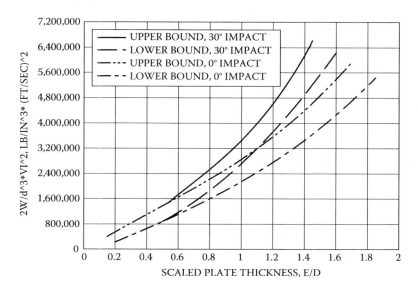

FIGURE 3.37 Perforation of homogeneous armor (BHN 250 to 300) by uncapped AP projectiles (TM 5-855-1, 1986).

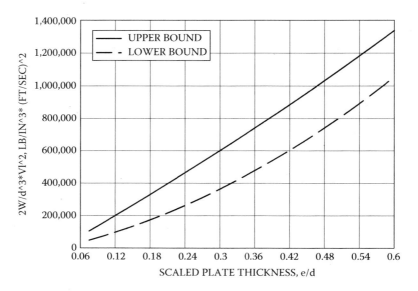

FIGURE 3.38 Perforation of homogeneous armor by bombs (TM 5-855-1, 1986).

TABLE 3.10
Armor Penetration Multiplication Factors for Various Materials

Material	Multiplication Factor
Steel armor	1.00
Mild steel	1.25
Aluminum	1.75
Lead	0.84
Concrete	2.00
Earth	6.00
Granite	1.50
Rock	1.75
Water	2.80
Green wood	3.60
Kiln-dried white oak (12% moisture)	6.70

Source: TM 5-855-1, 1986.

Conventional and Nuclear Environments

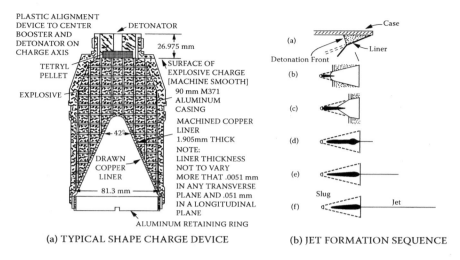

FIGURE 3.39 Shape charge device (Walters and Zukas, 1998).

3.3.6 Shaped Charges

A shaped charge is an explosive device that produces a high velocity molten metallic jet whose tip may travel at 10,000 m/s (Walters and Zukas, 1998). A typical shape charge device is made of an explosive charge cast against a metal (typically copper) conical liner as shown in Figure 3.39(a). Its action is initiated by a detonator located at the rear of the device. The resulting detonation wave travels in the explosive material toward the conical liner at a high velocity that, depending on the type of explosive used, could be about 8,000 m/s. The liner collapses under the high detonation pressure and its material is transformed into a molten metallic jet directed toward the front of the device.

Figure 3.39(b) shows the jet formation sequence. The time from charge initiation to full jet formation could be on the order of 15 to 20 μs. When the jet interacts with a target, the continuous high-pressure material steam will displace target material to penetrate it (similarly to the action of a water jet against soil). Shape charge penetration effectiveness will depend, among other parameters, on the amount of material in the jet flow into the target material, which is a function of the charge design. Because of velocity differences between the jet tip and the material that follows it, the jet tends to break up after some time, which is also a function of the distance to the liner's apex, and its penetration effectiveness diminishes. This means that shape charges must be detonated at optimal distances from their targets to achieve desired outcomes.

The traditional approach for protection is to provide thicknesses adequate to defeat the jet. Typical performances are shown in Table 3.11. Figure 3.40 can be used for estimating the penetrations of various shaped charges into steel. If other materials are considered, the factors in Table 3.10 can be used to multiply the

TABLE 3.11
Armor Penetration Data for Various Shaped Charge Munitions

Country	Weapons	Warhead Diameter (mm)	Armor Penetration (mm)
Argentina	Recoilless Gin Model 1968	105	200
Belgium	RL-83 Rocket Launcher	83	300
	Rocket Launcher MPA75	75	270
	PRB 415 Antitank Rocket	89	200
	MECAR Light Gun	90	350
China	Type 56 Antitank Grenade Launcher	80	265
	Type 69 Antitank Grenade Launcher	85	320
	Recoilless Rifle Type 36	57	63.5
	Recoilless Rifle Type 52	75	228
	Antitank Rocket Launcher Type 51	90	267
	Recoilless Gun Type 65	82	240
Czechoslovakia	Antitank Grenade Launcher Type P-27	120	250
	Recoilless Gun Type T-21	82	228
	Recoilless Guns M59 and M59A	82	250
Finland	Recoilless Antitank Grenade Launcher M-55	55	200
	Recoilless Antitank Gun M58	95	300
France	SARPAC Antitank Missile	68	500
	STRIM Antitank Rocket Launcher	89	400
	ENTAC Antitank Missile	150	650
	SS11 Antitank Missile	164	600
Federal Republic of Germany	PZF44 Portable Antitank Weapon	67	370
	B810 Cobra 2000 Antitank Missile	100	500
	Mamba Portable Antitank Weapon System	120	475
International	HOT Vehicle-Mounted Antitank Weapon System	136	800
	Milan Portable Antitank Weapon	90	352
Spain	Antitank Rocket Launcher M-65	88.7	330
Sweden	Miniman Light Antiarmor Weapon	74	300
	Carl Gustaf M2RCL Gun and Carl Gustaf System 550	84	400
	Recoilless Rifle PV-11110	90	380
USSR	RPG-2 Portable Rocket Launcher	82	180
	RPG-7 Portable Rocket Launcher	85	320
	SPG-9 Recoilless Gun	73	>390
	SPG-82 Rocket Launcher	82	230
	B-10 Recoilless Gun	82	240
	B-11 Recoilless Gun	107	380
		120	>400

TABLE 3.11 (CONTINUED)
Armor Penetration Data for Various Shaped Charge Munitions

Country	Weapons	Warhead Diameter (mm)	Armor Penetration (mm)
USA	HEAT Rocket Launchers M72, M72AI, M72A2	66	305
Yugoslavia	M60 Rifle Grenade	60	200
	M57 Antitank Launcher	90	300
	M60 Recoilless Gun	82	220
	M65 Recoilless Gun	105	330

Source: TM 5-855-1, 1986.

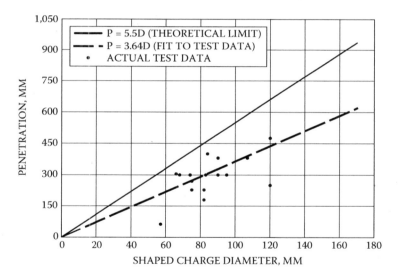

FIGURE 3.40 Penetration of steel armor versus shaped charge diameter (TM 5-855-1, 1986).

thicknesses obtained from Figure 3.40. Generally, it is accepted that the penetration depth of shaped charges could be up to about six times the warhead diameter. Other approaches to defend against shape charges might include forcing a premature detonation by placing an obstacle in front of the target (e.g., a wire screen, plywood, etc.) that will cause the jet to break down before it reaches the target or the use of reactive armor devices.

3.4 HE-INDUCED GROUND SHOCK, CRATERING, AND EJECTA

3.4.1 HE Charges and Conventional Weapons

Up to this point we have been concerned with above-ground facilities. When protected structures are buried in soil or rock, the damaging effects are related to the transfer of detonation energy to the structure or a delayed burst following the weapon's penetration into the overlay (or soil). The parameters to be considered are (1) weapon type and size, (2) properties of medium in which the structure is buried, and (3) depth of burst. Two principal types of events must be considered, as illustrated in Figure 3.41.

- Overhead burst on, or an explosive within protective layers of concrete or rock rubble, loading the roof slab
- Side or slant burst, or weapon in the soil loading the walls and floor

The phenomena associated with these conditions include cratering, ejecta, and wave propagation, as will be discussed in the following sections.

3.4.2 Three-Dimensional Stress Wave Propagation

One of the most important topics that must be understood is the propagation of shock waves through various types of geological media. For this purpose, wave propagation must be reviewed based on available previous work (Achenbach, 1973; Davis, 1988; Kolsky, 1963; Whitham, 1974). Consider a small element in a large elastic medium whose dimensions are δx, δy, and δz, along the x, y, and

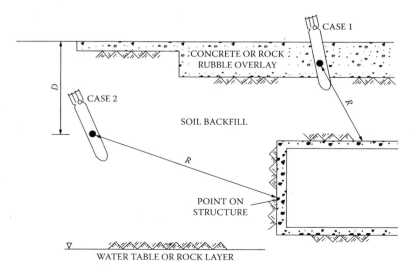

FIGURE 3.41 Geometry for explosion against a buried facility (TM 5-855-1, 1986).

Conventional and Nuclear Environments

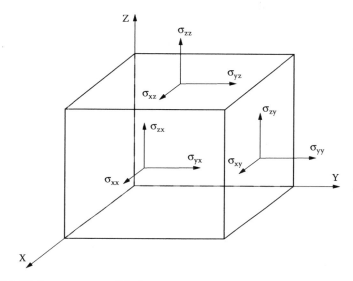

FIGURE 3.42 Stresses on a solid element.

z axes, respectively, as shown in Figure 3.42. In order for the cube to be in equilibrium, one can show that the stress components $\sigma_{ij} = \sigma_{ji}(i \neq j)$. Sometimes, the $\sigma_{ij}(i \neq j)$ terms are defined as τ_{ij}. Because of equilibrium, there are six components of stress that define the state of stress of the element: σ_{xx}, σ_{yy}, σ_{zz}, σ_{xy}, σ_{xz}, and σ_{yz}. The displacement of a point in the element will have the components u, v, and w that correspond to the directions x, y, and z. The point will be displaced from (x, y, z) to (x + u, y + v, and z + w).

We can employ the following definitions of strain and body rotations (Timoshenko and Goodier, 1970):

$$\varepsilon_{xx} = \partial u/\partial x, \; \varepsilon_{yy} = \partial v/\partial y, \; \varepsilon_{zz} = \partial w/\partial z$$

$$\varepsilon_{xy} = \partial u/\partial y + \partial v/\partial x, \; \varepsilon_{xz} = \partial u/\partial z + \partial w/\partial x, \; \varepsilon_{yz} = \partial v/\partial z + \partial w/\partial y \quad (3.15)$$

$$2\omega_x = \partial w/\partial y - \partial u/\partial z, \; 2\omega_y = \partial u/\partial z - \partial w/\partial x, \; 2\omega_z = \partial u/\partial y - \partial v/\partial x$$

where ε_{xx}, ε_{yy}, and ε_{zz} are longitudinal strains, ε_{xy}, ε_{xz}, and ε_{yz} are shear strains (sometimes defined as γ_{ij}), and ω_x, ω_y, and ω_z are rigid body rotations. One can also employ the following stress–strain relationships:

$$\sigma_{xx} = \lambda\Delta + 2\mu\,\varepsilon_{xx}$$

$$\sigma_{yy} = \lambda\Delta + 2\mu\,\varepsilon_{yy}$$

$$\sigma_{zz} = \lambda\Delta + 2\mu\,\varepsilon_{zz} \quad (3.16)$$

$$\sigma_{yz} = \mu \, \varepsilon_{yz}$$

$$\sigma_{zx} = \mu \, \varepsilon_{zx}$$

$$\sigma_{xy} = \mu \, \varepsilon_{xy}$$

where:

$$\Delta = \varepsilon_{xx} + \varepsilon_{yy} + \varepsilon_{zz} \tag{3.17}$$

and λ, and μ are Lamé's constants that have the following relationships:

$$E = \frac{\mu(3\lambda + 2\mu)}{\lambda + \mu}$$

$$\nu = \frac{\lambda}{2(\lambda + \mu)} \tag{3.18}$$

$$\kappa = \lambda + \frac{2\mu}{3}$$

where μ is the shear modulus, sometimes called G, E is Young's modulus, ν is Poisson's ratio, and κ is the bulk modulus. Considering our element and Newton's second law, we can write the dynamic equilibrium equation:

$$\Sigma F = ma \tag{3.19}$$

One can show that applying this equation in the x direction leads to the following relationship:

$$\frac{\partial \sigma_{xx}}{\partial x} + \frac{\partial \sigma_{xy}}{\partial y} + \frac{\partial \sigma_{xz}}{\partial z} = \rho \frac{\partial^2 u}{\partial t^2} \tag{3.20}$$

where ρ is the material density and u is displacement in the x direction.

One can derive two similar equations for the y and z directions, then employ the strain definitions and their relationships with displacements to obtain the following equations of motion of an elastic isotropic solid without body forces:

$$\rho\frac{\partial^2 u}{\partial t^2} = (\lambda+\mu)\frac{\partial \Delta}{\partial x} + \mu\nabla^2 u$$

$$\rho\frac{\partial^2 v}{\partial t^2} = (\lambda+\mu)\frac{\partial \Delta}{\partial y} + \mu\nabla^2 v \qquad (3.21)$$

$$\rho\frac{\partial^2 w}{\partial t^2} = (\lambda+\mu)\frac{\partial \Delta}{\partial z} + \mu\nabla^2 w$$

There are two principal types of waves (Archenbach, 1973; Kolsky, 1963; Rinehart, 1975). The first type has the following form:

$$\rho\frac{\partial^2 \Delta}{\partial t^2} = (\lambda + 2\mu)\nabla^2 \Delta \qquad (3.22)$$

This equation shows that the dilatation Δ propagates through the medium with a velocity:

$$c_1 = \sqrt{\frac{\lambda + 2\mu}{\rho}} \qquad (3.23)$$

These waves have no rotational components and are termed irrotational waves, longitudinal waves, or primary (P) waves.

The second type of wave equation has the following format:

$$\rho\frac{\partial^2 \omega_x}{\partial t^2} = \mu\nabla^2 \omega_x \qquad (3.24)$$

in which ω_x is the rotation around the x axis. One can obtain similar equations for y and z directions. Equation (3.24) shows that the rotation propagates with a velocity:

$$c_2 = \sqrt{\frac{\mu}{\rho}} \qquad (3.25)$$

These waves have no dilatational components and are termed equivoluminal waves or shear waves.

Based on these derivations, one notes that the following equation is a general form of a wave propagation equation:

$$\frac{\partial^2 \alpha}{\partial t^2} = c^2 \nabla^2 \alpha \qquad (3.26)$$

When an unbounded elastic isotropic solid is considered, only two types of waves can be propagated, as discussed earlier. However, elastic surface waves can occur when a surface is introduced into a solid (e.g., considering a semi-infinite elastic isotropic solid). These waves are similar to gravitational surface waves in liquids and their effects decrease rapidly with depth. Their propagation velocity is less than the body wave velocity. These waves were first investigated by Lord Rayleigh in 1887 (Kolsky, 1963), and are named after him. The derivation of their equations of motion is more involved than for the previous cases, and it will not be shown herein. Nevertheless, Rayleigh waves will not disperse and the shapes of such plane waves remain unchanged. Also, these waves attenuate very quickly with depth and Rayleigh waves with higher frequencies will alternate faster than similar waves with lower frequencies.

This phenomenon is analogous to skin effect in high frequency alternating current (AC) current in conductors. The path of any particle moving along the wave is an ellipse with its major axis normal to the free surface. Since these waves are found close to the free surface and do not exhibit rapid attenuation with distance from their source, it was suggested by Lord Rayleigh that their measurement can be useful for seismic purposes.

Sometimes, surface waves have no vertical component. Therefore, they cannot be Rayleigh waves that have both vertical and horizontal components. Furthermore, for Rayleigh waves, the direction of vibration of the horizontal component should be parallel to the direction of propagation, but the horizontal component of the measured seismic waves may vibrate parallel to the wave front. It was shown by Love (1967) that such waves can occur if there is an outer layer near the free surface having material properties different from those of the interior. Such waves can propagate in that layer without penetrating into the interior and are called Love waves. The order in which seismic waves arrive at a measuring station is: (1) longitudinal (P) waves, (2) shear (S) waves, and (3) surface waves with large amplitudes, as compared to longitudinal and shear waves.

Practical aspects of one-dimensional wave propagation will be addressed next.

3.4.3 One-Dimensional Elastic Wave Propagation

Several phenomena related to one-dimensional (1-D) wave propagation are illustrated in this section, as shown in Figure 3.43, given a small element of a rod through which an elastic dilatational wave (longitudinal or P wave) propagates. The state of stress on the element is longitudinal in the x direction. The element

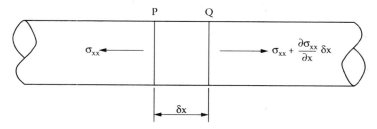

FIGURE 3.43 One-dimensional wave propagation.

length is δx. As noted in Section 3.4.2, one can employ the equation of dynamic equilibrium ΣF = ma to derive the following equation:

$$\rho A \delta x \frac{\partial^2 u}{\partial t^2} = A \frac{\partial \sigma_{xx}}{\partial x} \delta x \qquad (3.27)$$

in which ρ is the material density of the rod. Now, replace σ_{xx} by its relationship to strain that is a function of displacement to obtain:

$$\rho \frac{\partial^2 u}{\partial t^2} = E \frac{\partial^2 u}{\partial x^2} \qquad (3.28)$$

This is the equation of a plane dilatational wave propagating along the bar with a velocity:

$$c_o = \sqrt{\frac{E}{\rho}} \qquad (3.29)$$

The solution to this equation is:

$$u = f(c_o t - x) + F(c_o t + x) \qquad (3.30)$$

F and f are arbitrary functions, depending on initial conditions. Also, f is the solution for a wave traveling in the x+ direction while F represents a wave traveling in the x− direction. This is an approximate solution because it was assumed that plane cross-sections of the rod remain plane during the passage of the stress waves and that stresses act uniformly over each cross-section.

Longitudinal motions in the rod will cause lateral motions, and the lateral strains are related to longitudinal strains by Poisson's ratio ν. This will cause a nonuniform distribution of stresses over the bar cross-sections; hence, it will distort plane cross-sections so that they will not remain plane. Such lateral motions become extremely important when the wavelengths are of the same order of magnitude as the bar cross-sectional dimensions (i.e., diameter for a cylindrical bar). Therefore, in order to make our solution more accurate, we shall assume that the wavelengths are large compared to the cross-sectional dimension.

Considering a plane wave traveling in the −x direction, for which the solution is:

$$u = f(c_o t + x) \tag{3.31}$$

Its derivative is:

$$\frac{\partial u}{\partial x} = f'(c_o t + x) \tag{3.32}$$

in which:

$$f' = \frac{\partial f}{\partial (c_o t + x)} \tag{3.33}$$

Also, we know that:

$$\frac{\partial u}{\partial t} = c_o f'(c_o t + x) \tag{3.34}$$

Equation (3.32) and Equation (3.34) can be used to obtain the following:

$$\frac{\partial u}{\partial t} = c_o \frac{\partial u}{\partial x} \tag{3.35}$$

However, one can use the following relationships:

$$\sigma_{xx} = \left(\frac{E}{c_o}\right)\frac{\partial u}{\partial t} = \rho c_o \frac{\partial u}{\partial t} \tag{3.36}$$

Equation (3.36) shows a linear relationship between the stress at any point and the particle velocity, from which one obtains the following expression:

$$\frac{\sigma_{xx}}{\partial u/\partial t} = \rho c_0 \tag{3.37}$$

Conventional and Nuclear Environments

The product ρc_0 is defined as the characteristic impedance or acoustic impedance. This is the resistance to the material flow in wave propagation, similar to the resistance of a conductor to electric current. Equation (3.37) can be rewritten as follows:

$$\dot{u} = \frac{\partial u}{\partial t} = \frac{\sigma_{xx}}{\rho c} \qquad (3.38)$$

or in general:

$$\dot{u} = \frac{\sigma}{\rho c} \qquad (3.39)$$

The particle velocity is directly proportional to the acting stress and inversely proportional to the acoustic impedance.

One may plot the general 1-D wave solution in Equation (3.30) on a time versus location (t versus x) plane, as shown in Figure 3.44. This graphic representation is termed a solution by the method of characteristics, in which the lines F and f (characteristics) represent wave front locations as a function of time.

If a disturbance is introduced in the bar at $t_1 = 0$ and $x = x_1$, it will propagate along the bar in both directions with a speed c. The points x_2 and $-x_2$ will not be affected until $t = t_2$, when the wave will reach those locations.

The area between the f and F characteristics (Figure 3.44) defines the zone of influence on the behavior of a bar until the time t_2. If a disturbance occurs at $x = x_1$ and $t = t_1$, only locations between points $-x_2$ and x_2 will be affected by it at time $t = t_2$. The dashed area between the characteristics lines f and F shows

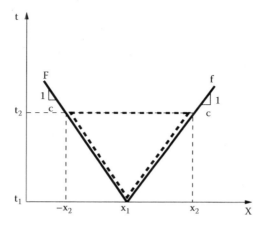

FIGURE 3.44 One-dimensional wave propagation in the t versus x plane.

the influenced zone that defines whether events will affect certain regions in the system during the time of interest, for example, determine whether a stress wave will reach the boundaries within a given time, whether it will reflect back into the domain of interest, and whether it will affect a specific location in that domain. Such issues are addressed next.

3.4.4 Reflection and Transmission of One-Dimensional D-Waves between Two Media

Consider a plane wave that travels along the bar in the +x direction and reaches a boundary between two materials, as shown in Figure 3.45. The wave is reflection and transmitted at that boundary.

Define the following transmission and reflection parameters:

$$t_\sigma \equiv \frac{\sigma_t}{\sigma_i} \qquad r_\sigma \equiv \frac{\sigma_r}{\sigma_i}$$

$$t_{\dot{u}} \equiv \frac{\dot{u}_t}{\dot{u}_i} \qquad r_{\dot{u}} \equiv \frac{\dot{u}_r}{\dot{u}_i}$$

(3.40)

One can show (Achenbach, 1973; Kolsky, 1963; Rinehart, 1975) that imposing the interface boundary conditions of $\sigma_1 = \sigma_2$ and $u_1 = u_2$ will lead to the following relationships:

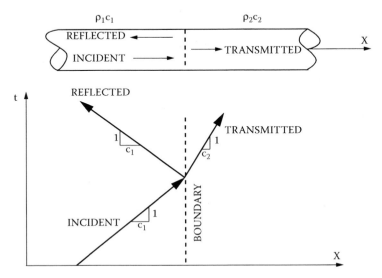

FIGURE 3.45 Reflection and transmission in one-dimensional wave propagation.

$$r_\sigma = \frac{I_2 - I_1}{\Sigma I} \tag{3.41a}$$

$$t_\sigma = \frac{2I_2}{\Sigma I} \tag{3.41b}$$

$$r_{\dot u} = \frac{I_1 - I_2}{\Sigma I} \tag{3.41c}$$

$$t_{\dot u} = \frac{2I_1}{\Sigma I} \tag{3.41d}$$

where

$$I_1 = \rho_1 c_1,\ I_2 = \rho_2 c_2 \text{ and } \Sigma I = I_1 + I_2 \tag{3.41e}$$

One can consider the following two special cases to illustrate the application of these relationships.

Case 1 — The interface is a free end (i.e., material 2 is a vacuum): $I_2 = \rho_2 c_2 = 0$.

$$r_\sigma = (0 - I_1)/(0 + I_1) = -1$$

The reflected stress is equal to the incident stress, but with an opposite sign.

$$t_\sigma = (2*0) / (0 + I_1) = 0 \text{ (no transmitted stress)}$$

$$r_{\dot u} = (I_1 - 0) / (0 + I_1) = 1$$

The reflected wave has the same velocity.

$$t_{\dot u} = (2\ I_1) / (0 + I_1) = 2$$

The transmitted velocity is twice the incident velocity.

Case 2 — The interface is a fixed end (i.e., material 2 is rigid) $I_2 = \rho_2 c_2 \to \infty$

$$r_\sigma = \frac{I_2 - I_1}{I_2 + I_1} = \frac{I_2\left(1 - \frac{I_1}{I_2}\right)}{I_2\left(1 + \frac{I_1}{I_2}\right)} = 1$$

The same stress is reflected.

$$t_\sigma = \frac{2I_2}{I_1 + I_2} = \frac{2I_2}{I_2\left(1 + \frac{I_1}{I_2}\right)} = \sim \frac{2I_2}{I_2} \sim 2$$

The velocity doubles upon transmission.

$$r_{\dot{u}} = \frac{I_1 + I_2}{I_1 + I_2} = \sim -\frac{I_2}{I_2} = \sim -1$$

An equal and opposite velocity is reflected.

$$t_{\dot{u}} = \frac{2I_1}{I_1 + I_2} = \sim \frac{2I_1}{I_2} \sim 0$$

There is no transmitted velocity.

3.4.5 Computational Aspects

In the case of a layered medium, each layer has its material properties (i.e., ρ and c) and therefore the slopes of the characteristics will change as one moves from one layer to the next, as illustrated in Figure 3.46.

It is necessary to draw lines with different slopes, representing the variation of the seismic speed in the layers, and it becomes difficult to prevent graphical mistakes. Another computational approach is to use the same slope for all lines. To enable this simplification, one material is chosen to represent the wave speeds in all layers, but the thicknesses of the other layers must be adjusted to ensure correct travel times. The adjustment is as follows:

$$y'_n = y_n \frac{c}{c_n} \quad (3.42)$$

where y_n is the thickness of layer n, y_n' is the "corrected" thickness of layer n, c_n is the actual wave speed in layer n, and c is the wave speed to be used for the

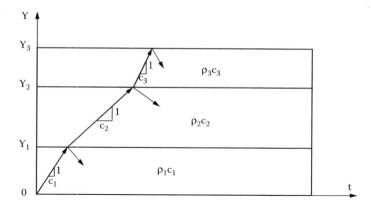

FIGURE 3.46 Characteristic path for one-dimensional wave propagation in layered medium.

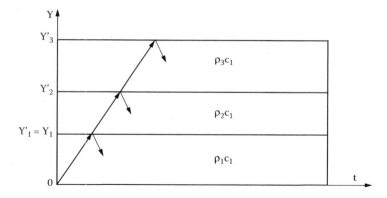

FIGURE 3.47 Characteristic path for one-dimensional wave propagation in corrected layered medium.

line slopes. Figure 3.47 shows the new configuration; the first layer was used to normalize the others.

The reflection and transmission coefficients are computed with the actual ρ and corrected c of each layer, but for the new geometric configuration.

3.4.6 Shock Waves in One-Dimensional Solids

Shock waves are directly associated with explosive phenomena in gases, liquids, and solids (Kolsky, 1963; Rinehart, 1975; Liberman and Velikovich, 1986; Ben Dor, 1992). It can be shown that the treatment of nonlinear (e.g., plastic) waves by the Eulerian method (a fixed reference in space is chosen and the motions are derived with respect to that region; the Lagrangian method is based on a moving reference region) will lead to deriving equations of motion and continuity that are similar to the equation of a wave of finite amplitude in a fluid, as shown below.

$$\frac{\partial \varepsilon}{\partial t} + (c+V)\frac{\partial \varepsilon}{\partial x} = 0 \qquad (3.43)$$

Here, $c + V$ is the propagation velocity of a disturbance in the medium. If the elastic modulus is constant, large compressive disturbances will travel faster than smaller ones. This will result in faster wave components overtaking the slower ones, and the finite compressive pulse will develop a steep front. The fundamental shock wave equations, known as the Rankine-Hugoniot equations (sometimes called only the Hugoniot equations), are derived from the equations for conservation of mass, momentum, and energy in the medium, as discussed in Section 2.3.5. Nevertheless, a more detailed treatment of this problem is presented next.

Consider a rod through which a plane shock wave travels at a constant velocity (Figure 3.48). The variables P, v, and ρ are the corresponding pressure, particle velocity, and density in the two parts of the rod. Part A represents the stressed conditions behind the shock front, while Part B is the unstressed material before the shock. The very narrow transition zone represents the region over which the conditions in the material change from unshocked to shocked. One can repeat the derivation procedure used previously to derive the equations of motion and formulate the conservation relationships for the transition zone as follows:
Conservation of mass per unit time:

$$m = \rho_A V_A = \rho_B V_B \qquad (3.44)$$

Conservation of momentum:

$$m(V_B - V_A) = P_A - P_B \qquad (3.45)$$

Conservation of rates of work and energy:

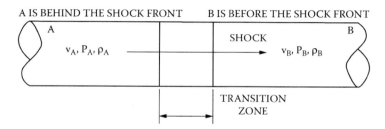

FIGURE 3.48 One-dimensional shock propagation.

Conventional and Nuclear Environments

$$P_A V_A - P_A V_B = m\left[\frac{1}{2}\left(V_B^2 - V_A^2\right) + \Delta U\right] \tag{3.46}$$

in which, ΔU is the change of internal energy per unit mass. These equations enable one to obtain the following relationship:

$$V_B = v_B \left[\frac{P_A - P_B}{v_B - v_A}\right]^{0.5} \tag{3.47}$$

in which v is $1/\rho$ and V_B is the wave speed in the undisturbed zone for a stationary transition zone. Therefore, $V_B = c$. Also, the relative velocity can be expressed as follows:

$$V = V_B - V_A = \left[(P_A - P_B)(v_B - v_A)\right]^{0.5} \tag{3.48}$$

The internal energy change is expressed as follows:

$$\Delta U = \frac{1}{2}(P_A + P_B)(v_B - v_A) \tag{3.49}$$

As noted in Chapter 2, Equation (3.46) through Equation (3.48) are the Rankine-Hugoniot equations. They are valid for chemical reactions and can be applied when explosively produced shocks are considered. It can be shown from Equation (3.46) and Equation (3.47) that the following relationship exists between the pressure difference and the particle velocity, that also can be derived for plane plastic waves:

$$\Delta P = \rho c V \tag{3.50}$$

Shock waves are formed when the stress wave behind the shock front has a higher speed than the wave before the front because the material behind the front is compressed by the shock and becomes more dense. The faster waves over take the slower ones, and thus form a shock front, as shown in Figure 3.49(a). As the shock weakens, the front will deteriorate as shown in Figure 3.49(b).

3.4.7 STRESS WAVE PROPAGATION IN SOILS

Consider several cases of high stresses propagating in soil according to the approach of Rakhmatulin and Dem'yanov (1966). It has been shown that the pressure behind the shock front is of the form:

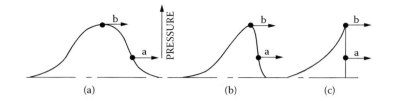

A. SHOCK FORMATION (RINEHART, 1975)

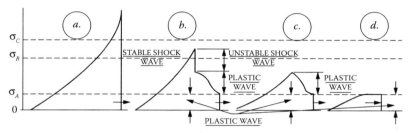

B. SHOCK DETERIORATION (HENRYCH, 1979)

FIGURE 3.49 Shock wave behavior. A. Shock formation (Rinehart, 1975). B. Shock deterioration (Henrych, 1979).

$$P^*(x^*) - P_a = \frac{b}{x^*} \int_0^{\frac{x^*}{b}} (P_o - P_a) dt \quad (3.51)$$

where P^* is the pressure behind the shock front, x^* is the location of the shock front with respect to the free boundary before application of the load ($t = 0$), P_o is the applied over-pressure (load), P_a is the atmospheric pressure (constant), and b is the shock velocity (assumed constant because of small changes in density induced by a weak shock wave).

The following two cases have been investigated using the above equation:

Case 1 — Shock induced by a nuclear detonation where the over-pressure is given by:

$$P(t) = P_{so}(1 - \tau)(ae^{-\alpha\tau} + be^{-\beta\tau} + ce^{-\gamma\tau}) \quad (3.52)$$

in which P_{so} is the peak over-pressure (taken as 200 psi) and a, b, c, α, β, and γ are constants defined in Crawford et al. (1974). Equation (3.52) can be introduced into Equation (3.51), and after integration one obtains the following expression of the pressure behind the shock front, as a function of distance x^*:

$$P^*(x^*) = 36e^{-2.5\frac{x^*}{b}} + 4.59e^{-18\frac{x^*}{b}} + 0.25e^{-80\frac{x^*}{b}} -$$
$$\frac{b}{x^*}\left[21.6e^{-2.5\frac{x^*}{b}} + 4.71e^{-18\frac{x^*}{b}} + 0.246e^{-80\frac{x^*}{b}} - 26.562\right] \quad (3.53)$$

Case 2 — The shock is induced by the following triangular over-pressure:

$$P(t) = P_{so} - \left(\frac{P_{so}}{\Delta t}\right)t \quad (3.54)$$

in which P_{so} is the peak over-pressure and Δt is the duration of pulse. Introducing Equation (3.54) into Equation (3.51) and integrating will yield:

$$P^*(x) = 200 - 666.67\frac{x^*}{b} \quad (3.55)$$

Results obtain with the above solutions compared well with measured data. Now consider the case of reflected shock waves from a rigid wall in a 1-D sand material. One assumes the following stress–strain curve:

$$\sigma = A\varepsilon^n \quad (3.56)$$

The shock front velocity will be:

$$c = \left[\frac{\sigma/\varepsilon}{\rho}\right]^{\frac{1}{2}} \quad (3.57)$$

and the particle velocity:

$$V_p = c\varepsilon \quad (3.58)$$

From which one obtains:

$$\sigma_\iota = k(A,\rho,n)\cdot V_{pi}^{\frac{2n}{n+1}} \quad (3.59)$$

Using the boundary conditions at the rigid wall, one derives:

$$\sigma_r = k(A,\rho,n) \cdot (2V_p^{(i)})^{\frac{2n}{n+1}} \tag{3.60}$$

where, $V_p^{(i)}$ is the incident particle velocity. Using the approach presented in Section 3.3.4, one defines the reflection factor as:

$$R \equiv \frac{\sigma_r}{\sigma_\iota} = 2^{\frac{2n}{n+1}} \tag{3.61}$$

from which it is seen that for $n > 1$, one obtains $R > 2$. Similarly, the following ratio of particle velocities has been found:

$$\frac{V_p^{(r)}}{V_p^{(i)}} = \left[R \cdot \frac{\varepsilon_r}{\varepsilon_i} \cdot \frac{\rho_1}{\rho_2} \right]^{\frac{1}{2}} \tag{3.62}$$

where $V_p^{(r)}$ is the reflected particle velocity, $V_p^{(i)}$ is the incident particle velocity, R is the reflection factor, ε_r is the strain in sand after reflection, ε_i is the strain in sand before reflection, and ρ_1 and ρ_2 represent the sand density in front of and behind the incident shock, respectively. Results derived with these solutions were in good agreement with experimental data obtained by Baker (1967).

Now, consider a shock wave reflection and transmission by an interface between two sand materials (Figure 3.50). In the first case, the shock travels from a softer material into a harder material. In the second, the shock travels from a

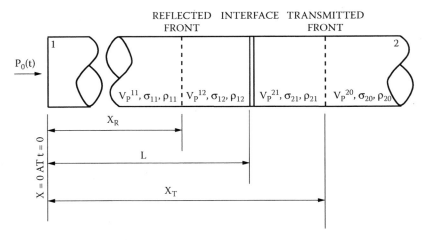

FIGURE 3.50 One-dimensional shock wave transmission and reflection model.

harder material into a softer one. In the first case, we find a transmitted shock in the harder material and a reflected shock in the softer material. In the second, we find a transmitted shock in the softer material and an "unloading stress" reflected into the harder material. These cases are studied in more detail below.

Case 1 (reflected shock) — On the reflected front use the following conservation equations:

Conservation of mass:

$$\rho_{11} \cdot \left(\dot{x}_r - V_p^{11} \right) = \rho_{12} \cdot \left(\dot{x}_r - V_p^{12} \right) \quad (3.63)$$

$$\dot{x}_r = \text{reflected shock velocity}$$

Conservation of momentum:

$$\rho_{12} \cdot \left(\dot{x}_r - V_p^{12} \right) \cdot V_p^{12} - \rho_{11} \cdot \left(\dot{x}_r - V_p^{11} \right) = \sigma_{12} - \sigma_{11} \quad (3.64)$$

Similarly, on the transmitted shock:

Conservation of mass:

$$\rho_{20} \cdot \left(\dot{x} - 0 \right) = \rho_{21} \cdot \left(\dot{x}_T - V_p^{21} \right) \quad (3.65)$$

$$\dot{x}_T = \text{transmitted shock velocity}$$

Conservation of momentum:

$$\rho_{21} \cdot \left(\dot{x}_T - V_p^{21} \right) \cdot V_p^{21} = \sigma_{21} \quad (3.66)$$

Boundary conditions on the interface:

$$V_p^{12} = V_p^{21} \quad (3.67a)$$

$$\sigma_{11} + \sigma_{12} = \sigma_{21} \quad (3.67b)$$

From Equation (3.62) one derives:

$$\left(1 - \frac{\rho_{12}}{\rho_{11}}\right) \cdot \dot{x}_r + \frac{\rho_{12}}{\rho_{11}} \cdot V_p^{12} = V_p^{11} \qquad (3.68)$$

Equation (3.65) leads to:

$$V_p^{21} = \dot{x}_T \cdot \left(1 - \frac{\rho_{20}}{\rho_{21}}\right) \qquad (3.69)$$

Now, define the following relationships:

$$\xi_1 \equiv 1 - \frac{\rho_{12}}{\rho_{11}} \qquad (3.70)$$

$$\xi_2 \equiv 1 - \frac{\rho_{20}}{\rho_{21}} \qquad (3.71)$$

Introducing Equation (3.69) into Equation (3.66) will provide:

$$\dot{x}_T^2 \cdot \rho_{20} \cdot \xi_2 = \sigma_{21} \qquad (3.72)$$

Similarly, introducing Equation (3.69) into Equation (3.68) results in the following:

$$\xi_1 \cdot \dot{x}_r + \frac{\rho_{12}}{\rho_{11}} \cdot \dot{x}_T \cdot \xi_2 = V_p^{11} \qquad (3.73)$$

Introducing Equation (3.69) into Equation (3.64) leads to the following:

$$\rho_{12} \cdot (\dot{x}_r - \dot{x}_T \cdot \xi_2) \cdot \dot{x}_T \cdot \xi_2 - \rho_{11} \cdot (\dot{x}_r - V_p^{11}) \cdot V_p^{11} = \sigma_{12} - \sigma_{11} \qquad (3.74)$$

Now, introduce Equation (3.67a) into Equation (3.71) and solve for σ_{12} which is introduced into Equation (3.73.) Similarly, solve for $\dot{x}_r$ from Equation (3.73) and introduce into Equation (3.74). These lead to the following:

$$\dot{x}_T^2 \cdot \rho_{20} \cdot \xi_2 = \sigma_{11} + \sigma_{12} \rightarrow \sigma_{12} = \dot{x}_T^2 \cdot \rho_{20} \cdot \xi_2 - \sigma_{11}$$

$$\rho_{12} \cdot (\dot{x}_r - \dot{x}_T + \xi_2) \cdot \dot{x}_T \cdot \xi_2 - \rho_{11} \cdot (\dot{x}_r - V_p^{11}) \cdot V_p^{11} = \dot{x}_T^2 \cdot \rho_{20} \cdot \xi_2 - 2\sigma_{11}$$

$$(3.75)$$

Conventional and Nuclear Environments

$$\rho_{12} \cdot \left(\frac{V_p^{11} - \frac{\rho_{12}}{\rho_{11}} \cdot \dot{x}_T \cdot \xi_2}{\xi_1} - \dot{x}_T \cdot \xi_2 \right) \cdot \dot{x}_T \cdot \xi_2 - \rho_{11} \cdot \left(\frac{V_p^{11} - \frac{\rho_{12}}{\rho_{11}} \cdot \dot{x}_T \cdot \xi_2}{\xi_1} - V_p^{11} \right) \cdot$$

$$V_p^{11} = \dot{x}_T^2 \cdot \rho_{20} \cdot \xi_2 - 2\sigma_{11}$$

From which one derives:

$$-\dot{x}_T^2 \left[\rho_{12} \cdot \xi_2^2 \cdot \left(\frac{\rho_{12}}{\rho_{11}} + \xi_1 \right) + \rho_{20} \cdot \xi_1 \cdot \xi_2 \right] + \dot{x}_T \left[2 \cdot \rho_{12} \cdot V_p^{11} \cdot \xi_2 \right] +$$

$$2 \cdot \xi_1 \sigma_{11} - \rho_{11} \cdot V_p^{11} = 0$$

(3.76)

Equation (3.74) is a quadratic equation in the form $Ay^2 + By + C = 0$ in which y is $\dot{x}_T$, from which one obtains $\dot{x}_T$. After $\dot{x}_T$ is known, one derives σ_{21} from Equation (3.72) and σ_{12} from Equation (3.67). V_p^{21} is found from Equation (3.67a) and Equation (3.67b) and $\dot{x}_r$ is obtained from Equation (3.73) along with values for σ_{12}, ρ_{12}, and ε_{12}. Similar derivations were achieved by Weidlinger and Matthews (1965).

3.4.7.1 Numerical Evaluation

This approach was evaluated using data obtained from Triandafilidis et al. (1968) as follows.

Case 1 (reflected shock) —

$$\rho_{10} = 2.83 \; \frac{lb \; sec^2}{ft^4} \quad \text{from} \quad \gamma_{10} \cong 91 \; pcf$$

$$\rho_{20} = 3.48 \; \frac{lb \; sec^2}{ft^4} \quad \text{from} \quad \gamma_{20} \cong 112 \; pcf$$

For 200 psi on sand (Material 1 in Figure 3.50), we find $\varepsilon \simeq 0.023$ from which:

$$\rho_{11} = \rho_{10} \left(\frac{1}{1-\varepsilon} \right) \cong 2.9 \; \frac{lb \; sec^2}{ft^4}$$

Assuming $\rho_{12} \approx \rho_{10}$ and sand (Material 2 in Figure 3.50) will give only $\varepsilon \simeq 0.01$. Therefore, assume:

$$\rho_{21} = 1.01\rho_{20} = 3.51 \frac{lb \; sec^2}{ft^4}$$

$$\xi_1 \cong -0.023 \quad \xi_2 \cong 0.01$$

$$c = u_{shock}^0 = \left[\frac{(P/\varepsilon)_{max}}{\rho}\right]^{\frac{1}{2}} = \left[\frac{\frac{200 \cdot 144}{0.023}}{2.83}\right]^{\frac{1}{2}} = 665 \; fps$$

$$V_p^{11} = c \cdot \varepsilon = 665 \cdot 0.023 = 15.3 \; fps$$

These are the shock wave and particle velocities for the incident front. We assume that the interface is close to the free boundary (i.e., $\sigma_i = 200$ psi). Therefore, from Equation (3.75), A = 0.0004, B = 0.90576, and C = −1830.861. Solving the quadratic equation will provide:

$$\dot{x}_T = \frac{-0.90576 \pm \sqrt{0.90576^2 - 4 \cdot 0.0004 \cdot (-1830.861)}}{0.0008}$$

$\dot{x}_T = 1288.35$ ft/sec., or −3552.75 ft/sec. One selects the positive solution to find:

$$\sigma_{21} = \dot{x}_T^2 \cdot \rho_{20} \cdot \xi_2 = \frac{1288.35^2 \cdot 3.48 \cdot 0.01}{144} = 401 \; psi$$

Then, $\sigma_{12} = 201$ psi and

$$V_p^{21} = \dot{x}_T \cdot \xi_2 \cong 1288.35 \cdot 0.01 = 12.88 \; fps = V_p^{12}$$

$$\dot{x}_r = \left(\frac{\sigma_{12}/\varepsilon}{\rho_{11}}\right)^{\frac{1}{2}} = \left(\frac{\frac{200 \cdot 144}{0.02}}{2.9}\right)^{\frac{1}{2}} = 706.4 \; fps$$

Case 2 (no reflected shock) — Use the conservation equations on the transmitted front:

Conservation of mass:

Conventional and Nuclear Environments

$$\rho_{20} \cdot (\dot{x}_T - 0) = \rho_{21} \cdot (\dot{x}_T - V_p^{21}) \tag{3.77}$$

Conservation of momentum:

$$\rho_{21} \cdot (\dot{x}_T - V_p^{21}) \cdot V_p^{21} = \sigma_{21} \tag{3.78}$$

On the interface:

$$V_p^{11} = V_p^{21} \tag{3.79}$$

From Equation (3.76), one obtains the following:

$$\dot{x}_T \cdot \left(1 - \frac{\rho_{20}}{\rho_{21}}\right) = V_p^{21} \tag{3.80}$$

Introducing Equation (3.78) into Equation (3.79) and $\xi_2 \cong \left(1 - \dfrac{\rho_{20}}{\rho_{21}}\right)$ leads to the following:

$$\dot{x}_T \cdot \xi_2 = V_p^{11} \quad \dot{x}_T = \frac{V_p^{11}}{\xi_2} \; [\tag{3.81}$$

When $\dot{x}_T$ is known, one finds σ_{21} and can compute:

$$\sigma_{12} = \sigma_{11} - \sigma_{21} \tag{3.82}$$

Now, one can evaluate this case by using the same test data, as for Case 1:

For 200 psi:

$$c = u_{shock}^0 = \left[\frac{(P/\varepsilon)_{max}}{\rho_{10}}\right]^{\frac{1}{2}} = \left[\frac{\frac{200 \cdot 144}{0.01}}{3.48}\right]^{\frac{1}{2}} = 910 \, fps$$

$$\left(V_p^{11}\right)_{max} = c \cdot \varepsilon_{max} = 9.1 \, fps$$

$$\xi_2 \cong 0.02$$

$$\dot{x}_T \cong \frac{9.1}{0.02} = 455 \, fps$$

$$\sigma_{21} = 109 \, psi$$

$$\sigma_{12} = 200 - 109 = 91 \, psi \, (unloading)$$

One can compare these results with the those obtained by using the approach proposed by Rakhmatulin and Dem'yanov (1966) as follows.

Assumed constant shock velocity — Rakhmatulin and Dem'yanov (1966) showed that the shock front velocity is almost constant because of small changes in density. Furthermore, it is assumed that the changes in density in the pressure range of 140 ÷ 210 psi will be 4 ÷ 6%. The previous example indicated density changes that are even smaller. Since the example is in the above-noted stress range, one can assume a constant shock velocity.

Pressure attenuation — The pressure behind the shock front will attenuate with distance from the free boundary as shown clearly by Triandafilidis (1968). However, to simplify the procedure it was assumed that the interface is close to the free boundary and the amount of attenuation is small (e.g., at a distance of 5 ft we have over 90% of the over-pressure). It is seen that the results are in good agreement with the theory.

HE-induced shock waves in soils — As the degree of soil saturation increases, especially if it is 95% or higher, the soil exhibits a stiffer behavior. This will cause increases in the peak stresses and accelerations. A sharp jump in the soil seismic velocity to over 5000 fps is an indication of a saturated layer. These observations are true mainly for cohesive soils. In granular soils, the stiffness is provided by granular contact and one can observe similar behavior only when such soils are at low densities as the granular skeleton collapses. The seismic velocity will be computed from:

$$c = \sqrt{\frac{M}{\rho_o}} \tag{3.83}$$

in which M is the stiffness (modulus) of the soil and ρ_o is its mass density. Low seismic velocities indicate poor quality of ground shock transmission. However, cemented desert dry alluvium may exhibit high seismic velocities (close to 4000 fps), but because of high porosity they transmit shock poorly. Soils with high relative densities (low volume air voids) attenuate ground shock more slowly than do low density soils. Some results for contained explosions are presented in Figure 3.51. The time of arrival t_a is the time for a ground shock to reach a given location and is computed as follows:

$$t_a = R / c \tag{3.84}$$

Conventional and Nuclear Environments 129

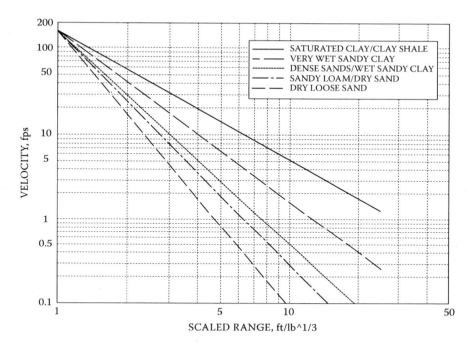

FIGURE 3.51 Peak stress and particle velocity from contained explosions in various soils (TM 5-855-1, 1986).

where R is the range and c is the average seismic velocity over that range. The approaches presented above may not be directly applicable for design-related analysis. Therefore, one should treat cases similarly to the approaches in available design manuals (e.g., TM 5-855-1, 1986), as presented next.

3.4.8 Application to Protective Design

The rise time t_r from the ground shock arrival to the peak value is approximated (TM 5-855-1, 1986) as follows:

$$t_r = 0.1\, t_a \quad (3.85)$$

Beyond the peak, the pulse decays to the ambient pressure or velocity (assumed to be zero) are estimated according the following expressions:

$$P(t) = P_o e^{\alpha t/t_a} \quad t \geq 0 \quad (3.86a)$$

$$V(t) = V_o(1 - \beta t/t_a)e^{-\beta t/t_a} \quad t \geq 0 \quad (3.86b)$$

in which P(t) is the pressure, V(t) is the particle velocity, P_o and V_o are the corresponding peak pressure and velocity as shown in Figure 3.51, and α and β are the following time constants:

$$\alpha = 1.0 \quad \beta = \frac{1}{2.5} \tag{3.87}$$

In stiff soils, the pulses are shorter, are at higher frequencies and accelerations, and have lower displacements than in softer (looser) soils. The relationship between peak particle velocity and peak pressure (stress) is based on the previous derivation for wave propagation as follows:

$$P_o = \rho c V_o \tag{3.88}$$

in which P_o is the peak soil stress, ρ is the soil mass density, c is the seismic velocity derived based on the 1D wave propagation model $c = (M/\rho)^{0.5}$, M is the soil modulus, and V_0 is the peak particle velocity.

The following relationships can be employed for bombs detonating on or in burster slabs, or in the soil near a target (Department of the Army, 1986):

$$P_0 = f \cdot (\rho c) \cdot 160 \, \lambda^{-n} \tag{3.89}$$

$$V_o = f \cdot 160 \cdot \lambda^{-n} \tag{3.90}$$

$$W^{1/3} \cdot a_o = f \cdot 50 \cdot c \cdot \lambda^{(-n-1)} \tag{3.91}$$

$$W^{1/3} \cdot d_o = f \cdot \frac{500}{c} \cdot \lambda^{(-n+1)} \tag{3.92}$$

$$W^{-1/3} \cdot I_o = f \cdot \rho_o \cdot 1.1 \cdot \lambda^{(-n+1)} \tag{3.93}$$

where λ equals $R/W^{1/3}$, f is the coupling factor, ρc is the acoustic impedance (pounds per square inch/feet per second), R is the range (feet), W is the charge weight (pounds), n is the attenuation coefficient, V_o is the peak particle velocity (feet per second), a_o is the peak (gravitational or g) acceleration, c is the seismic velocity (feet per second), d_o equals peak displacements, I_o is the impulse (pounds per second per square inch), and ρ_o is mass density (pounds per second per foot[4]; compute from ρc values in Table 3.12 or γ/g). Some typical values for various materials are presented in Table 3.12. Users should obtain actual values for a site under consideration. A more detailed list of parameters has been developed based on explosive test programs; see Table 3.13.

TABLE 3.12
Soil Properties for Calculating Ground Shock Parameters

Material Description	Seismic Velocity c (fps)	Acoustic Impedance c (psi/fps)	Attenuation Coefficient (n)
Loose, dry sands and gravels with low relative density	600	12	3–3.25
Sandy loam, loess, dry sands, and backfill	1,000	22	2.75
Dense sand with high relative density	1,600	44	2.5
Wet sandy clay with air voids (less than 1%)	1,800	48	2.5
Saturated sandy clays and sands with small number of air voids (less than 1%)	5,000	130	2.25–2.5
Heavy saturated clays and clay shales	>5,000	150–180	1.5

Source: TM 5-855-1, 1986.

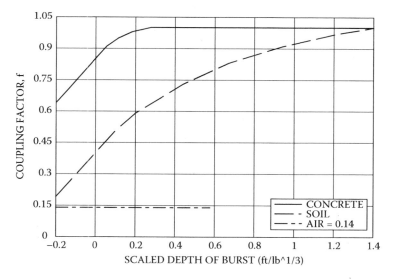

FIGURE 3.52 Ground shock coupling factor as function of scaled depth of burst for air, soil, and concrete (TM 5-855-1, 1986).

In all cases, an appropriate coupling factor f must be chosen. This factor represents the degree of energy transfer into the medium, as compared to the case of a fully contained explosion. For explosions in air, it is assumed that f = 0.14 and is assumed to be constant for detonations near the ground surface. Figure 3.52 should be used to select the coupling factor for explosives in soil and concrete.

TABLE 3.13
Soil Properties from Explosion Tests

Soil Description	Dry Unit Weight (pcf)	Total Unit Weight (pcf)	Air-Filled Voids (%)	Seismic Velocity c (fps)	Acoustic Impedance c (psi/fps)	Attenuation Coefficient (n)
Dry desert alluvium and playa, partially cemented	87	93–100	>25	2,100–4,200[a]	40	3–3.25
Loose, dry poorly graded sand	80	90	>30	600	11.6	3–3.5
Loose, wet, poorly graded sand with freestanding water	97	116	10	500–600	12.5–15	3
Dense dry sand, poorly graded	99	104	32	900–1,300	25	2.5–2.75
Dense wet sand, poorly, graded with freestanding water	108	124	9	1,000	22	2.75
Very dense dry sand, relative density 100%	105	109	30	1,600	44	2.5
Silty clay, wet	95–100	120–125	9	700–900	18–25	2.75–3
Moist loess, clayey sand	100	122	5–10	1,000	28	2.75–3
Wet sandy clay, above water table	95	120–125	4	1,800	48	2.5
Saturated sand, below water table in marsh	–	–	1–4[b]	4,900	125	2.25–2.5
Saturated sandy clay, below water table	78–100	110–124	1–2	5,000–6,000	130	2–2.5
Saturated sandy clay, below water table	100	125	<1	5,000–6,000	130–180	1.5
Saturated stiff clay, saturated clay-shale	–	120–130	0	>5,000	135	1.5

[a] High because of cementation.
[b] Estimated.

Source: TM 5-855-1, 1986.

Conventional and Nuclear Environments

For layered systems (combinations of different soils, or concrete–soil, etc.), one may compute a coupling factor for explosions when a bomb detonates in more than one layer with the following expression:

$$f = \sum f_i \left(\frac{W_i}{W}\right) \tag{3.94}$$

in which f_i equals f for a given layer I, W_i is part of the charge in layer I, and W is the total charge. The following expression can be used for cylindrical bombs (assuming a uniform charge density distribution):

$$f = \sum f_i \left(\frac{L_i}{L}\right) \tag{3.95}$$

where L_i is the length of weapon in layer i and L is the total weapon length.

In a layered system, the effects of reflections from various interfaces and how they combine to modulate the actual load/pressure on a structure must be considered. Consider the schematic configuration in Figure 3.53. Now, define several path lengths as follows:

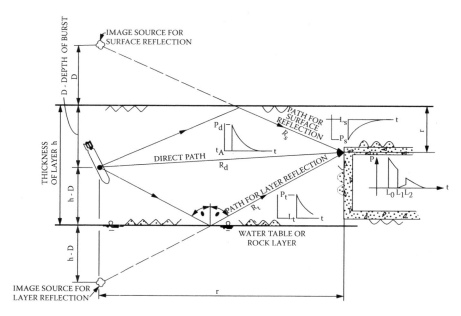

FIGURE 3.53 Ray path for reflections from surface and lower layers (TM 5-855-1, 1986).

Direct path from explosion to structure:

$$R_d = \sqrt{(D-z)^2 + r^2} \tag{3.96a}$$

Total path length of reflection from free surface:

$$R_s = \sqrt{(D+z)^2 + r^2} \tag{3.96b}$$

Total path length of reflection from deeper layer:

$$R_l = \sqrt{(2h-D-z)^2 + r^2} \tag{3.96c}$$

where h is the layer thickness, D is the depth of explosion center below the free surface, z is the depth of target point below the free surface, and r is the horizontal explosion center to target point distance.

The total pressure–time histories at the target point are calculated, as discussed in the section on wave propagation, to obtain the following (TM 5-855-1, 1986):

$$P_d = P_o(R_d)e^{-\alpha t/t_d} \qquad t \geq t_d \tag{3.97a}$$

$$P_s = -P_o(R_s)e^{-\alpha t/t_s} \qquad t \geq t_s \tag{3.97b}$$

$$P_l = KP_o(R_l)e^{-\alpha t/t_L} \qquad t \geq t_l \tag{3.97c}$$

Also, the peak stress at each distance R_i is calculated with the following equation:

$$P_o(R_i) = \rho c \cdot 160 \cdot \left(\frac{R_i}{W^{\frac{1}{3}}}\right)^{-n} \tag{3.98}$$

where P_d is the directly transmitted stress, P_s is the surface reflected stress, unloading, P is the lower layer reflected stress, K is the reflection coefficient from the lower layer, and n is the attenuation coefficient. The times of arrival for each pulse are given by:

$$t_d = \frac{R_d}{c_1} \tag{3.99a}$$

$$t_s = \frac{R_s}{c_1} \qquad (3.99b)$$

$$t_l = \frac{R_l}{c_1} \qquad (3.99c)$$

and

$$K = (\cos\theta - K_0)/(\cos\theta + K_0) \text{ for } [(c_2/c_1)\sin\theta]^2 > 1$$
$$K = 1 \qquad \text{otherwise} \qquad (3.100)$$

$$K_0 = (\rho_1 c_1/\rho_2 c_2)\{1 - [(c_2/c_1)\sin\theta]^2\}^{0.5} \qquad (3.101)$$

in which ρ_1 is the mass density of layer 1 (upper layer), c_1 is the seismic velocity in layer 1, ρ_2 is the mass density of layer 2 (lower layer), and c_2 is the seismic velocity in layer 2. The trigonometric functions are defined as follows:

$$\sin\theta = \frac{r}{R_d}$$
$$\cos\theta = \frac{2h - D - z}{R_d} \qquad (3.102)$$

Once the calculations are done for the three propagating stresses, their values are combined (with their corresponding times of arrival considered) to obtain the total stress history of the shock load at the target point as follows:

$$P(t) = P_d + P_s + P_l \qquad (3.103)$$

This computation must be modified for events in which more than two layers are considered; more complicated cases may require a wave propagation computer code.

3.4.9 CRATERING

When an explosion occurs near a ground surface, it excavates a large hole (crater). The geologic material thrown out of the crater is termed ejecta, and part of it will fall back into the crater. The visible crater is an apparent crater because part of the true crater has been filled with fallback (ejecta that fell back into it). One can describe the main parameters of craters, as shown in Figure 3.54. Note the following four regions:

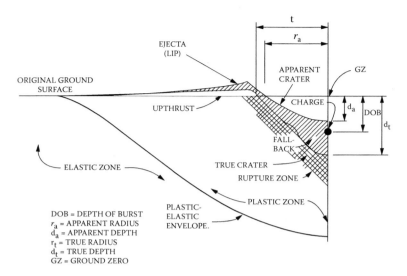

FIGURE 3.54 Half-crater profile taken about a vertical centerline through ground zero and showing crater nomenclature and notation (TM 5-855-1, 1986).

- The crater and the debris that fell back into and around it
- The rapture zone in which the material is severely damaged
- The plastic zone in which the material underwent plastic deformation
- The elastic zone in which the material returned to its original state

The parameters that control the shape of a crater and are employed for predicting it are (1) the type and amount of explosive, (2) the type of medium in which a crater is formed, and (3) the depth of burst as illustrated in Figure 3.55. The depth of the crater will increase as the depth of burst (DOB) increases up to an optimum depth, beyond which it will decrease. When a burst is fully contained, a crater will be formed underground, but it will not be seen on the surface. This case is termed a camouflet. The crater dimensions can be estimated from Figure 3.56 for uncased charges and different soil types. These values should be corrected for actual cased weapons and nonhorizontal surfaces.

Similar curves can be used for craters in concrete (Figure 3.57). When contact detonations occur on finite thickness concrete elements, shallow craters will result. It is assumed that breaching occurs when the penetration depth exceeds the material thickness. However, when the front-face crater depth (true depth) is about two-thirds the thickness of the element, the outcome could be breaching due to the combined effect of cratering and rear-face scabbing (Figure 3.58). Although reinforcing bars tend to improve the structural behavior for buried structural concrete systems (but not for contact explosions), they have virtually no effect on crater dimensions.

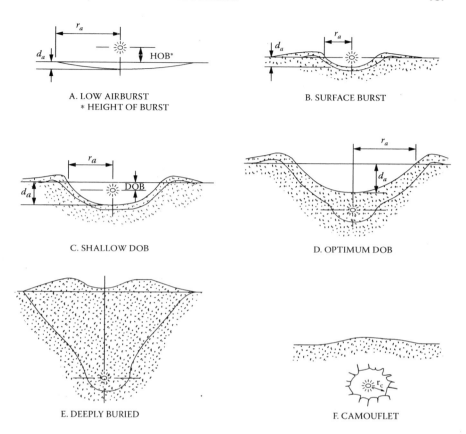

FIGURE 3.55 Variation in crater size and shape with depth of burst. Upper profile in each figure indicates apparent crater; lower profile indicates true crater (TM 5-855-1, 1986).

3.4.10 Ejecta

Large explosive charges detonated underground cause massive amounts of material to be ejected from a crater and deposited within a certain radius from the explosion center (ground zero or GZ). The amount of ejected material and the maximum size to be expected can be estimated from the charts in Figure 3.59 through Figure 3.61. For geological materials, 40 to 90% (by weight) of the ejecta will be deposited within a distance two to four times the apparent radius r_a from GZ. For soils, the ejecta is confined totally to within 30 r_a, but in rock the distance may increase up to 75 r_a.

Ejecta particles, especially for rock, may perforate structural elements (e.g., steel plates) that they impact as shown in Figure 3.62. One may use such estimates for selecting the thickness of steel plates to be placed over a structure for its protection from such debris. Hard rock is assumed to be seven times more capable

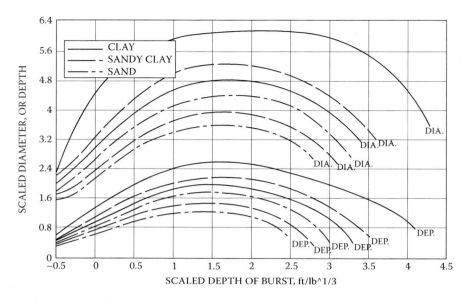

FIGURE 3.56 Apparent crater dimensions from cased and uncased high explosives in various soils (TM 5-855-1, 1986).

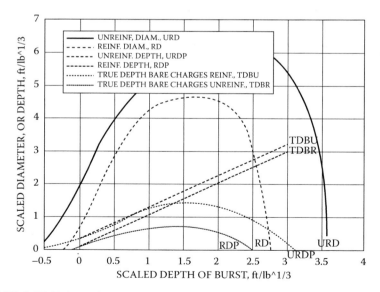

FIGURE 3.57 Estimated crater dimensions in massive concrete (TM 5-855-1, 1986).

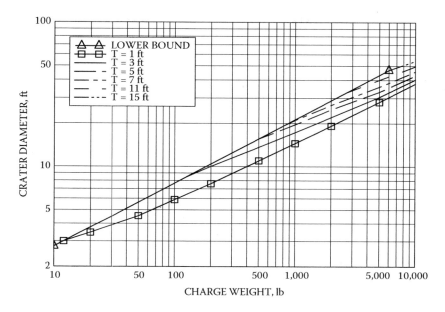

FIGURE 3.58 Charge weight for breach of bridge pier or similar reinforced concrete structure. Dashed line indicates minimum breach; solid line, the desired crater diameter (TM 5-855-1, 1986).

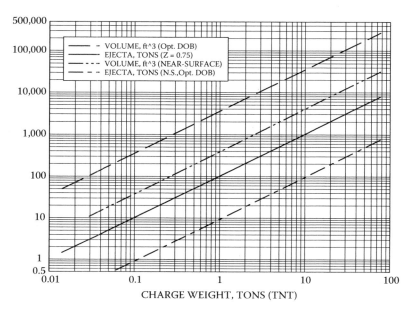

FIGURE 3.59 Crater ejecta weight and volume relations for hard rock (TM 5-855-1, 1986).

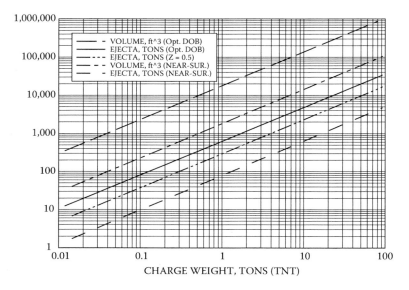

FIGURE 3.60 Crater ejecta weight and volume relations for soft rock and cohesive soils (TM 5-855-1, 1986).

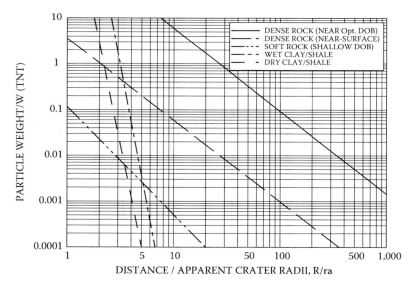

FIGURE 3.61 Maximum expected ejecta particle size versus range (TM 5-855-1, 1986).

Conventional and Nuclear Environments

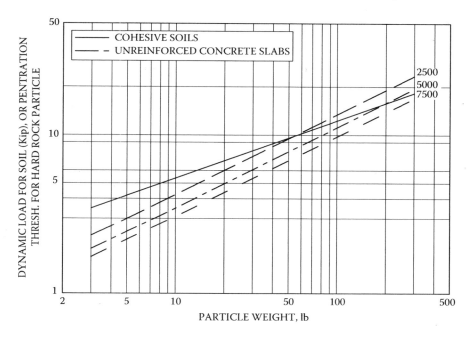

FIGURE 3.62 Ejecta impact parameters (TM 5-855-1, 1986).

of perforating a structure than soft rock. Similar figures can be used to estimate perforation of unreinforced concrete. Some simplified equations for estimating dynamic loads of soft particles are illustrated in Figure 3.63.

3.5 CRATERING, EJECTA, AND GROUND SHOCK FROM NUCLEAR DEVICES

The order of discussion in this section is different from the section on HE explosions. Here we shall start with the effects of cratering, then move to ejecta, and finally ground shock. The reason for this difference is because the direct-induced ground shock of a nuclear event must be combined with other sources of ground shock (such as airblast). These effects are coupled also for HE explosions, but because of the relatively smaller amounts of energy released by conventional weapons, some of the related effects are secondary. Additional information on this topic is found in ASCE (1985).

3.5.1 CRATERING

The definition of crater parameters is identical to definition in the HE case as illustrated in Figure 3.64. Here too a weapon's yield, height of burst (HOB), depth of burst (DOB), material properties, and geological structure determine crater geometry.

142   Modern Protective Structures

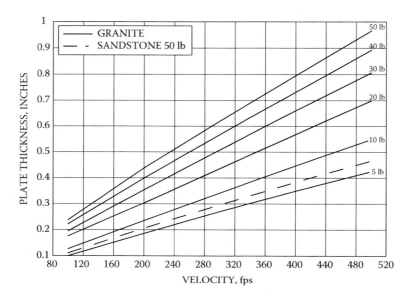

FIGURE 3.63 Perforation of mild steel plate by rock particles (TM 5-855-1, 1986).

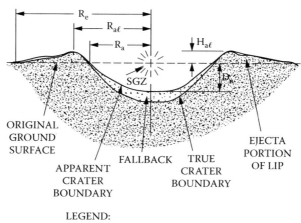

LEGEND:
D_a = APPARENT CRATER DEPTH
$H_{a\ell}$ = HEIGHT OF APPARENT LIP
R_a = APPARENT CRATER RADIUS
$R_{a\ell}$ = CRATER RADIUS TO APPARENT LIP
R_e = EJECTA RADIUS
SGZ = SURFACE GROUND ZERO

FIGURE 3.64 Crater dimensions of interest (ASCE 1985).

Conventional and Nuclear Environments

Predictions of crater dimensions are empirical and are based on very few experiments at the Nevada Test Site (two near-surface bursts) and several high yield surface bursts at the Pacific Proving Ground. The data have been studied and extrapolated to various other conditions. The data have all been scaled to a standard yield of 1 kt and the most significant burst for crater formation is the low air burst when HOB $\leq$10 ft/kt$^{1/3}$. A contact burst has been defined for an HOB of 1.6 ft (0.5 m) for all yields. A shallow-buried burst is defined for DOB $\leq$ 6 ft/kt$^{1/3}$ (5 m/kt$^{1/3}$).

The best estimate for the apparent crater volume as a function of HOB is presented in Figure 3.65 for near-surface explosions of yields in the range of W $\geq$10 kt in. Since most available information is for W = 1 kt, scaling should be employed for extrapolation purposes as follows:

$$\frac{V_a}{V_{a_1}} = W^{0.882} \tag{3.104}$$

$$\frac{HOB}{HOB_1} = W^{0.294} \tag{3.105}$$

in which V_a is the apparent crater volume for 1 kt, V_{a_1} is the apparent crater volume for W kt, HOB_1 is the height of burst for 1 kt, and HOB is the height of burst for W kt. The shapes and other dimensions as functions of various parameters are given in Table 3.14 and Table 3.15.

High yield explosions on saturated media create dish-shaped craters for HOB cases and bowl-shaped craters for deep-buried bursts. However, there are no data for the intermediate region defined by the following range: $0 < DOB/W^{0.294} \leq 16$ ft/kt$^{0.294}$. The following empirical relationships may be employed for crater assessment:

$$\frac{R_a}{V_a^{\frac{1}{3}}} = 1.2W^b \tag{3.106}$$

$$\frac{D_a}{V_a^{\frac{1}{3}}} = 0.5W^{-c} \tag{3.107}$$

R_a and D_a were defined in Figure 3.64, V_a is the volume for a yield W, and the constants b and c are obtained from Figure 3.65 for large yields. Some other relationships that might be used are shown below:

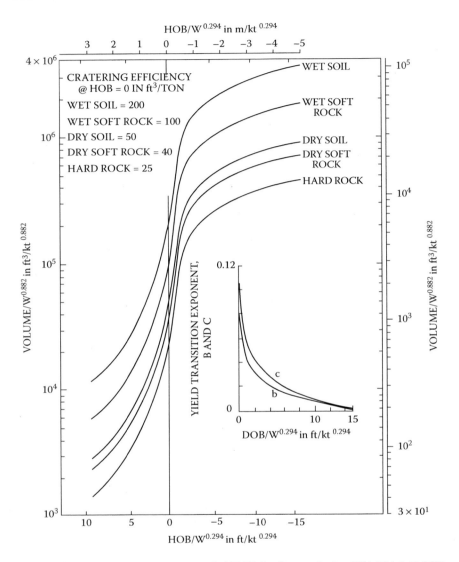

FIGURE 3.65 Crater volume versus scaled HOB for five geologies (W ≥10 kt) (ASCE, 1985).

$$R_{a\ell} = 1.25 R_a \tag{3.108}$$

$$0.25 D_a \leq H_{a\ell} \leq 0.33 D_a \tag{3.109}$$

where R_a is the radius to apparent crater lip and H_a is the height of apparent crater lip.

TABLE 3.14
Crater Shapes Resulting from Specified Heights of Bursts and Geologies

Scaled Height of Burst (HOB)	Geology	Crater Shape
$HOB/kt^{1/3} > 0$	Unsaturated	Bowl
	Saturated	Dish
-16 ft $\leq HOB/kt^{1/3} < 0$	Unsaturated	Bowl
(-5 m)	Saturated	Dish/bowl

Source: ASCE, 1985.

TABLE 3.15
Crater Dimensions Resulting from Near-Surface Bursts

Dimension of Apparent Crater	Crater Shape	
	Bowl	Dish
Radius	$1.1 V_a^{\frac{1}{3}} \quad R_a \quad 1.4 V_a^{\frac{1}{3}}$	$1.1 W^{0.08} \quad \dfrac{R_a}{V_a^{\frac{1}{3}}} \quad 1.4 W^{0.08}$
Best estimate	$R_a = 1.2 V_a^{\frac{1}{3}}$	$\dfrac{R_a}{V_a^{\frac{1}{3}}} = 1.2 W^{0.08}$
Depth	$0.35 V_a^{\frac{1}{3}} \quad D_a \quad 0.7 V_a^{\frac{1}{3}}$	$0.35 W^{-0.12} \quad \dfrac{D_a}{V_a^{\frac{1}{3}}} \quad 0.7 W^{-0.12}$
Best estimate	$D_a = 0.5 V_a^{\frac{1}{3}}$	$\dfrac{D_a}{V_a^{\frac{1}{3}}} = 0.5 W^{-0.12}$

Source: ASCE, 1985.

3.5.2 Ejecta

The same definitions given for HE explosions are applicable here, but some additional empirical relationships were compiled from various tests. Only a summary of the main parameters and their estimations are given here, based on ASCE (1985). The depth of the ejecta layer as a function of range from GZ is provided in Figure 3.66.

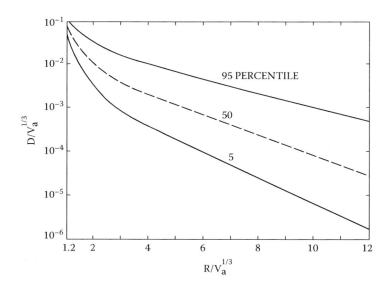

FIGURE 3.66 Ejecta depth as a function of range for nuclear bursts (ASCE, 1985).

The areal weight density can be estimated from:

$$\delta = \gamma D \qquad (3.110)$$

where δ is areal density, γ is the bulk unit weight of ejecta (100 lb/cu ft, 1600 kg/cu m), and D is the ejector depth from Figure 3.66. Other relationships can be used in place of or with Figure 3.66, as shown in Figure 3.67 through Figure 3.69.

One may employ other relationships for estimating ejecta size distributions, number of impacts per unit area and impact probability, and ejecta impact velocity and angles. Such relationships are found in ASCE (1985) and Crawford et al. (1974). Geological factors (such as water tables and bedrock) influence crater formation, ejecta, and ground shock. For surface explosions, the influence of a water table is not significant if it is deeper than 11 ft/kt$^{0.294}$; for shallower water tables, the apparent radius, volume, and lip height of the crater increase exponentially as it approaches the free surface.

For shallow-buried depths of burst (DOB $\leq$ 15 ft/kt$^{0.294}$), the final crater radius may be about 50% greater and the crater depth will be about 30% shallower. One can use Figure 3.70 for estimating crater volume V_a as a function of a shallow water table. This figure requires some iteration but the solution converges rapidly. There are similar techniques for three-layer systems and corresponding references on this subject should be employed. Bedrock has a similar influence on cratering as a water table. There will be a small increase in crater radius (5 to 10%) and the crater will be about 30% shallower.

Conventional and Nuclear Environments

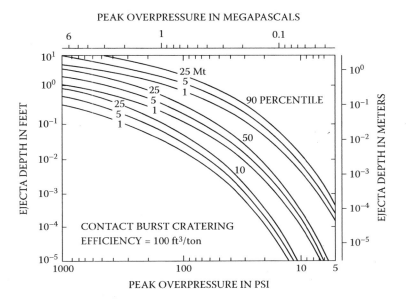

FIGURE 3.67 Ejecta depth as function of over-pressure (ASCE, 1985).

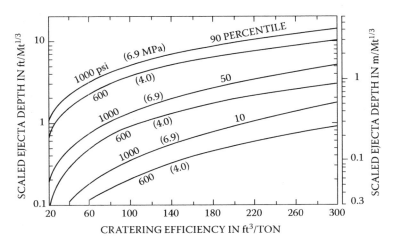

FIGURE 3.68 Scaled ejecta depths at 1,000 psi (6.9 MPa) and 600 psi (4 MPa) levels as functions of cratering efficiency (ASCE, 1985).

3.5.3 Ground Shock

The terminology used in this discussion is illustrated in Figure 3.71. The condition shown in the figure is termed *superseismic* if the airblast shock velocity is greater than the velocity of shock waves in the upper layer of the soil medium. The initial

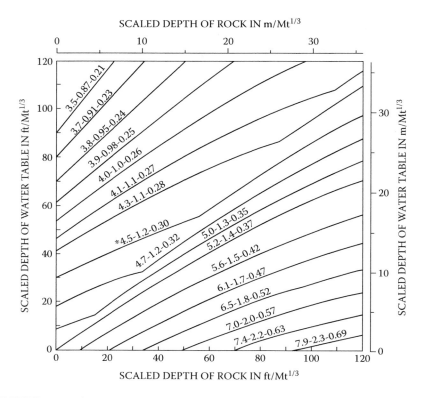

FIGURE 3.69 Ejecta thickness at 1,000 psi (6.9 MPa) level from high-yield burst in layered geology (ASCE, 1985).

component of ground shock is induced by airblast and is termed *airslap-induced*. As the airblast slows, a region is reached where the airblast velocity and the ground shock velocity are equal; this is the *trans-seismic* region. Further slowing of the airblast creates a situation of outrunning ground shock, in which a point is loaded by the ground shock before the airblast arrives. If the lower layer is stiffer than the upper layer, shock waves transmitted into the lower layer outrun those in the upper layer and load the upper layer ahead of the shock waves (*head waves*) traveling in the upper layer. The following waves can be considered:

- Airslap-induced waves and their reflection from the interface
- Direct-induced waves originating from direct coupling of energy into the ground (from cratering) and from upstream airblast
- Head waves from either direct-induced or upstream airblast-induced

The airslap-induced shock environment is illustrated in Figure 3.72. The compression wave front is inclined at an angle θ_p, while the shear wave at an angle θ_s, as defined in the figure, where U is the airblast shock front velocity, C_p is the

Conventional and Nuclear Environments

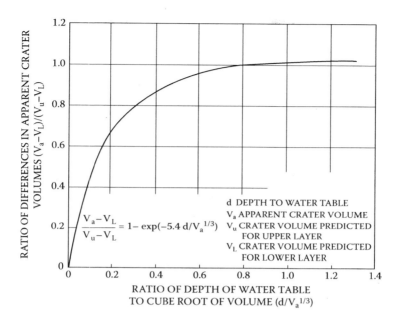

FIGURE 3.70 Effect of shallow water table on crater volume (ASCE, 1985).

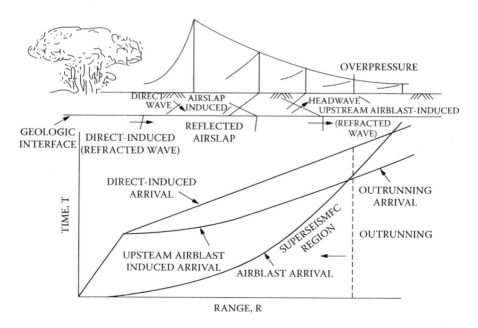

FIGURE 3.71 Surface-burst ground shock phenomenology for layered geology (ASCE, 1985).

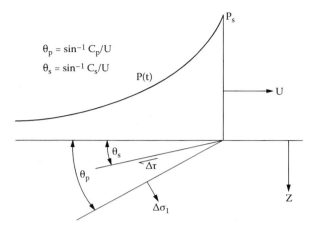

FIGURE 3.72 Superseismic airblast load P(t) applied to surface of elastic half-plane (ASCE, 1985).

compression wave propagation velocity, and C_s is the shear wave propagation velocity. Because of the compression front, there will be a change in $\Delta\sigma_1$ in the principal stress perpendicular to the front, causing the following change in particle velocity:

$$\Delta V_p = \frac{\Delta\sigma_1}{\rho C_p} = \frac{\Delta\sigma_1}{\rho U \sin\theta_p} \qquad (3.111)$$

ρ is the mass density, and the ratio between horizontal to vertical velocity change is defined as:

$$\frac{\Delta V_u}{\Delta V_u} = \tan\left[\sin^{-1}\left(\frac{C_p}{U}\right)\right] \qquad (3.112)$$

If θ_p is a small angle (very superseismic wave), $\theta_p \approx \sin\theta_p$, and $\Delta V_v \approx \Delta_{vp}$. Near the surface, the change in horizontal particle velocity will be approximately:

$$\Delta V_u \quad \frac{P_s}{\rho U} \qquad (3.113)$$

At the shear front, the principal stresses are equal and opposite and inclined at 45° and 135° to the wavefront. The motion is parallel to the front, causing a change in the particle velocity, due to a change in shear stress $\Delta\tau$.

Conventional and Nuclear Environments

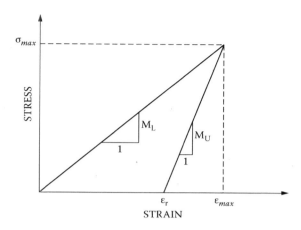

FIGURE 3.73 Typical bilinear stress–strain curve (ASCE, 1985).

$$\Delta V_s = \frac{\Delta \tau}{\rho C_s} = \frac{\Delta \tau}{\rho U \sin \theta_s} \qquad (3.114)$$

For vertical airslap-induced ground shock, 1-D wave propagation techniques with simple (bilinear) stress–strain curves for soil may be applied as shown in Figure 3.73. As a result of such computations, attenuation curves can be derived as shown in Figure 3.74. From these and the peak stress propagation velocity, one can obtain attenuation factors for accelerations as shown in Figure 3.75 ($100 < P_o < 1000$ psi), and displacement attenuation as a function of a reflective layer at depth H in Figure 3.76. Advanced computer codes must be employed for deriving complete stress–time histories. Similar techniques can be employed for horizontal airslap-induced ground shock. None of these methods is accurate and significant uncertainties are involved, as shown in Table 3.16.

TABLE 3.16
Uncertainty in Airslap-Induced Motion Predictions

Component	Uncertainty Factor F
Vertical velocity	± 1.25
Horizontal velocity	± 2.0
Vertical acceleration	± 2.5
Horizontal acceleration	± 5.0

Source: ASCE, 1985.

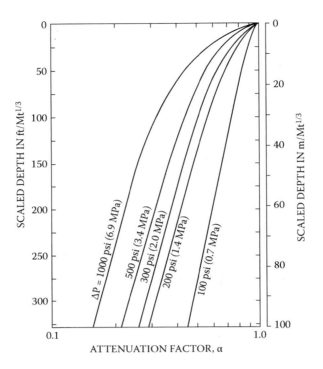

FIGURE 3.74 Attenuation factor α versus scaled depth ξ (ASCE, 1985).

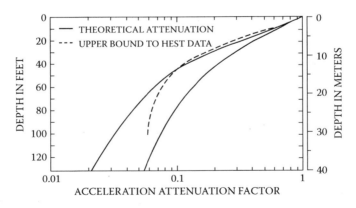

FIGURE 3.75 Attenuation of airslap-induced peak acceleration (ASCE, 1985).

Studies were performed on upstream-induced ground shock and various empirical models were proposed; such models from cited literature should be utilized. Finally, all the ground shock components are combined with their corresponding times of arrival to obtain an estimate of ground motions. This topic is very complex and usually requires a major computational effort. Much additional research is necessary to improve the accuracy of existing models.

Conventional and Nuclear Environments

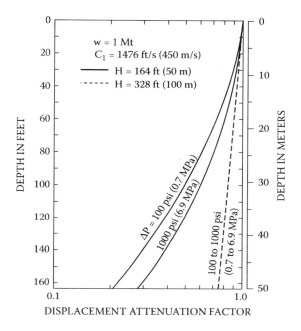

FIGURE 3.76 Attenuation of airslap-induced peak vertical displacement (ASCE, 1985).

3.6 FRAGMENTATION

Fragment loading on structures is a critical issue. Unlike blast, fragments may perforate a structure and cause significant damage. Furthermore, multiple-fragment impacts must be considered to derive a safe design. This section discusses design approaches and findings from recent studies. The discussion is based on an approach proposed by Gurney in the early 1940s (Zukas and Walters, 1998) and focuses on fragments from naturally fragmenting munitions. The problem of munitions designed to produce predetermined fragments can be handled similarly.

The weight distribution of fragments resulting from the detonation of an evenly distributed explosive in a uniform thickness, cylindrical metal case is given by the following expression (TM 5-855-1, 1986):

$$N_m = \left(\frac{W_C}{2M}\right) e^{-\sqrt{\frac{m}{M}}} \quad (3.115)$$

in which N_m is the number of fragments with weights larger than m, W_c is the total casing weight (ounces), M equals $B_x^2 t_c^{5/3} d_i^{2/3}(1 + t_c/d_i)^2$, B_x is the explosive constant from Table 3.17, t_c is the average casing thickness (inches), and d_i is the average inside casing diameter (inches). When m = 0, the total number of generated fragments N_T can be estimated as follows.

TABLE 3.17
Explosive Constants

Explosive	B_x [for Equation (2-205), Figure 3.77)] $(Oz)^{1/2}/(in.)^{7/6}$	G^* [for Equation (3.119)] 10^3 fps
AMATOL	0.35	6.190
BARATOL	0.51	5.200
COMP. A-3	0.22	
COMP. B	0.22	8.800
COMP. C-4		8.300
CYCLONITE (RDX)		9.300
CYCLOTOL (75/25)	0.20	8.900
CYCLOTOL (20/80)		8.380
CYCLOTOL (60/40)	0.27	7.880
H-6	0.28	8.600
HBX-1	0.26	8.100
HBX-3	0.32	
HMX		10.200
HTA-3		8.500
OCTOL (75/25)		9.500
PENTOLITE (50/50)	0.25	8.100
PTX-2	0.23	
TNT	0.30	7.600
TORPEX		7.450
TRITONAL (80/20)		7.600

* G = Gurney explosive energy constant.

Source: TM 5-855-1, 1986.

$$N_T = \frac{W_c}{2M} \quad (3.116)$$

The parameter $\sqrt{M}/B_x$ as a function of d_i and t_c is obtained from Figure 3.77.

The choice of a design fragment weight W_f as a function of a particular confidence level CL is obtained from Figure 3.78. Usually, CL $\geq$ 0.95 is recommended. CL is defined as the probability that W_f is the largest fragment where:

$$W_f = M[\ln(1-CL)]^2 \quad (3.117)$$

and the number of fragments heavier than W_f is found from the following expression:

$$N_f = N_T(1-CL) \quad (3.118)$$

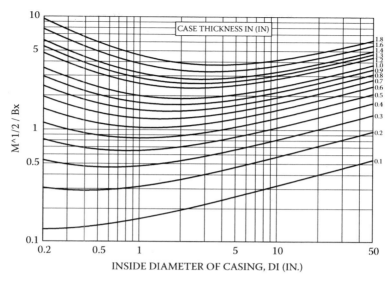

FIGURE 3.77 $\sqrt{M}/B_x$ versus casing geometry (TM 5-855-1, 1986).

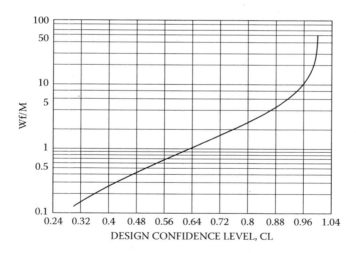

FIGURE 3.78 Design fragment weight versus design confidence level (TM 5-855-1, 1986).

The quantity $B_x^2 N_T / W_c$ for a given casing geometry is obtained from Figure 3.79.

The preceding approach applies to cylindrical cased charges. There are no models for actual munitions; average cross-sectional dimensions of the weapon are recommended for creating an "equivalent" cylindrical shape.

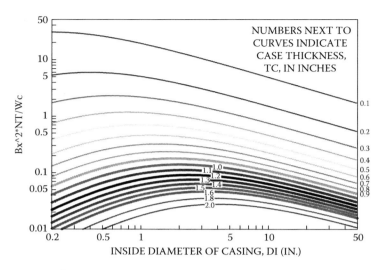

FIGURE 3.79 $B_x^2 \, N_T \, / \, W_c$ versus casing geometry (TM 5-855-1, 1986).

The initial fragment velocity can be estimated from the Gurney equation as follows:

$$V_{0I} = G \left[\frac{\left(\dfrac{W}{W_C}\right)}{1 + 0.5 \dfrac{W}{W_C}} \right]^{0.5} \tag{3.119}$$

in which W is the weight of explosive, W_C is the casing weight, G is the Gurney explosive energy constant (see Table 3.17 for constants for different explosives), and V_{0I} is the fragment initial velocity (10^3 feet per second). Figure 3.80 can be used for the computation. The design characteristics for some weapons are presented in Table 3.18.

For targets of up to 20 feet from an explosion, one may use the initial fragment velocity as the striking velocity. Beyond 20 feet, the velocity will decrease because of air drag. The striking velocity V_{sf} in 10^3 feet per second as a function of the distance R_f in feet and the air fragment weight W_f is computed from Equation (3.120).

$$V_{sf} = V_{0I} e^{-0.004 \left(\dfrac{R_f}{W_f^{1/3}} \right)} \tag{3.120}$$

V_{sf} can also be obtained from Figure 3.81.

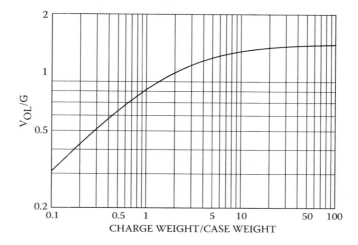

FIGURE 3.80 Initial velocity of primary fragments for cylindrical casing (TM 5-855-1, 1986).

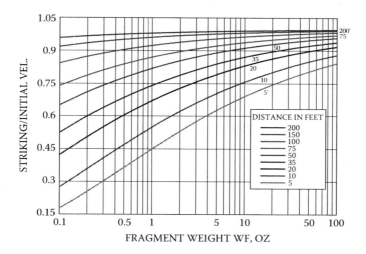

FIGURE 3.81 Variation of primary fragment velocity with distance (TM 5-855-1, 1986).

An important parameter is the fragment shape to be used for penetration calculations. The assumed shape is rather blunt and a normal impact of 90° to the target surface is employed.

Caliber density D_d is defined as:

$$D_d = \frac{W_f}{d^3} \tag{3.121}$$

TABLE 3.18
Design Fragment Characteristics of Selected Munitions

Munition Type	Country/Size	Model No. or Type	M [Equation (3.115)] (oz)	W_f (CL = 0.95) (oz)	V_{ol} (10^3 fp)
Mortar round	USSR/82 mm	0-832D	0.030	0.27	4.34
	USSR/120 mm	OF 8434	0.192	1.73	2.66
	USSR/160 mm	F-853A	0.210	1.89	3.28
	USSR/240 mm	F-864	0.277	2.49	4.74
	US/60 mm	M49	0.009	0.084	4.82
	US/81 mm	M362A1	0.0075	0.068	6.34
	US/4.2 in.	M327A2	0.024	0.219	5.55
Artillery round	USSR/122 mm	OF 472	0.163	1.47	3.30
	USSR/130 mm	OF 482M	0.282	2.54	2.78
	USSR/152 mm	OF 540	0.249	2.24	3.45
	US/105 mm	M1	0.051	0.463	4.06
	US/155 mm	M107	0.253	2.27	3.38
	US/175	M437A2	0.146	1.32	4.55
	US/8 in.	M106	0.383	3.44	3.78
Tank round	USSR/115 mm	OF-18	0.195	0.86	3.85
Rocket round	USSR/140 mm	M-14-OF	0.134	1.21	3.93
	USSR/240 mm	9	0.058	0.52	6.93
Bomb	US/100 lb	GP	0.0175	0.16	8.03
	US/250 lb	GP	0.05	0.45	7.86
	US/500 lb	GP	0.072	0.65	7.88
	US/1000 lb	GP	0.204	1.84	7.41
	US/2000 lb	GP	0.25	2.25	7.76
	US/1600 lb	AP	1.80	16.18	3.46
	US/2000 lb	SAP	1.05	9.49	5.16

Source: TM 5-855-1, 1986.

And one assumes the general fragment cylindrical shape with a spherical nose, as shown in Figure 3.82. The fragment's nose shape factor is defined as follows:

$$N_s = 0.72 + 0.25\sqrt{n - 0.25} \qquad (3.122)$$

The above shape will lead to $D_d = 0.186$ lb/in^3, $n = 0.5$, and $N_s = 0.845$. It is suggested to design for normal (90° impact) conditions.

3.6.1 Fragment Penetration

The general topic of penetration was discussed previously and some additional information will be provided here for the "design" fragment. Fragment penetration into mild steel (BH 150) in inches is obtained as follows:

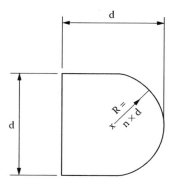

FIGURE 3.82 Primary fragment shape (TM 5-855-1, 1986).

$$X = 0.21 W_f^{0.33} V_{sf}^{1.22} \tag{3.123}$$

For other steels, one can use the following expression:

$$X' = X \exp[8.77 \cdot 10^{-6}(B'^2 - B^2) - 5.41 \cdot 10^{-3}(B' - B)] \tag{3.124}$$

in which W_f is fragment weight (ounces), V_{sf} is fragment striking velocity (in 10^3 feet per second), B is the BH for mild steel (150), and B' is the BH for other steel. Perforation occurs if the values of X and X' are larger than the plate thickness t_s.

The penetration depth can be obtained also from Figure 3.83 and Figure 3.84. The residual velocity after penetration is derived from Figure 3.85 or from the following empirical equation:

$$V_r = \frac{\left[V_{sf}^2 - 12.92\left(\dfrac{t_s}{W_f^{1/3}}\right)^{1.64}\right]^{1/2}}{\left(1 + 1.44\dfrac{t_s}{W_f^{1/3}}\right)} \tag{3.125}$$

Fragment penetration into concrete is computed from the following expression:

$$X = \frac{0.95 W_f^{0.37} V_{sf}^{0.9}}{f_c'^{0.25}} \quad \text{for} \quad X \leq 1.4 W_f^{1/3} \tag{3.126a}$$

or

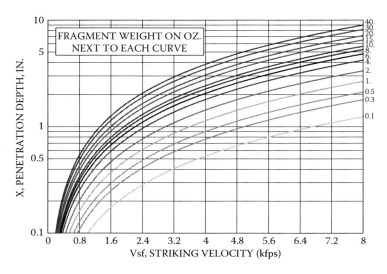

FIGURE 3.83 Steel penetration design chart; mild steel fragments penetrating mild steel plates (TM 5-855-1, 1986).

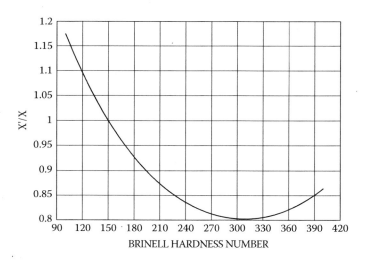

FIGURE 3.84 Variation of steel penetration with Brinell hardness (TM 5-855-1, 1986).

$$X = \frac{0.464 W_f^{0.4} V_{sf}^{1.8}}{f_c'^{0.5}} + 0.487 W_f^{1/3} \quad \text{for} \quad X > 1.4\, W_f^{1/3} \quad (3.126b)$$

Conventional and Nuclear Environments

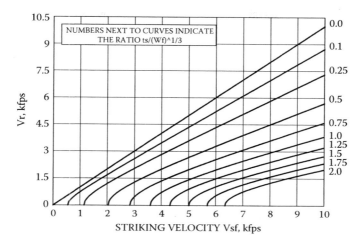

FIGURE 3.85 Residual velocity after perforation of steel (TM 5-855-1, 1986).

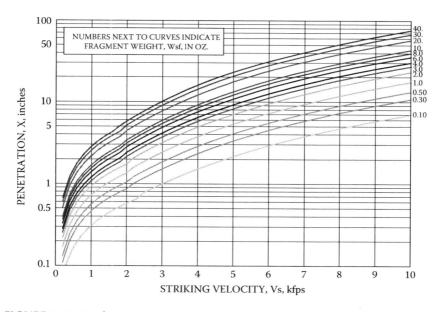

FIGURE 3.86 Penetration of mild steel fragments into massive 3,000 psi concrete (TM 5-855-1, 1986).

Figure 3.86 can be used to estimate such penetration. The thickness of concrete (in inches) that the fragment is just capable of perforating is calculated from Equation (3.127):

$$t_{pf} = 1.09\, XW_f^{0.033} + 0.91\, W_f^{0.33} \tag{3.127}$$

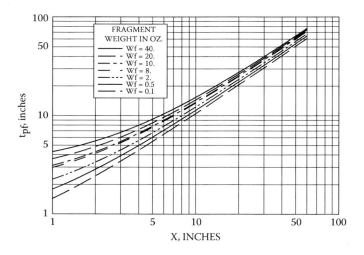

FIGURE 3.87 Thickness of concrete perforated versus penetration into massive concrete (TM 5-855-1, 1986).

This equation is presented graphically in Figure 3.87. If the actual thickness of the design wall is greater tan t_{pf}, the fragment will come to rest in the wall. However, a designer should note that the rear face may still spall and that secondary spall fragments may cause additional damage to equipment and/or injuries to people inside the protected space. If spall protection is desired, the design wall must be thicker than the thickness computed from Equation (3.128):

$$t_{sp} = 1.17\, XW_f^{0.033} + 1.47 W_f^{0.33} \tag{3.128}$$

This equation is plotted in Figure 3.88.

If the fragment perforates the wall, its residual velocity is given by the following expressions:

$$V_r = V_{sp}\left[1 - \left(\frac{t_s}{t_{pf}}\right)^2\right]^{0.555} \quad \text{for } X \le 1.4\, W_f^{1/3} \tag{3.129a}$$

$$V_r = V_{sp}\left(1 - \frac{t_s}{t_{pf}}\right)^{0.555} \quad \text{for } X > 1.4\, W_f^{1/3} \tag{3.129b}$$

where t_s is the thickness of the concrete section (inches). These equations are plotted in Figure 3.89 and Figure 3.90. The thickness of wood perforated by

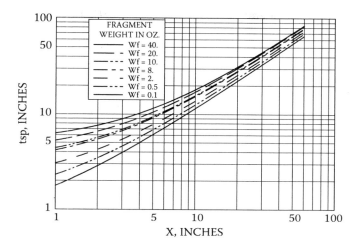

FIGURE 3.88 Thickness of concrete that will spall versus penetration into massive concrete (TM 5-855-1, 1986).

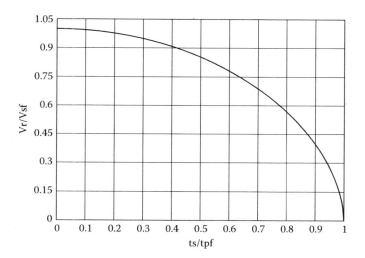

FIGURE 3.89 Residual fragment velocity upon perforation of concrete barriers for cases where $X \leq 1.4\, W_f^{1/3}$ (TM 5-855-1, 1986).

fragments of the shape presented earlier herein, as defined in TM 5-8551 (1986), and with normal impact conditions is given by the following expression:

$$t_p = \frac{10041 V_{sf}^{0.41}\, W_f^{0.59}}{\rho H^{0.54}} \quad (3.130)$$

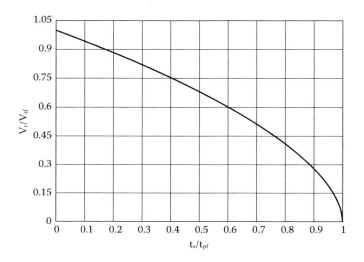

FIGURE 3.90 Residual fragment velocity upon perforation of concrete barriers for cases where $X \geq 1.4 \, W_f^{1/3}$ and for sand barriers (TM 5-855-1, 1986).

This equation is plotted in Figure 3.91 for dry fir plywood. t_p is the perforation thickness (inches), V_{sf} is the striking velocity (10^3 feet per second), W_f is the fragment weight (ounces), ρ is the wood density (pounds per cubic foot) shown in Table 3.19, and H is wood hardness (pounds) shown in Table 3.19. If the thickness of the wood is less than or equal to t_p, the fragment will perforate it and the residual velocity after perforation is given by the following equation and shown in Figure 3.92:

$$V_r = V_{sp} \left[1 - \left(\frac{t_1}{t_p} \right)^{0.5735} \right] \quad (3.131)$$

where V_r is residual velocity (10^3 feet per second) and t_1 is the actual thickness of the wood (inches).

The penetration into soil by fragments with the design shape given in Figure 3.93 and with normal impact conditions is given by the following expression:

$$t_p = 1.975 \, W_f^{1/3} K_p \log_{10} \left(1 + 4.65 \, V_{sf}^2 \right) \quad (3.132)$$

where t_p is the thickness perforated (inches), W_f is the fragment weight (ounces), K_p is the soil penetration constant (see Table 3.20), and V_{sf} is the striking velocity (10^3 feet per second). The fragment residual velocity V_r after soil perforation of a thickness t_1 that is less than t_p is obtained from the following expression:

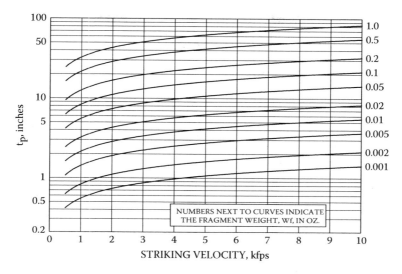

FIGURE 3.91 Penetration of dry fir plywood ($\rho = 30$ pcf, H = 75 lb) (TM 5-855-1, 1986).

TABLE 3.19
Densities and Hardnesses of Wood Targets

Type of Wood	Sample Density (pcf)	Hardness (lb)
Pine, dry	22–25	38.7
Pine, wet	30	51.1
Maple, dry	35	76.9
Maple, wet	40	72.0
Green oak, dry	51–59	88.1
Green oak, wet		72.1
Marine plywood, dry	37	68.9
Marine plywood, wet		58.8
Balsa, dry	6	21.0
Balsa, wet	6	61.5
Fir plywood, dry	30	75.0
Fir plywood, wet		68.9
Corrisa	27	
Hickory, dry	50	74.3
Hickory, wet	55	63.5

Source: TM 5-855-1, 1986.

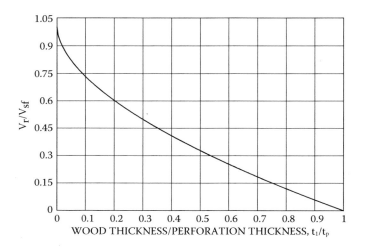

FIGURE 3.92 Residual velocity after perforation of wood (TM 5-855-1, 1986).

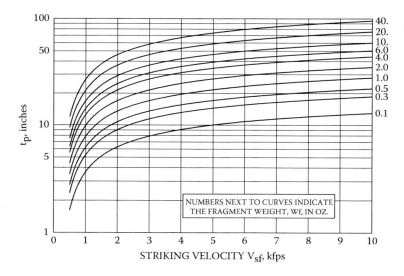

FIGURE 3.93 Penetration of sand ($K_p = 5.29$) (TM 5-855-1, 1986).

$$V_r = V_{sf}(1-t_1/t_p)^{0.555} \tag{3.133}$$

The design and analysis of composite barriers are based on the fairly accurate, if not conservative, assumption that there is no interaction between the various materials a composite so that each material's penetration resistance is not affected by the other materials' presence in the composite. This means that one can trace the path of the fragment through each successive layer, calculating whether or

TABLE 3.20
Soil Penetration Constants

Soil type	K_p (inches/oz$^{1/3}$)
Limestone	0.775
Sandy soil	5.29
Soil containing vegetation	6.95
Clay soil	10.6

Source: TM 5-855-1, 1986.

not it perforates that layer and, if it does, using the appropriate residual velocity equation to obtain a striking velocity for the next layer. Of course, if the fragment comes to rest in any layer, the wall of barrier has successfully defeated perforation. However, if the last layer is concrete and if the fragment enters this layer, the possibility of spalling should be considered by using the appropriate computational model.

3.7 FIRE, CHEMICAL, BACTERIOLOGICAL, AND RADIOLOGICAL ENVIRONMENTS

Fire, chemical, bacteriological, and radiological attacks have been used by both military and irregular forces. Such environments can be created with commercially available materials and with materials that may be obtained through other means. This is a very specialized area in protective construction that will not be addressed in detail here. For more information, consult special publications on protection against such environments.

Military forces have a wide range of incendiary weapon systems available for use against different types of targets and some of the systems have been appropriated also by terrorist organizations. Irregular forces and terrorist organizations also use flammable liquids and accelerators, as typically used by arsonists. Furthermore, it is well documented that various types of commercially available chemicals can be used for such attacks.

In general, ensuring fire protection is relatively simpler for reinforced concrete structures than for other types of structures. Care must be given to design details that can help prevent fires from spreading (such as choice of materials, fire suppression measures, and pressurized ventilation). The protection against chemical, bacteriological, and radiological effects requires special features. For chemical and bacteriological protection, one must ensure an airtight facility with positive pressure capabilities and decontamination facilities.

4 Conventional and Nuclear Loads on Structures

The significant differences between conventional and nuclear loads must be considered in the design approach. There are also significant differences between existing design procedures and state-of-the-art information from recent research. This section addresses both topics.

4.1 CONVENTIONAL LOADS ON STRUCTURES

4.1.1 BURIED STRUCTURES

Existing design procedures are for cases with a scaled range $\lambda > 1.0$. There are several tests for $\lambda \leq 1.0$ but the severity of such loading environments makes it very difficult to resist the applied loads. Therefore, the "milder" cases are discussed here and the severe cases will be addressed together with other test data.

First, one must estimate the reflected pressure on the structure and the duration of the pressure. Since this was discussed earlier, only a brief summary is provided here. For estimates, $P_r = 1.5\, P_o$ may be used. Then one needs an estimated duration based on the element thickness (e.g., about six transit times through a slab), or the distance ($t_{\text{free edge}} - t_{\text{point}} + t_{\text{travel point}} - {}_{\text{free edge}}$). The shorter value is used. For arrival time, the soil loading wave velocity is used and the travel time is based on twice this velocity.

Another approach is to employ a single-degree-of-freedom (SDOF) computation similar to the one proposed by Weidlinger and Hinman (July 1988). It is based on obtaining the load on a structure that moves relative to the soil, and then using simple structural materials to obtain the response. Other techniques employ more advanced structural models (TM 5-855-1, 1998 and later) and some of these are discussed later. Another approach is to use Figure 4.1 for obtaining an equivalent uniform load on the roof of a buried structure due to an overhead explosion. One may also employ the following formulation for obtaining the equivalent pressure at other points:

$$P_R = P_{or} \left(\frac{D}{R_s} \right)^3 \qquad (4.1)$$

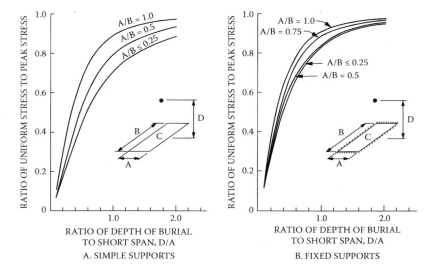

FIGURE 4.1 Equivalent uniform load in flexure (TM 5-855-1, 1986).

in which P_R is the pressure on the roof at distance R_s from the explosive charge, P_{or} is the pressure on the roof directly below the explosive charge, D is the depth from the explosive charge to the roof, and R_s is the distance from the charge to the specific point on the structure.

As for the pulse duration, it is recommended to use the duration of the free-field pressure pulse at the quarter point of the short span along a section at the center of the long span (TM 5-855-1, 1986). From that, one may develop an equivalent triangular pulse. The analysis of wall loads is performed similarly. More accurate load estimates can be obtained with advanced computer codes that can be used to perform shock wave propagation in geologic media, and medium-structure interaction.

4.1.2 Above-Ground Structures

The airblast components on above-ground structures arise from the free-field incident pressure, the dynamic pressure, and reflected pressures. The effects of fragments, discussed earlier, should be included in the design. The assumptions of a rectangular shape structure and loading by the Mach reflection region will be maintained, as discussed in Chapter 3. Figure 4.2 illustrates the pressure derivation on the front wall.

The clearing time to remove the reflected pressure P_r is defined as follows:

$$t_c = \frac{3S}{U} \qquad (4.2)$$

Conventional and Nuclear Loads on Structures

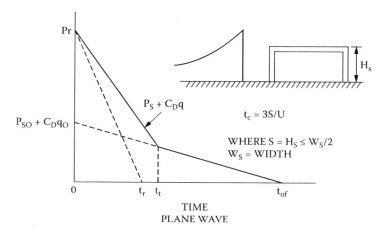

FIGURE 4.2 Front-wall loading (TM 5-855-1, 1986).

U is the shock front velocity and S is the structure's height or one-half of its width, whichever is smaller. t_c represents the time it takes for the shock wave to travel to the closest edge of the wall and for a relief wave to travel back to the point of interest.

After the time t_c, the acting pressure is defined as follows:

$$P_{tc} = P_s + C_D q \qquad (4.3)$$

in which P_s is the incident pressure, C_D is the drag coefficient, taken as 1 for the present ranges, and q is the dynamic pressure. For higher pressure ranges, the above procedure may lead to unrealistic pressure–time histories, and the dotted line in Figure 4.2 should be used. This is based on the total reflected pressure impulse i_r obtained from shock wave parameter charts (please see Chapter 3). The fictitious duration t_r is derived based on a triangular pressure pulse assumption, as follows:

$$t_r = \frac{2i_r}{P_r} \qquad (4.4)$$

in which P_r is the peak reflected pressure.

The loading on the wall is the curve that gives the smallest impulse (i.e., the smallest area under the curve). It should be noted that the reflected pressure impulse includes the effects of incident and dynamic pressures.

Here it was assumed that the load results from the shock wave part in the Mach system (i.e., the triple point is above the structure). For close-in detonations, the pressure and impulse vary along the wall and an average pressure should be assumed over a length of the structure that equals 1.3 times the normal distance

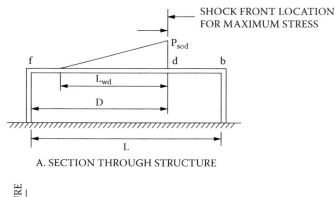

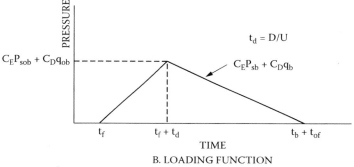

FIGURE 4.3 Roof and side-wall loading: span direction perpendicular to shock front (TM 5-855-1, 1986).

to the explosion (but shall not be larger than 2S). The average pressure and impulse should be determined at the mid-height front wall. For determining the loads on other surfaces such as sidewalls and roofs, it may be necessary to employ a computer code for hydrodynamic computations. An appropriate procedure is presented as follows.

From Figure 4.3, note that the maximum stresses in the member occur when the front reaches point d. For that case, the equivalent uniform pressure–time history shown in Figure 4.3b can be used and the coefficients C_E and D/L are obtained from Figure 4.4 where:

$$P_{or} = C_E P_{sob} + C_D q_{ob} \tag{4.5}$$

t_f is the time for the wave to reach the element (point f) and t_d is the time to go from f to d. The peak pressure P_{or} is the contribution of equivalent incident and drag pressures. P_{sob} is the peak over-pressure at point b and q_{ob} is $C_E P_{sob}$. The pressure decays to zero at time $t_b + t_{of}$, where t_b is the time for the front to reach point b and t_{of} is the fictitious positive phase duration. A similar procedure may be employed for the negative phase (if required), and it is assumed also at point

Conventional and Nuclear Loads on Structures

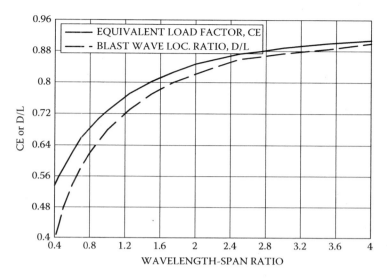

FIGURE 4.4 Equivalent load factor and blast wave location ratio versus wave length–span ratio (TM 5-855-1, 1986).

TABLE 4.1
Drag Coefficients

Peak Dynamic Pressure (psi)	Drag Coefficient
0–25	−0.40
25–50	−0.30
50–130	−0.20

Source: TM 5-855-1, 1986.

b. Values of C_D can be obtained from Table 4.1. For span directions parallel to the shock front, use the time history in Figure 4.5.

The loads on the rear wall result from the waves traveling around the structure and over the roof. The procedure is similar and it is outlined in Figure 4.6. The peak pressure (Figure 4.6b) is obtained with the pressure at point 2 (Figure 4.6a), that is, at a distance of H_S past the rear edge of the roof. The equivalent load factor C_E is based on the wavelength of the peak pressure above the unsupported length of the rear wall, and so are the rise time and duration. The effects of dynamic pressure are similar to the previous cases for the roof and side walls. Deriving these loads is discussed further in Section 4.3.

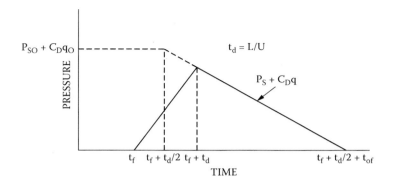

FIGURE 4.5 Roof and sidewall loading: span direction parallel to shock front (TM 5-855-1, 1986).

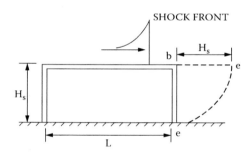

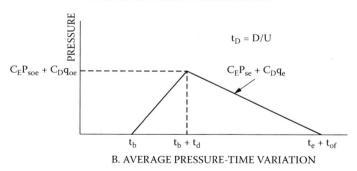

FIGURE 4.6 Rear-wall loading (TM 5-855-1, 1986).

Conventional and Nuclear Loads on Structures   175

4.1.3 Mounded Structures

Mounded structures should be considered above-ground structures. Nevertheless, recent tests have shown that their behavior is much better because soil berms modify the incident pressure waves and absorb a significant amount of airblast-induced and fragment-induced impulse. These phenomena have not been studied enough to be included in current design manuals.

4.1.4 Surface-Flush Structures

These structures must be designed for a combination of airblast- and ground shock-induced loads, according to the procedures outlined above.

4.1.5 Blast Fragment Coupling

Recent studies on the effects of load coupling where airblast and fragment effects were considered found that for close-in explosions, these effects tend to load a structure beyond what was expected. One of those studies conducted at the Ernst Mach Institute (Koos, 1987) demonstrated that for explosions of cased charges at relatively small distances from a structure or target, the impulses of airblast and fragments are combined to load the structure. These effects should be incorporated in the analysis, assessment, and eventual design. That study showed that for very close detonations, the airblast arrives first (Figure 4.7a), while fragments arrive first for farther detonations (Figure 4.7b).

Obviously, there is a range between those two where both blast and fragments will arrive at about the same time. The airblast fragment load coupling is illustrated in Figure 4.8. Note that the impulse delivered to the target increases with the weight of the charge casing. For example, at a detonation range of 3 m, an uncased charged will produce about 25 kg-m/s, a lightly cased charge 40 kg-m/s, and a heavier cased charge 100 kg-m/s. This shows clearly that one may have to consider much larger impulses when designing protective measures to resist cased explosive devices.

McCarthy (2006) also studied the combined effects of blasts and fragments. He employed a novel armored blast and impact sensor developed and experimentally tested under both laboratory and field conditions. The sensor permits the determination of the pressure–time history of an applied loading and also yields the spatial distribution of the loading. Field experimental trials were performed using an explosive blast and fragment generator, and for reference, a bare charge identical in construction to the fragment generator but without the preformed spherical fragments. The airblast and fragment field velocity characteristics of the charges were determined experimentally. The sensor package signals resulting from both blast and combined blast and fragment impact were compared to the applied loadings. A reasonable correlation was possible for the airblast loading, but difficulties in predicting the force arising from a fragment impact precluded meaningful comparison with gauge signals.

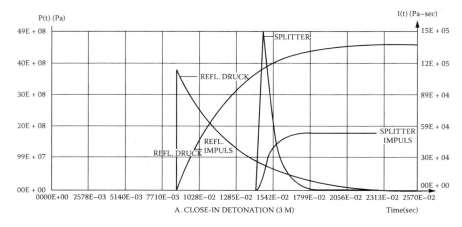
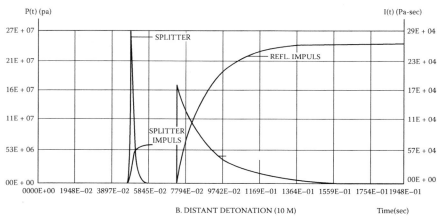

FIGURE 4.7 Detonation distance effects on blast and fragment arrival (Koos, 1987).

It was also difficult to discern the blast components of combined loadings from the gauge signal. Finally, the global structural responses of the structural target under both types of loading were considered. The combined blast and fragment loading proved to be much more severe than the airblast-only loading and fragment impact loading cannot be ignored. Field tests confirmed that the blast wave lagged behind the arrival of the fragments at a 10 m standoff and it began to catch up with the fragment field at a 5 m standoff. It may be concluded that as the standoff decreases, the arrival times of the fragments and blast wave will converge. Eventually, for even shorter standoff distances, the fragments may arrive before the blast wave as shown by Koos (1987). Clearly, this area needs further research to define accurately the combined blast and fragment effects.

Conventional and Nuclear Loads on Structures

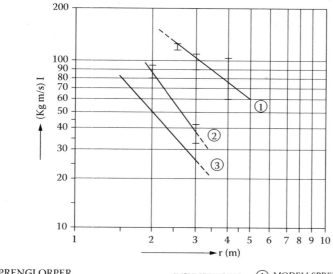

5: KOMBINIERTER LUFTSTOB-SPLITTER-IMPULS UND LUFTSTOB-IMPULS DER REINEN LADUNG, GEMESSEN MIT DEM BALLISTISCHEN ROLLPENDEL

① MODELLSPRENGKORPER MK 4 (1 Kg COMP B, 4 Kg STANIMANTEL)
② MODELLSPRENGKORPER MK 1 (1 Kg COMP B, 1 KG STANIMANTEL)
③ MODELLSPRENGKORPER MK 0 (1 Kg COMP B, OHNE MANTEL)

FIGURE 4.8 Impulse versus range for three charge types (Koos, 1987). Line 1: 1 kg Comp B in 4 kg case. Line 2: 1 kg Comp B in 1 kg case. Line 3: bare 1 kg Comp B.

4.2 NUCLEAR LOADS ON STRUCTURES

4.2.1 Above-Ground Structures

Airblast effects on above-ground structures in the nuclear domain are essentially the same as for conventional cases, as discussed in Chapter 3. The main difference for nuclear weapons is in the relative dimension of shock front to structure sizes and the much longer durations. One should recall the discussion on defining incident, reflected, and dynamic pressures. Since the total force is a combined effect of shock reflection and dynamic pressure, one must employ drag or lift coefficients for computing the additional effect as follows:

$$F = (C_d \text{ or } C_L) q A \tag{4.6}$$

in which F is the total drag or lift force on the exposed structure, C_d and C_L are the drag or lift coefficients (see Table 4.2), q is the dynamic pressure, and A is the projected area (perpendicular to airflow for drag and parallel to airflow for lift as shown in Figure 4.9). Similarly to the load definitions for HE devices, the

TABLE 4.2
Object Drag-and-Lift Coefficients for Structural Shapes of Infinite Length at Low Over-Pressures[a]

Profile and Wind Direction	C_d	C_L
→ □	2.0	0
→ I	2.0	0
→ I	2.0	0
→ ⊢⊣	1.8	0
→ L	2.0	0.3
→ ⌐	1.8	2.1
→ C	2.0	−0.1
→ ⌈	1.6	−0.5
→ ⊢	2.0	0
→ T	2.0	−1.2
→ ⌷	2.2	0

Note: Coefficients of References 7 through 17 in (ASCE 1985) rounded to nearest tenth.

[a] Below 20 psi or 138 kPa.

Source: ASCE, 1985.

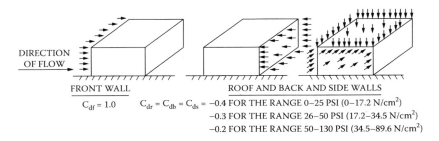

FRONT WALL
$C_{df} = 1.0$

ROOF AND BACK AND SIDE WALLS
$C_{dr} = C_{db} = C_{ds} =$ −0.4 FOR THE RANGE 0–25 PSI (0–17.2 N/cm²)
−0.3 FOR THE RANGE 26–50 PSI (17.2–34.5 N/cm²)
−0.2 FOR THE RANGE 50–130 PSI (34.5–89.6 N/cm²)

FIGURE 4.9 Average surface drag coefficients for rectangular structure (ASCE, 1985).

information in Figure 4.10 can be used for the computational process to define the load–time histories on various building surfaces.

In the previous system, no information was provided for airblast loads on arches and domes. Because of the curve geometries, drag and lift forces act simultaneously on these structures. Figure 4.11 through Figure 4.13 provide some information on such effects for the 25 psi (170 kPa) range.

Conventional and Nuclear Loads on Structures

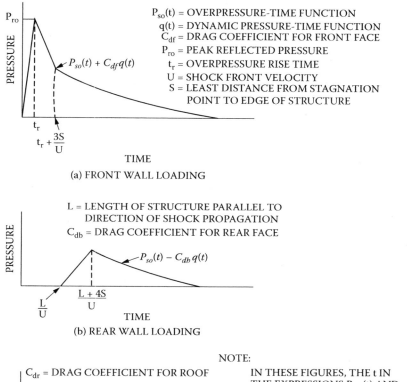

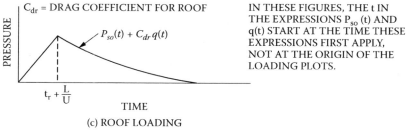

FIGURE 4.10 Loading on above-ground closed rectangular surfaces (ASCE, 1985).

4.2.2 Buried Structures

One of the important advantages of buried structures is the contribution of soil arching to lead resistance, as will be addressed later. In the dynamic domain, the effect of soil arching is more significant, especially for the late-time response (4 to 10 msec after load arrival). The problem is further complicated by stress wave propagation phenomena that combine with rigid body motions, structural response, and soil arching to modulate the loads. In other words, there is a dynamic soil–structure interaction (SSI) problem that must be considered for analysis and design.

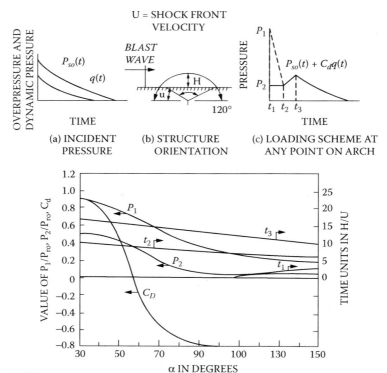

NOTES
1. P_1 AND P_2 ARE RELATED TO P_{ro}. THE IDEAL 0° REFLECTION COEFFICIENT. THESE COULD BE RELATED TO P_{ro} BUT THE RELATIONSHIP TO P_{ro} WAS FELT TO BE MORE APPROPRIATE FOR FRONT FACE LOADING. ON THE BACK SIDE, THE SAME RELATIONSHIP WAS USED FOR CONSISTENCY.
2. TIME UNITS ARE EXPRESSED AS A RATIO OF HEIGHT OF ARCH, H. TO VELOCITY OF SHOCK FRONT, U.

FIGURE 4.11 Ideal loading scheme for 120° arch for peak incident over-pressures of 25 psi (170 kPa) or less (ASCE 1985).

For shallow-buried rectangular structures (defined by the depth of burial range $0.2L \leq DOB \leq 1.5 L$; L is the clear span of roof or wall), one may use the following:

$$\sigma_r(t) = \sigma_{ff}(t)\left(2 - \frac{t}{t_d}\right) \quad \text{for} \quad t \leq t_d \qquad (4.7a)$$

and

$$\sigma_r(t) = \sigma_{ff}(t) \quad \text{for} \quad t > t_d \qquad (4.7b)$$

Conventional and Nuclear Loads on Structures

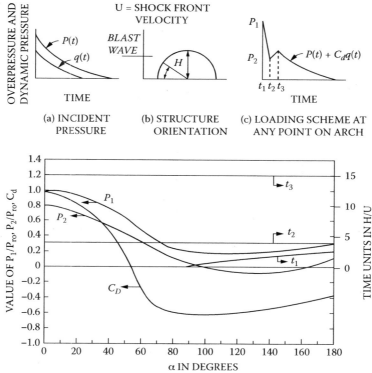

NOTES
1. P_1 AND P_2 ARE RELATED TO P_{ro}. THE IDEAL 0° REFLECTION COEFFICIENT. THESE COULD BE RELATED TO P_{ro} BUT THE RELATIONSHIP TO P_{ro} WAS FELT TO BE MORE APPROPRIATE FOR FRONT FACE LOADING. ON THE BACK SIDE, THE SAME RELATIONSHIP WAS USED FOR CONSISTENCY.
2. TIME UNITS ARE EXPRESSED AS A RATIO OF HEIGHT OF ARCH, H. TO VELOCITY OF SHOCK FRONT, U.

FIGURE 4.12 Ideal loading scheme for 180° arch for peak incident over-pressures of 25 psi (170 kPa) or less (ASCE, 1985).

in which $\sigma_r(t)$ is the stress acting on the roof or wall at time t, $\sigma_{ff}(t)$ is the incidental free-field stress at structure location, and t_d equals 12 times the travel time through the element thickness (1 to 5 msec). If soil arching is considered, it will affect Equation (4.7b) only (late time effect) as follows:

$$\sigma'_r(t) = C_a \sigma_{ff}(t) \quad \text{for} \quad t > t_d \tag{4.7c}$$

C_a is obtained from Figure 4.17 (see Section 4.2.3). The stress at the structure base can be estimated as follows:

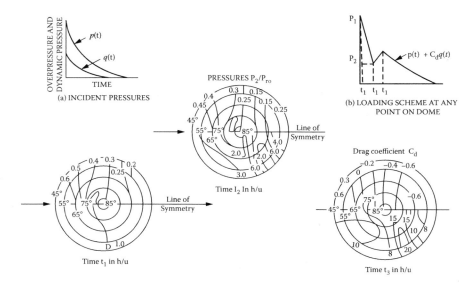

FIGURE 4.13 Ideal loading scheme for 45° dome for peak incident over-pressures of 25 psi (170 kPa) or less (ASCE, 1985).

$$\sigma_b(t) = \rho C_L v(t) \tag{4.8}$$

where ρ is the soil mass density, C_L is the loading wave velocity in soil, and $v(t)$ is the structural velocity. An approximation for such a stress could be:

$$\sigma_b(t) = \sigma_{f\!f}(t) \tag{4.9}$$

For the stress on the sides of the structures, one may use Equation (4.10):

$$\sigma_s(t) = K_o \sigma_{f\!f}(t) \tag{4.10}$$

where K_o is the coefficient of lateral soil pressure. The stress $\sigma_s(t)$ will introduce a membrane-compressive force in the roof, floor, etc., to increase their capacities.

When surface-flush structures are concerned (O ≤ DOB < 0.2 L), one may use the following expressions:

$$\sigma_r(t) = \sigma_{f\!f}(t)\left(2 - \frac{t}{t_d^1}\right) \quad \text{for} \quad t \le t_d' \tag{4.11a}$$

$$\sigma_r(t) = \sigma_{f\!f}(t) \quad \text{for} \quad t > t_d' \tag{4.11b}$$

$$\sigma_r(t) = P_{so}(t) \quad \text{for} \quad \text{DOB} = 0 \tag{4.11c}$$

Conventional and Nuclear Loads on Structures

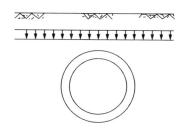

(a) BURIED HORIZONTAL CYLINDER

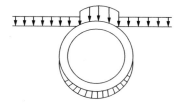

(b) VERTICAL STRESS COMPONENTS ON HORIZONTAL CYLINDER SHORTLY AFTER ARRIVAL OF THE INCIDENT WAVE

FIGURE 4.14 Buried horizontal cylinder subjected to incident plane wave (ASCE, 1985).

in which t'_d is the smallest value of t_d (defined earlier) or the time for the loading wave to travel from the soil surface to the structure roof. When arches and cylinders are considered, the loading conditions are complicated by the geometry and the structural motion in the soil, as illustrated in Figure 4.14. An approximation has been proposed by Crawford et al. (1974), as illustrated in Figure 4.15 for a horizontal octagonal cylinder:

$$F_A(t) = 0.8\sigma_A(t)RL \qquad (4.12)$$

$$F_B(t) = 0 \quad \text{for} \quad 0 \le t < \frac{0.3R}{C_L} \qquad (4.13a)$$

$$F_B(t) = 1.2\sigma_B(t)RL \quad \text{for} \quad t \ge \frac{0.3R}{C_L} \qquad (4.13b)$$

$\sigma_A(t)$ is obtained from Figure 4.15b and $\sigma_B(t)$ from Figure 4.15c. R is the cylinder's outer radius, and L is its length. The load resisted by rigid body motion can be estimated from the following expression (ignoring shear forces):

$$F_B(t) = 2\rho C_L v(t)RL \qquad (4.14)$$

and from dynamic equilibrium (Newton's Second Law):

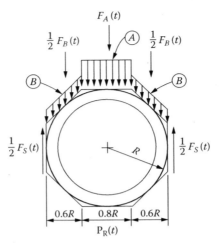

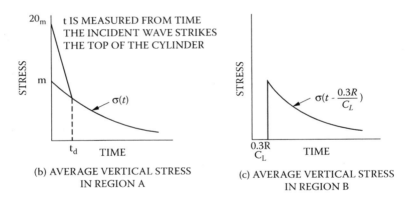

FIGURE 4.15 Approximate vertical loads for estimating vertical rigid body motions of horizontal octagonal cylinder (Crawford et al., 1974).

$$M\dot{v}(t) = F_A(t) + F_B(t) - F_R(t) \qquad (4.15)$$

The above approach is a very simple procedure for estimating loads on buried structures; it is based on deriving equivalent uniformly distributed loads as illustrated in Figure 4.16.

Other approximate methods for buried arches and cylinders were proposed (see ASCE, 1985) and for many cases they provide attractive alternatives to the above method. For more accurate assessments, it may be necessary to employ advanced computer codes.

Conventional and Nuclear Loads on Structures

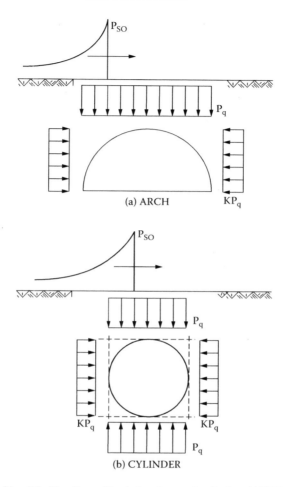

FIGURE 4.16 Simplified loading of buried arches and cylinders (ASCE, 1985).

Mounded structures can be evaluated in a similar way. First, the reflected and dynamic pressures are computed, and then the loads are transferred to the structure (as if a buried structure). Because of the complex geometry, these approaches are not very accurate and therefore adequate computer codes should be utilized.

4.2.3 Soil Arching

Soil arching occurs when there is a relative motion between structure and soil, especially if the soil can provide high shear strengths (e.g., sand). The classical approach for computing soil arching is use of the "trapdoor" mechanism (Terzaghi and Peck, 1948; Lamb and Whiteman, 1969; Proctor and White, 1977). The two types of soil arching are:

1. Passive arching: the structure moves away from the loading soil and the soil cannot follow it due to shear resistance.
2. Active arching: the structure is pushed into the soil.

Such models can lead to the following equation (ASCE, 1985):

$$\frac{P_q}{P_{so}} = \exp\left[\frac{-2K_o \tan\phi (a+1)bL^2}{aL^2}\right] \quad (4.16)$$

in which P_q is the stress applied to the structure, P_{so} is the surface pressure, K_o is the lateral earth pressure coefficient, ϕ is the angle of internal friction, a is the structural length-to-span length ratio, b is the DOB-to-span length ratio, and L is the structural span length. Typical arching factors for rectangular and arch structures are presented in Figure 4.17, and additional information on applying soil arching to protective design is found in ASCE (1985).

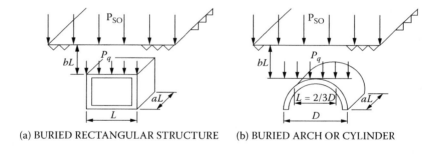

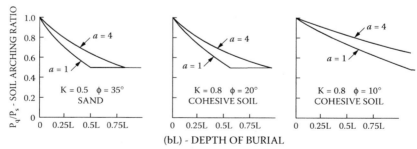

FIGURE 4.17 Soil arching as function of depth of burial (ASCE, 1985).

4.3 STEP-BY-STEP PROCEDURES FOR DERIVING BLAST DESIGN LOADS

4.3.1 Constructing Load–Time History on Buried Wall or Roof

1. Compute the free-field pressure at the center of the wall or roof (see Section 3.4.8) as follows:

$$P_o = f(\rho c)(160)(R/W^{1/3})^{-n}$$

2. Compute the free-field pressure decay from the following expression:

$$P(t) = P_0 e^{-\alpha t/t_a}$$

3. If using equivalent triangular loading pulse, compute duration as follows:

$$t_d = 2I_0/P_0$$

4. Compute the peak combined incident and reflected pressures; assume that it is equal to 1.5 P_0.
5. Compute the arrival time t_a and the rise time to peak stress t_0, $t_a = R/c$, $t_0 = 1.1\, t_a$, where R is the range and c is the seismic wave speed.
6. Compute the reflected pressure duration. $t_r = 12T/10{,}000$ fps. T is the wall thickness; if the roof or wall is very thick compared to its span, check the duration associated with clearing time and use the lowest duration.
7. Reduce the entire pressure–time history by the factor in Figure 4.18 for deriving equivalent uniform loading as shown in Figure 4.19.

4.3.2 Computing Pressure–Time Curve on Front Wall from External Explosion (Surface Burst)

1. Determine equivalent TNT charge weight and ground distance, R_g (see Section 3.1).
2. Calculate the scaled ground distance $Z_g = R_g/W^{1/3}$.
3. Determine free-field blast wave parameters from Figure 3.4 or Figure 3.7 for spherical or hemispherical conditions, respectively, and the corresponding scaled ground distance Z_g e.g., as shown in Figure 4.20. P_{so} is peak positive incident pressure, P_r is peak normal reflected pressure, U is shock front velocity, $i_s/W^{1/3}$ is scaled unit positive incident impulse, $i_r/W^{1/3}$ is scaled unit positive normal reflected impulse, and $t_a/W^{1/3}$ is scaled arrival time. Multiply scaled values by $W^{1/3}$ to obtain absolute values.

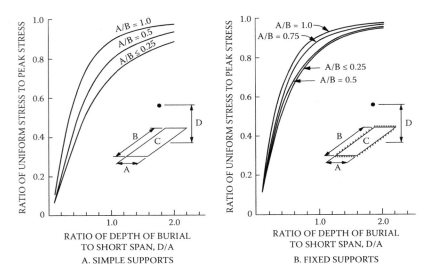

FIGURE 4.18 Equivalent uniform loads in flexure of buried structure (TM 5-855-1, 1986).

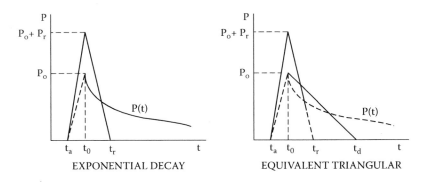

FIGURE 4.19 Time histories for loads on buried structure (TM 5-855-1, 1986).

4. Calculate clearing time t_c: $t_c = 3S/U$ where S is the height of front wall or one-half its width, whichever is smaller.
5. Calculate a fictitious positive phase duration t_{of}: $t_{of} = 2\, i_s/P_{so}$.
6. Determine peak dynamic pressure q_0 from Figure 4.21 for the given P_{so}.
7. Calculate $P_{so} + C_D q_0$ where $C_D = 1$.
8. Calculate a fictitious reflected pressure duration $t_r = 2i_r/P_r$.
9. Construct pressure–time curves as shown in Figure 4.22. The correct curve to use is whichever one (reflected pressure or reflected pressure plus incident pressure) that gives the smallest value of impulse (area under curve).

Conventional and Nuclear Loads on Structures

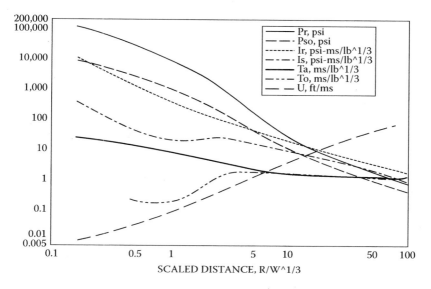

FIGURE 4.20 Shock wave parameters for hemispherical TNT surface bursts at sea level (TM 5-855-1, 1986).

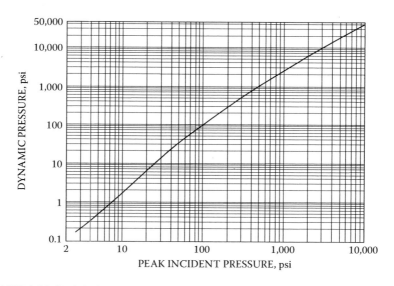

FIGURE 4.21 Peak incident pressure versus peak dynamic pressure (TM 5-855-1, 1986).

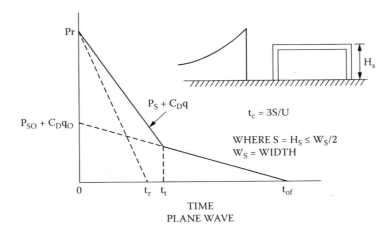

FIGURE 4.22 Front-wall loading (TM 5-855-1, 1986).

4.3.3 Computing Pressure–Time Curve on Roof or Sidewall (Span Perpendicular to Shock Front)

1. Determine shock wave parameters for hemispherical or spherical detonations from Figure 3.4 and Figure 3.7 for point b (back of structure) as shown in Section 4.3.2.
2. Calculate wave length-span ratio L_{wb}/L, and determine equivalent load factor C_E and blast wave location ratio D/L from Figure 4.4 (shown here again as Figure 4.23).
3. Calculate equivalent uniform load $C_E P_{sob}$ and wave location d.
4. Determine arrival times at front t_f and back t_b of span.
5. Calculate rise time t_d: $t_d = D/U$.
6. Calculate fictitious positive phase duration of t_{of}: $t_{of} = 2i_{sb}/P_{sob}$.
7. Determine the peak dynamic pressure q_{ob} from Figure 4.21 as shown in Section 4.3.2 for $C_E P_{sob}$.
8. Calculate $C_E P_{sob} + C_D q_{ob}$. Obtain C_D from Table 4.1.
9. Construct pressure–time curves as shown in Figure 4.24 and in Section 4.1.2.

4.3.4 Computing Pressure–Time Curve on Roof or Sidewall (Span Parallel to Shock Front)

1. Determine shock wave parameters for the point of interest from Figure 3.4 and Figure 3.7 as previously (e.g., see also Figure 4.20).
2. Calculate rise time t_d: $t_d = L/U$, where L is the width of the element considered.
3. Calculate fictitious positive phase duration t_{of}: $t_{of} = 2i_s/P_{so}$.

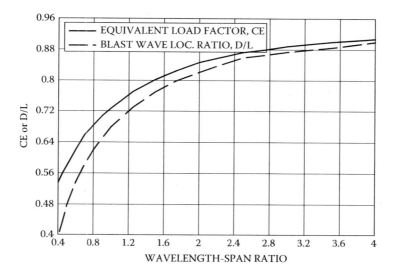

FIGURE 4.23 Equivalent load factor and blast wave location ratio versus wavelength-span ratio (TM 5-855-1, 1986).

4. Determine dynamic pressure q_0 from Figure 4.21 for the appropriate P_{so} as previously.
5. Calculate $P_{so}+ + C_D q_{ob}$. Obtain C_D from Table 4.1 as previously.
6. Determine arrival time t_f.
7. Construct the pressure–time curve as shown in Figure 4.25.

4.3.5 Blast Load on Rear Wall

The uniformly distributed pressure on the rear wall is computed as defined in Figure 4.26. Basically, one considers an imaginary extended roof for a distance that equals the structure height H_s, then computes the load on segment b–e shown in Figure 4.26 as follows:

1. The time at which the load on the rear wall starts is t_b, the time the blast load reaches the edge of the roof (see Section 4.1.2).
2. The peak load is defined by the combination of direct load and drag force: $C_E P_{soe} + C_D q_{oe}$ where C_E is obtained from Figure 4.23. It is based on the wavelength of the blast pressure above the fictitious length of the extended roof H_s. C_D is the drag coefficient from Table 4.1. P_{soe} is the peak over-pressure at point e and $q_{oe} = C_E P_{soe}$.
3. The time for the peak load is defined by $t_b + t_d$, and $t_d = D/U$. As defined in Section 4.1.2, U is the blast wave velocity and D is the location of the shock front (distance of the front from beginning of the roof as shown in Figure 4.24).

192   Modern Protective Structures

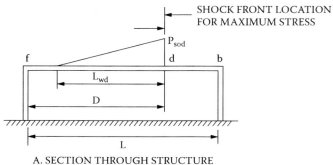

A. SECTION THROUGH STRUCTURE

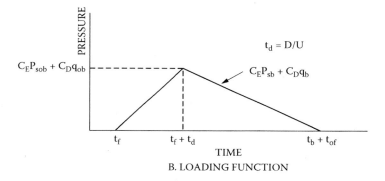

B. LOADING FUNCTION

FIGURE 4.24 Roof and side-wall loading: span direction perpendicular to shock front (TM 5-855-1, 1986).

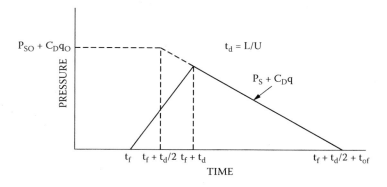

FIGURE 4.25 Roof and sidewall loading: span direction parallel to shock front (TM 5-855-1, 1986).

Conventional and Nuclear Loads on Structures

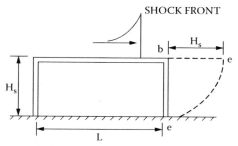

A. SECTION THROUGH STRUCTURE

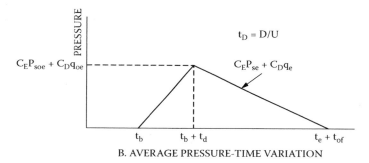

B. AVERAGE PRESSURE-TIME VARIATION

FIGURE 4.26 Rear-wall loading (TM 5-855-1, 1986).

4. The time at which the load reaches the ambient pressure is $t_e + t_{of}$, in which t_e is the time for the blast wave to reach point e and t_{of} is a fictitious duration of the positive phase over the extended part of the roof.
5. The linear blast load decay is defined by $C_E P_{so} + C_D q_0$, and these parameters were defined previously.

5 Behaviors of Structural Elements

5.1 INTRODUCTION

Traditionally, U.S. government agencies have developed and maintained manuals for the design of structures to resist blast effects. Such manuals have been directed primarily toward military structures, paying little attention to the design of nonmilitary buildings because until recently the threats to such buildings were minimal. Some design guidance for blast resistance is available to the general public; the primary users are industries that are aware of potential accidental explosions related to their normal operations (e.g., petrochemical plants). In addition, general design guidance such as that of the American Concrete Institute's Committee 318 (ACI, 2005) has served the public well. However, recent events such as the bombings of the World Trade Center in 1993 and the Alfred P. Murrah Federal Building (FEMA, 1996; Hinman and Hammond, 1997) and the events of September 11, 2001 (ASCE, 2003; FEMA, 2002) have heightened awareness in the U.S. of the need to consider potential blast effects in the designs of some buildings.

This chapter summarizes existing blast-resistant design approaches and addresses issues that are critical to the development of buildings with improved resistance to severe dynamic loads. Emphasis is given to the design and behavior of reinforced concrete structures, with particular attention to slabs, walls, columns, and connections.

This chapter provides brief reviews of materials available in modern texts on the subject. Although most of the information covers reinforced concrete structures (the most common types of protective structures), information on protective construction using steel is becoming more readily available and both types of structures are addressed.

Numerous reports and publications on weapons effects and structural responses to blasts are available in the open literature, but no single document covering all aspects of blast-resistant design exists. *Technical Manual 5-855-1* (Department of the Army, 1986 and 1998) serves as the Army's manual on protective construction. *Technical Manual 5-1300* (Department of the Army, 1990) is the tri-service manual for design against accidental explosions. The most widely used non-government design manual on blast resistance is the *ASCE Manual 42* published by the American Society of Civil Engineers (ASCE, 1985). Two more recent sources of relevant information from ASCE (1997 and 1999) contain extensive and useful information on the behavior, planning, design, and retrofit of blast-resistant structures. Summaries of the contents of key manuals and descriptions of the general guidance they provide appear in the following sections.

5.2 GOVERNMENT AND NON-GOVERNMENT MANUALS AND CRITERIA

5.2.1 Tri-Service Manual TM 5-1300 (Department of the Army, 1990)

Intended primarily for explosives safety applications, TM 5-1300 (Army designation) is the most widely used manual for designing structures to resist blast effects and is currently under revision. One reason for its widespread use by industry is that it is approved for public release with unlimited distribution. Its stated purpose is "to present methods of design for protective construction used in facilities for development, testing, production, storage, maintenance, modification, inspection, demilitarization, and disposal of explosive materials."

The manual is divided into six chapters:

1 Introduction
2 Blast, fragment, and shock loads
3 Principles of dynamic analysis
4 Reinforced conrete design
5 Structural steel design
6 Special considerations in explosive facility design

TM 5-1300 distinguishes between "close-in" and "far" design ranges for purposes of predicting modes of response. Taking into account the purpose of a structure and the design range, the allowable design response limits for the structural elements (primarily roofs and wall slabs) are given in terms of support rotations computed simply by taking the arc-tangent of the quantity given by the predicted midspan deflection, and dividing this by one-half the clear span length, i.e., a three-hinge mechanism is assumed.

One unique requirement of TM 5-1300 is the use of lacing reinforcement under certain conditions, particularly for very close-in explosions and for larger values of predicted support rotations. Lacing bars are reinforcing bars that extend in the direction parallel to the principal reinforcement and are bent into a diagonal pattern, binding together the two mats of principal reinforcement. It is obvious that the cost of using lacing is considerably greater than that of using stirrups because of the more complicated fabrication and installation procedures. Furthermore, it has been shown that one can achieve equivalent structural resistance with conventional shear reinforcement.

Behaviors of Structural Elements   **197**

5.2.2 Army Technical Manuals 5-855-1 (Department of the Army, 1986 and 1998), and UFC 3-340-01 (Department of Defense, June 2002)

TM 5-855-1 is intended for use by engineers involved in designing hardened facilities to resist the effects of conventional weapons. The manual includes design criteria for protection against the effects of penetrating weapons, contact detonation, and the blast and fragmentation from a standoff detonation. TM 5-855-1 does not call for the use of lacing, but does require minimum quantities of stirrups in all slabs subjected to blast. For beams, one-way slabs, and two-way slabs, the manual recommends a design ductility ratio (ratio of maximum deflection to yield deflection) of 5.0 to 10.0 for flexural design. The recommended response limits are given only in terms of ductility ratios, not support rotations. However, a more recent supplement to TM 5-855-1, *Engineer Technical Letter 1110-9-7* (Department of the Army, 1990), provides response limit criteria based on support rotations. TM 5-855-1 has been replaced by Unified Facilities Criteria issued by the Department of Defense (UFC 3-340-01, June 2002). The 1998 version of TM 5-855-1 and UFC 3-340-01 (2002) are for official use only.

5.2.3 ASCE Manual 42 (ASCE, 1985)

This manual was prepared to provide guidance in the design of facilities intended to resist nuclear weapons effects. It attempts to extract concepts from other documents such as the limited distribution design manual. However, the intent is to provide a more general approach. *ASCE Manual 42* presents conservative design ductility ratios for flexural response. Although the manual is an excellent source for general blast-resistant design concepts, it lacks specific guidelines on structural details.

5.2.3.1 Design of Blast Resistant Buildings in Petrochemical Facilities (ASCE, 1997)

This publication was developed to support the design and construction of blast-resistant industrial structures, with emphasis on the petrochemical industry. It provides a sound background on safety requirements and other relevant considerations such as load determination, types of construction, material behaviors under dynamic loads, dynamic analysis procedures, design approaches, typical details both for structural steel and structural concrete, architectural considerations, suggestions on upgrades, and examples.

5.2.3.2 Structural Design for Physical Security (ASCE, 1999)

Although not a design manual, this state-of-the-practice report contains detailed information on threat determination, load definition, structural behavior and design, design of windows and doors and utility openings, and retrofit of existing buildings.

5.2.4 DoD and GSA Criteria

The Department of Defense provided guidelines (UFC 4-010-01) to meet minimum antiterrorism standards for buildings (October 2003). Similarly, the General Services Administration (GSA) issued guidelines on how to assess and prevent progressive collapse (June 2003). Both publications recognize that progressive collapse may be the primary cause of casualties in facilities attacked with explosive devices. The DoD publication addresses these terrorism-related hazards and includes guidelines to mitigate them.

5.3 DISTANCES FROM EXPLOSION AND DYNAMIC LOADS

It is generally known that the greatest protection against blast effects is distance. The vehicle bomb has become a concern for buildings in the U.S. in regard to blast effects. Security and traffic route conditions that prevent close access by an explosives vehicle are significant protection measures. However, when distance cannot be reasonably guaranteed, specific design issues must be considered.

Generally, a prime location for a vehicle bomb is in a basement parking garage or in an exterior parking area. Structural protection against detonation in a basement parking area is rarely feasible. Major structural components may be vulnerable to very close-in blast effects. In such cases, good structural design may be successful only at limiting catastrophic progressive collapse of the building. In addition, loading is worsened by partial confinement of the interior explosion. By contrast, external detonations are vented to the outdoor environment and the potential distance to a major structural element is usually greater.

The manuals listed above and other available literature provide means to establish a reasonable approximation of loading (peak pressure and duration applied to exposed surfaces) to be expected for a given explosive threat. As openings are created (windows, walls, and floor slabs are breached), the blast loading on subsequent surfaces becomes very difficult to define and research in this area is continuing. Failed components or sections of structural elements may be propelled into other elements, thereby complicating the loading process. Additionally, heavy failed sections of floor and wall slabs may simply fall by gravity, causing a "pancake" effect on lower floors. Thus, structural details are necessary components for preventing progressive collapse.

5.4 MATERIAL PROPERTIES OF STEEL AND CONCRETE

5.4.1 Steel

There are various types of steel; some exhibit yield plateaus, while others do not, as illustrated in Figure 5.1. Most structural steels exhibit some sort of yielding at about 0.002 strain or lower. For high-strength steel, an offset approach must

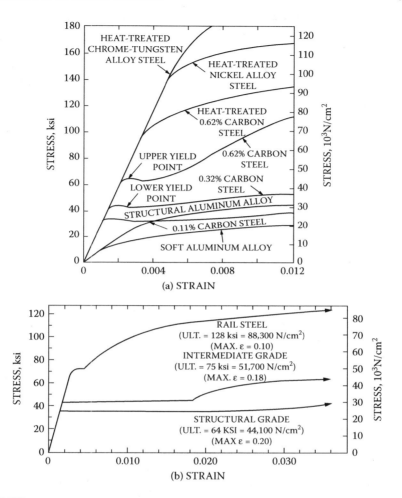

FIGURE 5.1 Typical stress–strain curves for various metals (ASCE, 1985).

be used to obtain an equivalent yield stress. The effect of loading rate on steel properties has been studied extensively and no clear answer yet exists. Several publications by Krauthammer et al. (1990, 1993, and 1994) address this issue for the analysis of reinforced concrete structures. Although for design applications a 25% strength increase is acceptable, the recommendations in TM 5-1300 should be followed.

5.4.2 Concrete

Typical uniaxial stress–strain curves for concrete in compression are presented in Figure 5.2 and the increase of strength with age is shown in Figure 5.3. The tensile capacity is about 10% of f'_c or six to seven times $(f'_c)^{0.5}$. The elastic

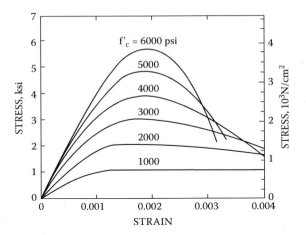

FIGURE 5.2 Typical stress–strain curves for concrete (ASCE, 1985).

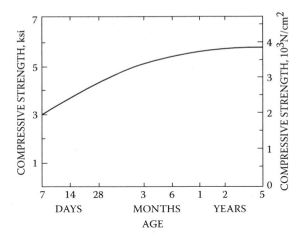

FIGURE 5.3 Effect of age on concrete compressive strength f'_c (ASCE, 1985).

modulus is proposed in ACI 318-05 (2005) as $E_c = 33w^{1.5}(f'_c)^{0.5}$, in which w is the unit weight (pounds per cubic foot) and f'_c is the uniaxial compressive strength (pounds per square inch). Also, one may use $\nu = 0.17$ for concrete. Concrete triaxial properties are presented in Figure 5.4 and Figure 5.5. One can observe that confining stresses have a significant positive effect on concrete behavior. The effect of transverse reinforcement must be assessed for each case, but it improves concrete strength. Clearly, concrete confined by spirals, steel tubes, or composite tubes or wraps is expected to outperform unconfined concrete.

Behaviors of Structural Elements

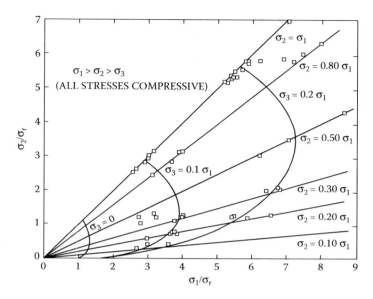

FIGURE 5.4 Normalized triaxial compression data (ASCE, 1985).

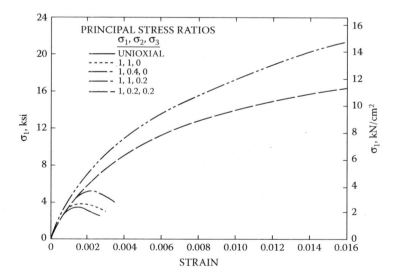

FIGURE 5.5 Effect of confining stresses on stress–strain properties of concrete (ASCE, 1985).

5.4.3 DYNAMIC EFFECTS

Generally, one adopts a design approach based on the concept that the reduced theoretical load carrying capacity of a structure should not be smaller than the magnified applied load combination. This way, one must use recommended load magnification factors and strength reduction factors to ensure that a structure will satisfy the anticipated demand imposed by the loading environment. Applying this approach to a severe dynamic loading domain requires one to account for two additional issues: (1) loading rate effects on material behavior and (2) inertia effects due to structural response. Therefore, the general design approach employed for protective structures is to the rate-enhanced material properties and perform a preliminary static design to obtain a tentative structural candidate. That structure is then analyzed under the dynamic loads to determine its ability to perform under design loads. The outcome of such an analysis will be used to determine required modifications that will lead to the final design.

These procedures are discussed in great detail in the previously cited design references and a brief summary of the key steps is presented next. Figure 5.6 shows the effect of dynamic loading and Figure 5.7 illustrates the effect of load cycling. Recommended enhancement values for steel and concrete are provided in a Department of the Army publication (1990).

5.5 FLEXURAL RESISTANCE

The general design approaches for reinforced concrete (RC) are based on ACI 318-05 (2005) procedures, as discussed extensively in various references (e.g., MacGregor and White, 2005). It is assumed for the ultimate state that one can use a rectangular stress block for concrete in compression, elasto-plastic steel,

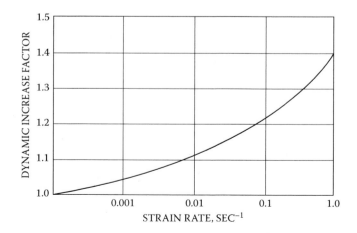

FIGURE 5.6 Dynamic increase factor for 28-day concrete in compression (ASCE, 1985).

Behaviors of Structural Elements

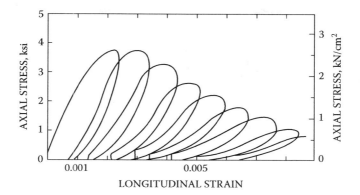

FIGURE 5.7 Repeated loading of plain concrete in compression (ASCE, 1985).

and concrete in tension is ignored. Here only the design formulations are presented and explained.

The ultimate theoretical moment capacity M_u of an RC beam on which an axial force is applied is derived as follows:

$$M_u = M_n + Pe \tag{5.1}$$

in which M_n is the pure-bending ultimate moment, P is the axial force, and e is the eccentricity of P with respect to the cross-sectional center. For cases where $P < P_b$ and P_b is the axial force to cause a balanced failure (crushing of top concrete fiber at the same time as the tension steel yields), the ultimate moment capacity is obtained as follows:

$$M_u = M_n = A_s f_y (d-d') + (A_s - A'_s) f_y (d - a/2) \tag{5.2a}$$

in which

$$a = [(A_s - A'_s) f_y]/0.85 f'_c b_w \tag{5.2b}$$

a equals the depth of the concrete compression block (also $\beta_1 k_u d$), A_s and A'_s represent tensile and compressive reinforcement areas, respectively, b_w is the beam width, t is the total thickness/depth, d' is the top compression face to center of compression steel distance, d is the effective depth (distance from top fiber to center of tensile reinforcement), f_y is the steel yield stress, f'_c is the concrete uniaxial strength, β_1 is a factor depending on concrete strength (0.85 for $f'_c \leq$ 4000 psi and reduced by 0.05 for each 1000 psi increase, but not lower than 0.65), and $k_u d$ is the depth of the neutral axis (NA) at ultimate moment.

Similar expressions can be computed for cases where $P > P_b$. Generally, the use of an "interaction diagram" is recommended for all cases where $P \neq 0$ and for finding the corresponding moment, thrust, and curvature relationships. These types of relationships also apply to one-way slabs; for these cases, the total flexural resistance per unit width (assuming symmetrical supports and restraints):

$$r_{yf} = 4\left(\frac{M_{el} + 2M_c + M_{e2}}{L^2}\right) \quad (5.3)$$

in which M_{e1}, M_{e2}, and M_c equal the corresponding values of M_n for ends 1, 2, and center, respectively, and L is the beam or slab-free span. For two-way slabs (Park and Gamble, 2000), the concept of the yield line theory and the requirement of a minimum load solution are usually employed. For a typical reinforced concrete slab (Figure 5.8 and Figure 5.9), the following is obtained:

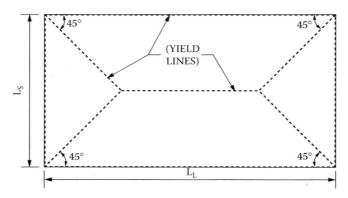

FIGURE 5.8 Idealized yield line pattern for two-way slabs (ASCE, 1985).

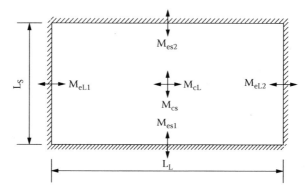

FIGURE 5.9 Definition of unit resisting moments in two-way slab (ASCE, 1985).

Behaviors of Structural Elements

$$r_{yf} = \frac{12(M_{es1} + 2M_{cs} + M_{es2})}{3L_S^2 - \frac{L_S^3}{L_L}} + \frac{12(M_{es1} + 2M_{cL} + M_{eL2})}{3L_S L_L - L_S^2} \tag{5.4}$$

It has been shown (Krauthammer et al., 1986; Park and Gamble, 2000) that these procedures tend to underestimate the structural load capacity. A more accurate approach is based on the inclusion of membrane effects, discussed later. Unfortunately, TM 5-1300 (Department of the Army, 1990) addresses only tensile membrane effects and ignores the significant benefits from compressive membrane effects.

When beam-slab systems are considered, these procedures should be combined in a rational manner. Some suggestions in design handbooks provide reasonable but conservative results. For flexural design of steel beams, one employs a similar approach according to the following expression:

$$M_p = f_y Z \tag{5.5}$$

in which M_p is the plastic moment capacity, f_y is the yield strength for the steel, and Z is the plastic modulus cited in steel design manuals (AISC, 2005).

5.6 SHEAR RESISTANCE

In general, two principal shear modes should be considered for structural concrete: the diagonal shear (punching, diagonal tension, or diagonal compression) and the direct shear (sometimes called pure or dynamic shear). This section reviews some of the proposed models for considering shear. Additional information is provided in the cited references. A diagonal shear case requires members to exhibit some flexural behavior. In the direct shear case, however, no such requirement is needed. For direct shear, the stresses are due to discontinuity (in geometry or in load) and the failure is local. For diagonal cases, the failure extends over a length about the member's depth. The general approach used for shear design is based on the following equation of equilibrium:

$$V_n = V_c + V_s \tag{5.6}$$

in which V_n is the theoretical (nominal) shear capacity, V_c is the concrete contribution, and V_s is the contribution of shear reinforcement. As for flexure, the factored shear capacity for a cross-section is equated to the factored shear force. The ACI Code (318-05, 2005) suggests the use of the following expression for concrete shear strength without shear reinforcement:

$$V_c = 2(f'_c)^{1/2} b_w d \text{ [lb]} \tag{5.7}$$

in which b_w is the width of the beam's web, and d is the beam's effective depth. Although another expression is provided in ACI 318-05, it is less rational than Equation (5.7), and can overestimate the value of V_c. Equation (5.7) should be adjusted for cases with acting axial forces, as discussed in the literature (ACI, 2005; MacGregor and White, 2005; Department of the Army, 1990). Once V_c is known, one can use Equation (5.6) to compute V_s, and the amount of shear reinforcement is determined as follows:

$$V_s = \frac{A_v f_y}{s} d \qquad (5.8)$$

in which A_v is the total area of steel per unit width over the spacing s of the shear reinforcement.

The coupling of shear and other effects such as flexure and axial forces has been presented in the literature (Krauthammer et al., 1990; Collins and Mitchell, 1991; MacGregor and White, 2005), and is also discussed later. For considering the effects of shear on deep members, one should use the ACI recommendations (2005). An alternative approach for computing V is based on the modified compression field theory (Collins and Mitchell, 1991) that employs a more rational attempt for describing shear effects. For slabs, the critical locations for shear are near supports and/or columns (Park and Gamble, 1999). The issue of direct shear is addressed in ASCE (1999), and discussed briefly later.

5.7 TENSILE AND COMPRESSIVE MEMBERS

A primary concern for buildings is the widespread use of columns. Columns (particularly exterior columns) should be avoided in protective structures; continuous walls are preferred. Columns are generally very stiff against lateral flexural response because of the relatively large quantities of longitudinal reinforcement. In addition, compressive forces from gravity loads enhance lateral strength at initial loading. The consequence is a structural element that will likely respond in a brittle mode such as shear unless specifically detailed with adequate confining reinforcement. Fortunately, columns are not generally vulnerable to "far-away" blast loadings that tend to have lower peak pressures and engulf all sides of a column. However, columns may be vulnerable to intense close-in blast loading, and it is imperative that the total structure be capable of redistributing gravity loads in case one or more columns are destroyed.

RC columns should be designed according to ACI (2005), as discussed in MacGregor and White (2005) and Department of the Army (1986 and 1990). The pure axial capacity of RC columns can be computed as follows:

$$P_0 = 0.85 \, f'_c(A_g - A_{st}) + f_y A_{st} \qquad (5.9)$$

Behaviors of Structural Elements

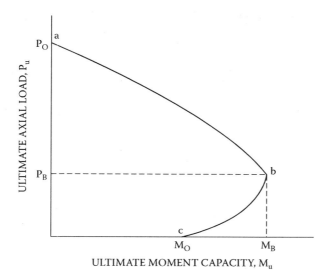

FIGURE 5.10 Typical interaction diagram for reinforced concrete column (ASCE, 1985).

Moment-thrust interaction diagrams are used for columns subjected to both flexure and axial forces, and discussed in MacGregor and White (2005) and shown in Figure 5.10. Also, as for flexure and shear, one should note the use of strength reduction factors Φ and overload factors α to modify the above relationships.

Similarly to structural concrete, structural steel design for protective applications should be performed according to Department of the Army (1990) specifications that refer also to the AISC manual. One should employ the latest AISC information (2005 and 2006) and appropriate books on steel design (McCormac and Nelson, 2003).

Tension and compressions members are analyzed as follows:

$$P_n = f_y A_g \tag{5.10}$$

P_n is the axial tensile force, f_y is the yield strength, and A_g is the gross cross-sectional area. In compression, however, one needs to ensure that a member will not buckle under the applied force.

5.8 PRINCIPAL REINFORCEMENT

Primary design issues include the placement (location) and continuity of principal reinforcement in slabs. In floor and roof slabs of protective structures, the principal reinforcement is usually equal in each face of the slab and continuous throughout the span. If lengths prohibit continuous bars, mechanical couplers or long splice lengths (prescribed in blast-resistant design manuals) are required. Additionally, all principal reinforcement is well anchored into the supporting elements. In contrast, floor slabs in buildings are primarily reinforced in tension zones and are vulnerable to "reversed" loading. Such reversed loading is likely when shock waves from an explosion project upward and load floor slabs located at levels higher than the explosive source. When conventional slabs are loaded underneath by blast pressure, the reinforcement is generally in the "wrong" place. Consequently, slabs are vulnerable to catastrophic failure and will probably come to rest in a pancake formation.

5.9 CYLINDERS, ARCHES, AND DOMES

The fundamental assumption is that these structures are rarely perfectly constructed and therefore the imposed pressures will be highly nonuniform, as shown in Figure 5.11. The internal thrust, shear, and moment (Crawford et al., 1974) are:

$$P_\theta = -qR + \frac{\overline{p}R}{3}\cos 2\theta \qquad (5.11a)$$

$$V_\theta = -\frac{2}{3}\overline{p}R \sin 2\theta \qquad (5.11b)$$

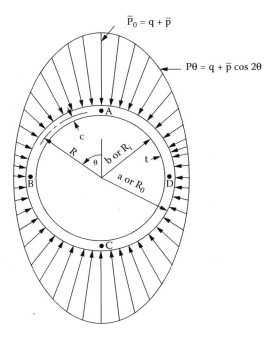

FIGURE 5.11 Circular ring subjected to nonuniform external pressure (Crawford, et al., 1974).

TABLE 5.1
Curvature Correction Factors for Straight Beam Formula

Section	$\frac{R}{C}$	Factor K Inside Fiber	Factor K Outside Fiber	y_o^*
	1.2	2.89	0.57	0.305R
	1.4	2.13	0.63	0.204R
	1.6	1.79	0.67	0.149R
	1.8	1.63	0.70	0.112R
	2.0	1.52	0.73	0.090R
	3.0	1.30	0.81	0.041R
	4.0	1.20	0.85	0.021R
	6.0	1.12	0.90	0.0093R
	8.0	1.09	0.92	0.0052R
	10.0	1.07	0.94	0.0033R

Note: K is independent of section dimensions.

Source: Crawford et al., 1974.

$$M_\theta = \frac{\bar{p}R^2}{3}\cos 2\theta \tag{5.11c}$$

in which q is the uniform pressure component and $\bar{p}$ is the maximum amplitude of the nonuniform component.

The beam theory may be employed for very slender cylinders (R/C > 10) to obtain the internal stresses. For thick cylinders, the following may be used. K (the Winkler–Bach correction factor) is provided in Table 5.1.

$$\sigma = K\left[\frac{Mc}{I} + \frac{P}{A}\right] \tag{5.12}$$

As for deflections, again based on elasticity assumptions, one can use the following expression:

$$\delta_{max} = \delta_{\bar{p}} + \delta_q = \frac{\bar{p}R}{3}\cos 2\theta\left(\frac{1}{q_{cr}-q}\right) \tag{5.13a}$$

in which

$$q_{cr} = \frac{3EI}{R^3} \tag{5.13b}$$

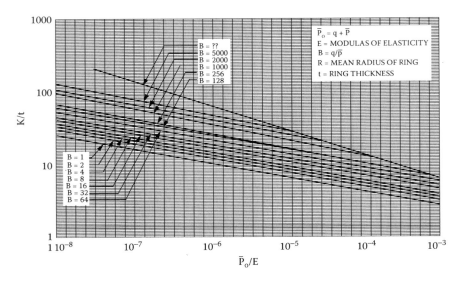

FIGURE 5.12 Critical buckling ratios for ring loaded with nonuniform pressure (Crawford, et al., 1974).

The maximum moment can be computed from the following expression:

$$M_{max} = M_{\bar{p}} + M_q = M_\theta + qR\delta_{max} = \frac{\bar{p}R^2}{3}\cos 2\theta \left(\frac{1}{1-q/q_{cr}}\right) \quad (5.14)$$

For steel rings, buckling must be introduced (see Figure 5.12); β is the ration p/q. The deflection of thin rings can be estimated from Figure 5.13.

For reinforced concrete rings, the concept of an interaction diagram should be employed, with the assumption that the amounts of tensile and compressive reinforcements are equal. It is rational to require a uniform cross-section, and the ultimate moment becomes:

$$M_n = \rho d^2 f_y \left(2 - \frac{t}{d}\right) \quad (5.15)$$

in which ρ is the tensile reinforcement ratio defined by A_s/bd, A_s is the cross-sectional area of the tensile steel, b is the cross-sectional width, and d is its effective depth.

It is advisable to design the cross-section such that the axial force is lower than the balance point P_b for the cross-section under combined moment and thrust (on the interaction diagram in Figure 5.10). At the collapse pressure, the ultimate deflection will have the following value:

Behaviors of Structural Elements

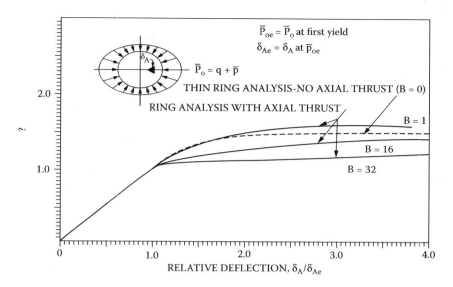

FIGURE 5.13 Load deformation relations for homogeneous ring with R/t = 10 (Crawford et al., 1974).

$$\frac{\delta_u}{R} = \frac{f_y/E_s}{2\left[1 - \frac{87,000}{87,000 + f_y}\left(\frac{3\beta-1}{3\beta+1}\right)\right]} \quad \text{[psi]} \qquad (5.16)$$

This information can be used for the design of horizontal and vertical cylinders, but one needs to consider the following differences between these two types:

1. Vertical cylinders carry a significant axial force from the closure area.
2. Vertical cylinders experience differential loading conditions between the top portion ($2R_o$ depth to the surface) and the rest of the cylinder (below $2R_o$ depth); R_o is the cylinder's outer radius (Figure 5.11).

Further information on this topic is available in Crawford et al. (1974). Circular arches can be treated as part of the cylinders and using the corresponding expressions on the basis of the following two response modes (ASCE, 1985):

1. Compression mode: the whole arch is compressed uniformly and the maximum structural resistance for a semicircular arch per unit curved area is derived, as follows:

$$r_{yc} = \left(0.85 f_c' + 0.9\rho_t f_y\right)\frac{A_c}{R} \qquad (5.17)$$

2. Flexural mode: end wall effects are neglected (this tends to underestimate the resistance), and for pin-supported semicircular cases, one may use the following expression:

$$r_{yf} = \frac{8 M_n}{(\beta R)^2}\left[1 - \left(\frac{\beta}{\pi}\right)^2\right]\left(\frac{b}{B}\right) \quad (5.18a)$$

For a fixed support, one can use the following expression:

$$r_{yf} = \frac{11.7 M_n}{(\beta R)^2}\left[1 - 0.6\left(\frac{\beta}{\pi}\right)^2\right]\left(\frac{b}{B}\right) \quad (5.18b)$$

in which β is the half central angle of arch (radians), R is the arch radius to the center of thickness, M'_u is the ultimate moment capacity of the concrete rib (also defined as M_n), b is the rib width, and B is the rib center–center-spacing; b/B = 1 for uniform thickness. The effects of end walls can be estimated based on information provided in ASCE (1985).

For domes, the approach is based on the elastic theory as for cylinders and arches, and the corresponding equations can be obtained. A summary of this approach can be found in Crawford et al. (1974, pp. 714–718).

5.10 SHEAR WALLS

Shear walls are designed to increase the lateral load capacity of a structure and their general configuration is illustrated in Figure 5.14. The load deformation relationship is shown in Figure 5.15. In the static domain, the following expressions were proposed in (ASCE, 1985). The horizontal static load resistance of a wall at the first cracking stage:

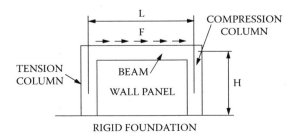

FIGURE 5.14 Principal elements of shear wall (ASCE, 1985).

Behaviors of Structural Elements

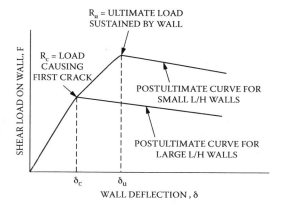

FIGURE 5.15 Characteristic load deflection curves for shear walls (ASCE, 1985).

$$R_c = 0.1 f'_c Lt \tag{5.19a}$$

in which f'_c is the concrete uniaxial compressive strength, L is the wall length between centers of columns, and t is the wall thickness. The ultimate static wall resistance R_u is computed from the following:

$$R_u = \frac{C}{1+\frac{10P}{C}} + \frac{2.1P}{\frac{P}{C}+0.6} \tag{5.19b}$$

The deflection at first cracking is obtained as follows:

$$\delta_c = \frac{R_c H}{E_c}\left(\frac{H^2}{3t} + \frac{2.2}{Lt}\right) \tag{5.19c}$$

while the deflection at the ultimate resistance is derived from the following expression:

$$\delta_u = \frac{24\delta_c H}{L} \quad (\delta_u \geq \delta_c) \tag{5.19d}$$

in which H is the wall height, A_s is the steel area in the column on the compression side, and

$$C = A_s f'_c \left[15 + 1.9(L/H)^2\right] \tag{5.19e}$$

$$P = f_y \rho t[H + L] \tag{5.19f}$$

5.11 FRAMES

Typical one-story frame structures are shown in Figure 5.16; the load-versus-response mode for uniformly distributed loads (nuclear blast loads or distant HE blast loads) is given in Figure 5.17. The total resistance can be computed from this assumed mechanism and by equating the total internal work done at the plastic hinges (sum of all plastic moments times their corresponding rotations) with the total work done by external forces through their corresponding displacement):

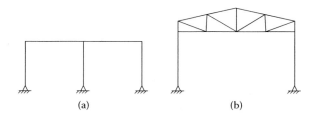

FIGURE 5.16 Typical framed structures (ASCE, 1985).

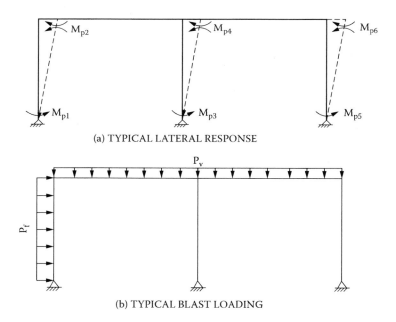

FIGURE 5.17 Nuclear or distant HE blast loading and response of typical framed structure (ASCE, 1985).

Behaviors of Structural Elements

$$P_F = 2 \Sigma M_{pn} / L^2 \text{ (force/length)} \quad (5.20)$$

where ΣM_{pn} is the summation of top and bottom moment capacities of columns and L is the frame height. It is assumed that the frame will fail in flexure. The possibilities of shear failures must be considered as well. For localized effects, the response may be more complicated and the critical members can be assessed according to the procedure proposed by Krauthammer et al. (1990, 1993a, 1993b, and 1994).

5.12 NATURAL PERIODS OF VIBRATION

As will be discussed in the next chapter, the dynamic response of a structural element is related to its natural period of vibration and to the duration of the applied loading pulse. Therefore, once a preliminary static design is obtained for the structural elements under consideration, one needs to also estimate their natural period. The information is found in the design manuals cited earlier and also in various handbooks (Harris and Piersol, 2002). Some of the typical cases based on ASCE (1985) are presented next.

Beams and one-way slabs (ρ is the tensile steel ratio):
Both ends simply supported:

$$T_f = \frac{L^2}{425,000\sqrt{\rho}} \quad [\text{sec}] \quad \text{(English units)} \quad (5.21a)$$

Both ends fixed:

$$T = \frac{L^2}{850,000\sqrt{\rho}} \quad (5.21b)$$

One end fixed, one simply supported:

$$T_f = \frac{L^2}{637,000\sqrt{\rho}} \quad (5.21c)$$

For thick slabs, one must employ a correction factor:

$$T_l = F_s T_f \quad (5.22a)$$

where

$$F_s = \sqrt{1 + \frac{2}{1-v}\left(\frac{d}{L}\right)^2} \qquad (5.22b)$$

in which v is Poisson's ratio.
Two-way slabs:

$$\frac{1}{T_2} = \frac{1}{T_{IS}} + \frac{1}{T_{IL}} \qquad (5.23)$$

T_2 is the two-way slab period and T_{IS} and T_{IL} are the corresponding one-way slab periods (seconds) in the short and long span directions, respectively. The effect of added masses can be considered as follows:

$$T = T'\sqrt{1 + \frac{m_a}{m}} \quad [\text{sec.}] \qquad (5.24)$$

in which T' equals T for the original element, m_a is the added mass, and m is the original mass. For steel beams, one may use the information in Table 5.2.
Arches:
Compression mode:
  Reinforced concrete:

$$T \approx \frac{R}{22,000}\sqrt{\frac{m_t}{m_a}} \qquad (5.25)$$

m_t is the total mass for the responding system per unit of arch surface area and m_a is the arch mass per unit of arch surface area.
  Steel:

$$T \approx \frac{R}{32,000}\sqrt{\frac{m_t}{m_a}} \qquad (5.26)$$

Antisymmetric flexural mode:
  Reinforced concrete:

$$T = \frac{C_t \theta' R^2}{270,000 \, d_c \sqrt{\rho_t}}\sqrt{\frac{m_t}{m_a}} \qquad (5.27)$$

TABLE 5.2
Natural Period of Vibration for Steel Beams

Member	Period
simply supported beam	$T = 0.64 L^2 \sqrt{\dfrac{W}{gEI}}$
simply supported beam with concentrated load at L/2	$T = 0.91 \sqrt{\dfrac{W_c}{g} \dfrac{L^3}{EI}}$
propped cantilever	$T = 0.42 L^2 \sqrt{\dfrac{W}{gEI}}$
propped cantilever with concentrated load at L/2	$T = 0.61 \sqrt{\dfrac{W_c}{g} \dfrac{L^3}{EI}}$
fixed-fixed beam	$T = 0.28 L^2 \sqrt{\dfrac{W}{gEI}}$
fixed-fixed beam with concentrated load at L/2	$T = 0.45 \sqrt{\dfrac{W_c}{g} \dfrac{L^3}{EI}}$

Source: ASCE, 1985.
T = period (seconds). W = supported weight (including beam) per unit length. W_c = total weight concentrated at midspan. E = modulus of elasticity. I = moment of inertia. g = gravitational constant.

Simply supported:

$$C_t = \frac{\left[\left(\dfrac{\pi}{\theta'}\right)^2 + 1.5\right]}{\left[\left(\dfrac{\pi}{\theta'}\right)^2 - 1\right]} \tag{5.28}$$

Fixed:

$$C_t = 0.64 \frac{\left[\left(\frac{\pi}{\theta'}\right)^2 + 1\right]}{\left[\left(\frac{\pi}{\theta'}\right) - 0.5\right]} \tag{5.29}$$

Other formulations (for different systems) can be found in previously cited manuals and handbooks.

5.13 ADVANCED CONSIDERATIONS

5.13.1 MEMBRANE BEHAVIOR

As discussed in structural dynamics textbooks such as the one by Biggs (1964), a fixed supported beam or one-way slab element under slowly applied uniform load initially undergoes elastic deflection. As loading continues, plastic hinges first form at the supports and later at midspan. The load deflection curve associated with the formation of the failure mechanism is referred to as *resistance function*. A common approach is to develop a single-degree-of-freedom model for dynamic analysis, using the resistance function to define the peak resistance (ultimate capacity) and response limit of the element.

The ultimate capacity and response limit can be significantly enhanced by membrane forces. The ultimate capacity may be enhanced by compressive membrane forces and response limit may be enhanced by tensile membrane forces as shown in Figure 5.18. As discussed by Park and Gamble (2000), the ultimate flexural capacity is enhanced by compressive membrane forces in slabs whose edges are restrained against lateral (outward) movement (segment OA in Figure 5.18). As the slab deflects, changes in geometry tend to cause its edges to move

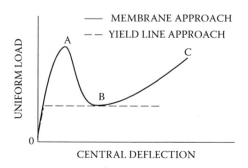

FIGURE 5.18 Slab resistance models.

Behaviors of Structural Elements

outward and react against stiff boundary elements. The membrane forces enhance the flexural strength of the slab sections at the yield lines.

Research has shown that compressive membrane forces can increase the ultimate capacities of both one-way and two-way slabs very significantly above the yield line (pure flexure) resistance, as discussed by Park and Gamble (2000). An ultimate capacity of 1.5 to 2 times the yield line value is quite common. Actually, past design manuals have not fully utilized compressive membrane theory in defining resistance functions, but criteria currently under development closely follow the theory presented by Park and Gamble (2000). Designers should be cautious in relying on compressive membrane behavior in buildings, but may find confidence in applying the theory to slab systems that include stiff supporting beams. In any case, knowledge that compressive membrane forces exist will provide a "hidden" safety factor.

Although compressive membrane behavior may be limited, tensile membrane behavior can be a significant factor in limiting catastrophic failure and progressive collapse. The tensile membrane region (segment BC in Figure 5.18) is where the load resistance increases as the deflection increases. As discussed by Park and Gamble (2000), after the ultimate load resistance has been reached, the load resistance decreases until membrane forces in the central region of the slab change from compression to tension (segment AB in Figure 5.18). In pure tensile membrane behavior, cracks penetrate the whole thickness, and yielding of the steel spreads throughout the central region of the slab. The load is carried mainly by reinforcing bars acting as a tensile net or membrane, and thus the resistance increases as the steel is strained until rupture occurs. The increase in load resistance accompanying this action is often called *reserve capacity*. The derivation of the membrane behavior is described next, based on the approach in Park and Paulay (2000).

Consider a reinforced concrete slab with a typical yield line pattern as shown in Figure 5.19. The yield lines represent locations of continuous plastic hinges that enable the slab to "fold" as a kinematic mechanism. The slab is then divided into strips in both the x and y directions and the locations of the plastic hinges are noted for each such strip. According to the yield line theory, one employs the virtual displacement approach, assuming a deflection δ for the segment EF, then assumes that the work done by the applied uniformly distributed load over the entire slab is equal to the work done by the internal yield moments (both positive and negative) along the plastic hinge lines. The deformed shape of one such strip is shown in Figure 5.20.

One can focus on part of the deformed strip, and redraw it as a free body diagram (Figure 5.21) in which C_c and C_s are the compressive concrete and steel forces, respectively, T is the tensile reinforcement force, and c is the neutral axis depth.

Park and Gamble (2000) showed that the virtual work consideration will lead to the derivation of Equation (5.30) that represents the peak compressive membrane capacity. The tensile membrane capacity is derived for the model shown in Figure 5.22 that leads to the tensile membrane Equation (5.32). Note, however,

220   Modern Protective Structures

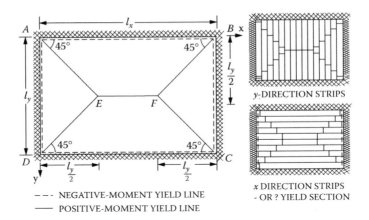

FIGURE 5.19 Typical yield line pattern for clamped RC slab and strips (Park and Gamble, 2000).

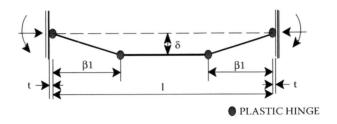

FIGURE 5.20 Deformed slab strip.

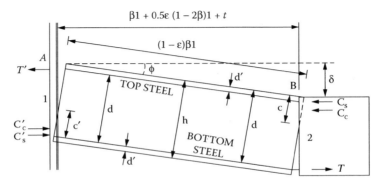

FIGURE 5.21 Free body diagram for deformed slab strip (Park and Gamble, 2000).

Behaviors of Structural Elements

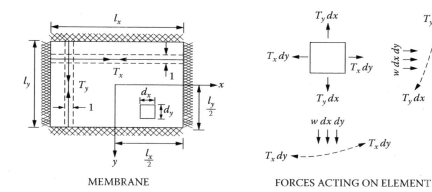

FIGURE 5.22 Tensile membrane model.

that Equation (5.32) is a straight line through the origin (zero load and deflection). In reality, the tensile membrane is activated not from the origin, but at Point B in Figure 5.18. The result will be a straight line parallel to Equation (5.32), but at a higher resistance as also found experimentally (Park and Gamble, 2000). Krauthammer (1984) and Krauthammer et al. (1986) modified both the compressive and tensile membrane formulations to correct for experimental findings and simplify the previously noted expressions.

$$w_u l_y^2 [3(\ell_x/\ell_y) - 1]/24 = 0.85 f'_c \beta_1 h^2 \{(l_x/l_y)(0.188 - 0.141\beta_1) + (0.479 - 0.245\beta_1) +$$
$$+ (\varepsilon'_y/16)(l_y/h)^2(l_x/l_y)(3.5\beta_1 - 3) + (\varepsilon'_y/16)(1_y/h)^2[2(l_x/l_y)(1.5\beta_1 - 1) + (0.5\beta_1 - 1)] +$$
$$- (\beta_1/16)(l_x/l_y)(l_y/h)^4[(\varepsilon'_x)^2(l_x/l_y) + (\varepsilon'_y)^2]\} - (3.4 f'_c)^{-1}[(T'_x - T_x - C'_{sx} + C_{sx})^2 +$$
$$+ (l_x/l_y)(T'_y - T_y - C'_{sy} + C_{sy})^2] + (C'_{sx} + C_{sx})(3h/8 - d'_x) +$$
$$+ (T'_x - T_x)(d_x - 3h/8) + (C'_{sx} + C_{sx})[(l_x/l_y)(h/4 - d'_y) + h/8] +$$
$$+ (T'_y - T_y)[(I_x/I_y)(d_y - h/4) + h/8] \quad (5.30)$$

in which

$$\varepsilon'_x = \varepsilon_x + 2t_x/1_x \text{ and } \varepsilon'_y = \varepsilon_y + 2t_y/l_y \quad (5.31)$$

$$\frac{w l_y^2}{T_y \delta} = \frac{\pi^3}{4 \sum_{n=1,3,5}^{\infty} (-1)^{(n-1)/2} \left\{ 1 - \frac{1}{\cosh\left[(n\pi l_x/2l_y)\sqrt{T_y/T_x}\right]} \right\}} \quad (5.32)$$

A reserve capacity is important in the design of protective structures since moderate to severe damage is often acceptable if collapse is avoided. It is possible for a slab's peak reserve capacity to equal or be greater than the ultimate capacity. Tensile membrane behavior can occur only if the principal reinforcement extends across the entire length of the slab (or is properly spliced) and is well anchored into supporting members or adjacent elements. Additionally, placing stirrups throughout the length of the slab helps to bind the top and bottom layers of principal steel, thus enhancing the integrity of the concrete element during the large deflections associated with tensile membrane behavior. Consideration also should be given to the strength of the supporting member for determining the extent of potential reserve capacity.

Although the information on membrane effects in structural concrete slabs has been known for more than three decades (Park and Gamble, 2000; Wood, 1971), its application for both numerical analysis and design has not yet matured. Regarding blast-resistant structural behavior, the elastic and elasto-plastic models in use since the early 1960s (Biggs, 1964) have been the most frequently used approaches until recently. However, as discussed earlier, these models considerably underestimated the load carrying capacities of slabs. These limitations, although acceptable for design, are not acceptable for the behavior analysis of slab-type structures subjected to severe dynamic loads. The durations of blast loads are about 1000 times shorter than those of earthquakes, and under such conditions the inertia effects dominate behavior. A slab that performs well under slow loading rates may collapse under faster loads. Although in statically loaded slabs, the capacity underestimation was a hidden safety factor, under severe dynamic conditions, the load–response relationship is much more complicated, and to determine the structural safety one needs to know how the loads are resisted. Obviously, analysts and designers could no longer accept "hidden" safety factors and attention had to be given to more precise definitions of structural resisting mechanisms.

The model presented by Park and Gamble (2000) included an expression for estimating the peak load carrying capacity w_{max} of one-way reinforced concrete slabs in the compression membrane domain. It was noted on the basis of test data obtained by various researchers that this capacity was associated with central deflections to slab thickness ratios (δ/h) between 0.1 and 0.89 (Point A in Figure 5.18). Woodson (September 1994) reported δ/h ratios near 0.3 for one-way slabs with length-to-effective-depth (L/d) ratios of 10. However, for deep one-way slabs (L/d 3 and 5), Woodson (November 1994) reported that the δ/h ratios varied between approximately 0.03 and 0.07. Obviously, the peak load capacity in deep slabs is reached at much smaller δ/h values than those in more slender slabs. Park and Gamble (2000) recommended that for slabs with L/h 20, the peak capacity w_{max} could be estimated at $\delta/h = 0.5$. For slabs with lower L/h ratios, the δ/h values are expected to be lower (peak capacity will be reached earlier). Furthermore, the peak load will be reached at lower δ/h values in strips (one-way action), as confirmed by Woodson (September and November 1994).

The transition into the tensile membrane domain was noted to occur in the range $1 \leq \delta/h \leq 2$, and it corresponded to the yield line capacity of the slab (Point B in Figure 5.18). Beyond that transition, the resistance is governed by the tensile strength of the steel and such a model was also discussed by Park and Gamble (2000) (segment BC in Figure 5.18). Employing these models for both the

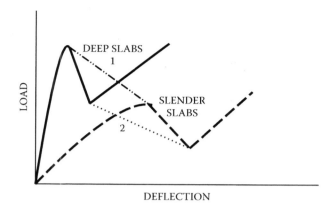

FIGURE 5.23 Generalized resistance function for slabs.

compression and tension membranes enabled one to describe a complete load–deflection relationship for structural concrete slabs subjected to uniformly distributed loads and for several typical support conditions.

Krauthammer et al. (2003) incorporated these observations into the previous model (Krauthammer et al., 1986) to define a general resistance function for slender, intermediate, and deep slabs, as shown in Figure 5.23. One notes two resistance functions: one for deep slabs and one for slender slabs. The peak compression membrane capacities are connected by line 1 and the transitions into the tensile membrane response are connected by line 2. These lines represent an assumed linear extrapolation between the deep and slender slab responses. Accordingly, the resistance function for an intermediate slab would be drawn such that its peak compressive membrane would be on line 1 and its transition to the tensile membrane would be on line 2.

Compressive membrane action can develop if sufficient resistance to both outward motion and rotation exists at the supports. Similarly, preventing inward motion at the supports is essential for the development of tensile membrane effects. Park and Gamble (2000) discussed the requirements of support restraint to enable compressive membrane action. They showed that a compressive membrane can be achieved when the lateral support stiffness has the value $S = 2E_c/(L/h)$, in which L/h is the span-to-slab-thickness ratio and E_c is the slab's modulus of elasticity. The ratio S/E_c defines the relative support stiffness, as compared with the modulus of elasticity. Although significant compressive membrane enhancement can be achieved even for low lateral support stiffness values, these support conditions must be adequately considered to ensure compressive membrane behavior.

Krauthammer (March 1984) showed that considering membrane effects (both in compression and tension) greatly improved the ability to explain some observations in slabs tested under explosive loads. These preliminary results motivated further attention to detailed structural models that included both improved membrane effects and direct shear resistance mechanisms (Krauthammer et al., April 1986; Krauthammer, August 1986). This improved model was a modified version of that presented by Park and Gamble (2000) that could be used for two-way slabs, and it included the effects of externally applied in-plane forces. In these later studies, such load–deflection relationships were adopted for dynamic structural analysis based on a single-degree-of-freedom (SDOF) approach.

Mass and stiffness parameters for the structural system under consideration are selected on the basis of the type of problem. These include the load source, type of structure, and general conditions for load application to the structure (localized load on a small part of the structure, distributed load over a large part of the structure). Obviously, the expected behavioral domain (linear elastic, elastic-perfectly plastic, nonlinear, etc.) will affect this relationship. The general approach for selecting such parameters was discussed in detail by Biggs (1964), and most design manuals contain similar procedures. Although neither Biggs nor the design manuals provide information on the treatment of fully nonlinear systems by SDOF simulations, such approaches were presented in the literature cited above and can be employed for analysis and design, as discussed by Krauthammer et al. (1990).

The effect of structural damping is usually small (typically 2 to 5%), but this should be assessed for the specific case under consideration. Such solutions are usually obtained by employing a numerical approach, as discussed in Biggs (1964) and Clough and Penzien (1993). The system response will depend not only on the magnitude of the force, but also on the relationship between the dynamic characteristics of the force and the frequency characteristics of the structure. These are defined by the ratio K/M and the effect of damping. A detailed discussion of these issues is presented in Biggs (1964), Clough and Penzien (1993), and Chapter 7 of ASCE (1985). The various design manuals (Department of the Army, 1986 and 1990; ASCE, 1985) contain dynamic response charts and tables based on SDOF considerations and these can be used for design.

The differences in the approaches discussed by Biggs (1964), the various design manuals cited above, and the modified approach proposed by Krauthammer (1984 and 1986) and Krauthammer et al. (April 1986) included the structural resistance function. Instead of employing the classical concept for a resistance function (elastic, elastic–perfectly plastic, or plastic), the term $R(x)$ was represented by the modified load displacement relationship that also included membrane effects as described above. That relationship included membrane effects (both in compression and tension for either one- or two-way slabs), the influence of externally induced in-plane forces, shear effects, and strain rate effects on the materials. Another modification extended the application of this approach to cases where the loads were localized rather than uniformly distributed.

When more advanced numerical approaches are considered important (finite element or finite difference methods), one needs to ensure that the approach can represent membrane effects correctly. Certain types of elements may include tension membrane effects but lack the ability to exhibit compression membrane behavior. Others may introduce numerical locking unrelated to the physical

Behaviors of Structural Elements

phenomenon of compression membrane behavior. Obviously, to ensure that the results are interpreted correctly, analysts and designers must understand the characteristics of the particular numerical tool they intend to use. Detailed discussion on such topics are available in various books on advanced structural analysis approaches (Bathe, 1996).

5.13.1.1 Membrane Application for Analysis and Design

The design of slabs in typical construction is often governed by serviceability requirements rather than by strength considerations. Therefore, the consideration of full compression and tension membrane enhancements may not be feasible. Since the peak compressive membrane capacity is expected to be associated with a central deflection of about 0.5 h, it is quite possible that such conditions would meet ACI deflection control requirements (ACI, 2005). Under certain conditions, considering some parts of membrane contribution could be advantageous. Furthermore, for special cases, the consideration of both compression and tension membrane effects could be justified.

Park and Gamble (2000) showed that by considering compression membrane effects, designers could reduce the total amount of steel in the slab to less than that required by the yield line theory, even though some additional steel must be added to the supports to ensure sufficient restraining of the slab. However, designers need to consider loading patterns on the floor slab to prevent over-stressing the tie reinforcements in the supporting beams. In summary, one can include consideration of compressive membrane effects, but the detailing of steel in the support regions must be carefully examined to ensure that the enhanced slab capacity will not jeopardize the support integrity.

The consideration of membrane effects in special structures could be much more attractive. In fortifications, one must rely on every possible contribution to structural resistance for surviving severe loading environments. Under such conditions, survivability rather than serviceability often governs the required performance criteria. Obviously, a slab that can resist higher loads, even if it is associated with large deflections, could be very useful. Test data discussed by Krauthammer (April 1986) and Woodson (September and November, 1994) can be used to show such advantages. However, very careful attention must be given to shear reinforcement and to the support design to ensure that the slab can reach the corresponding resistance and deformation levels.

Another type of structure that could benefit from membrane effect considerations is the culvert. Again, based on operational needs for such structures, one may limit the design to only compressive membrane enhancement if deflections must be controlled. Otherwise, both compressive and tensile membrane effects could be considered. In addition to the information presented by Park and Gamble (2000), two recent examples of how compressive membrane enhancement could be utilized in design may illustrate this approach.

Krauthammer et al. (1986) adopted the approach in Park and Gamble (2000) for the redesign of reinforced concrete culverts. A one-barrel box culvert 40 ft long with a 12 ft × 12 ft opening was considered. Slab thicknesses were 11.5 in. for the roof, 12 in. for the floor, and 8 in. for the walls. The findings confirmed the expectations in Park and Gamble as follows. First, in an existing slab, even small support stiffness can provide significant compressive membrane enhancements. When the ratio S/E_c was varied between 0.005 and 0.16 for the roof slab, the compressive membrane enhancements varied between 1.1 and 1.47. Second, when the slab was redesigned (the same steel was redistributed to accommodate compressive membrane action), its compressive membrane enhancement varied between 1.2 and 1.65. Third, a large S/E_c ratio would not ensure a much higher compressive membrane enhancement. When S/E_c was set to 50 in the original slab, the compressive membrane enhancement increased from 1.47 to 1.5. The slab was then checked for meeting serviceability requirements. It was found that cracking would be controlled by the given amount of steel and the δ/h ratio varied between 0.5 and 0.2 for the given range of S/E_c ratios. For the roof slab, these ratios corresponded to deflections between 0.57 and 2.3 in. For a span of 144 in., these deflections correspond to deflection to depth ratios (δ/h) in the range between 0.004 and 0.016 that are quite acceptable according to ACI deflection control limits (ACI, 2005).

Meamarian et al. (1994) continued the development of this approach by considering improved material and failure models and implemented them in an analysis and design computer code. In that approach, the general method presented by Park and Gamble (2000) was used to derive the ultimate load, deflections, and sectional forces. Then, the modified compression field theory (Collins and Mitchell, 1991) was used to compute the internal stresses, strains, crack angles, and total deflections. The proposed approach was used to simulate the behavior of ten slabs for which test data were available. The average ratios of computed results to test data for axial force, moment, load, and deflection were 1.07, 1.07, 1.04, and 1.0, respectively. However, the standard deviations for these comparisons were in the range between 0.13 and 0.38, showing that this approach required additional attention before it could be used in support of design activities. Interestingly, for these slabs the compression membrane enhancement was estimated in the range between 1.7 and 2.9, as compared to the yield line capacity. Furthermore, this enhancement was important for thinner slabs with moderate amounts of steel ($0.007 \leq \rho \leq 0.012$), but it did not change for slabs with lower reinforcement ratios ($\rho = 0.005$) in which ρ is the reinforcement ratio.

5.13.2 Direct Shear

Direct shear is a technical term that describes a localized shear response of a structural concrete element. It can appear at areas of discontinuity (geometrical or load). Examples include the support area and the edges of the loaded area (if the load is not applied over the entire element). This phenomenon was observed experimentally and studied extensively (Krauthammer et al., April 1986, April 1990, September 1993, and October 1994). A summary of this issue appeared in ASCE (1999) and contains a description of the problem and mathematical models

Behaviors of Structural Elements

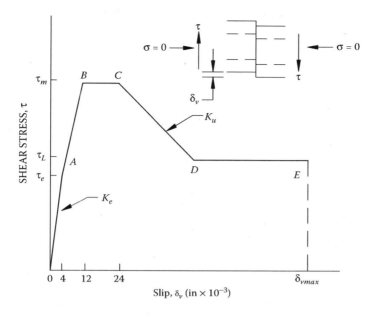

FIGURE 5.24 Direct shear resistance (Krauthammer et al., 1986).

that can be used, as shown in Figure 5.24. This resistance function based on previous studies (Hawkins, 1974) is defined by the following line segments (in pounds per square inch).

Segment OA — The response is elastic and the slope K_e of the curve is defined by the shear resistance τ_e for a slip of 0.004 in. The resistance is given by:

$$\tau_e = 165 + 0.157 f'_c \quad (5.33)$$

The initial response should be taken as elastic to not greater than $\tau_m/2$.

Segment AB — The slope of the curve decreases continuously with increasing displacements until a maximum strength τ_m is reached at a slip of 0.012 in. and the maximum strength is given by:

$$\tau_m = 8\sqrt{f'_c} + 0.8\rho_{vt} f_y \leq 0.35 f'_c \quad (5.34)$$

in which ρ_{vt} is the total reinforcement ratio of the steel crossing the shear plane and f_y and f'_c are the yield strength of reinforcement crossing the shear plane and the concrete uniaxial compressive strength, respectively.

Segment BC — The shear capacity remains constant with increasing slips and point C corresponds to a slip of 0.024 in.

Segment CD — The slope of the curve is negative, constant, and independent of the amount of reinforcement crossing the shear plane. The slope is given by:

$$K_u = 2{,}000 + 0.75\, f'_c \text{ (psi/in.)} \tag{5.35}$$

Segment DE — The shear capacity remains constant. The deformation at E varies with the level of damage, with a failure at a slip of Δ_{max} defined as follows for well anchored bars:

$$\Delta_{max} = C(e^x - 1)/120 \text{ (in.)} \tag{5.36}$$

in which

$$x = 900/[2.86(f'_c/d_b)^j] \tag{5.37}$$

and $C = 2.0$ and d_b is the bar diameter (inches). The limiting shear stress is defined as:

$$\tau_L = 0.85(A_{sb}/A_c)\, f'_s \tag{5.38}$$

in which A_{sb} is the area of bottom reinforcement (although this may be the total amount of steel crossing the shear plane), f'_s is the tensile strength of that steel, and A_c is the total cross-sectional area. Because of the nature of this type of response, the shear cracks are either vertical or close to vertical. Therefore, one needs to provide additional longitudinal reinforcement to enhance the direct shear strength of structural concrete elements.

Krauthammer et al. (1986) modified the original stress–slip relationship for direct shear. The original model describes the static interface shear transfer in RC members having well anchored main reinforcement in the absence of axial forces to account for effects of compressive stresses and loading rates on concrete shear strength. The authors considered both axial forces and rate effects and applied an enhancement factor of 1.4 to account also for test data, as shown in Figure 5.25.

5.13.3 Diagonal Shear Effects

It is well known (Park and Paulay, 1975) that diagonal shear will reduce the flexural strength of a reinforced concrete beam. Various studies of the interaction between flexure and shear were carried out (Kani, 1966; Krauthammer and Hall, 1982; Krauthammer et al., 1987 and 1990; Russo et al., 1991 and 1997; Kyung, 2004). Studies on beams without web reinforcement showed that the principal mechanisms of shear resistance are beam action and arch action (Park and Paulay, 1975). The influence of shear on the beam strength, as represented by the ratio between the actual moment capacity M_u and the theoretical moment capacity M_{fl} depends mainly on the shear span–depth ratio a/d and the reinforcement ratio $\rho = A_s/b_w d$, as shown by Kani (1966) and in Figure 5.26.

These test data show clearly that the largest effect of shear on the flexural moment capacity is observed at a/d ratios of about 2.5. The a/d ratio is considered

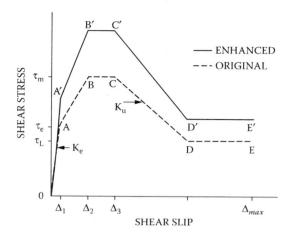

FIGURE 5.25 Original and modified direct shear models (Krauthammer et al., 1986).

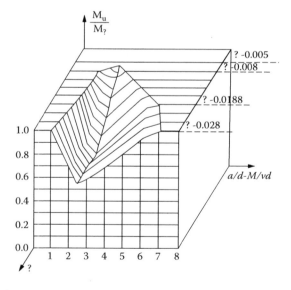

FIGURE 5.26 Flexure–shear interactions for RC beams without stirrups (Kani, 1966).

the transition from deep beam behavior (a/d <2.5) to slender beam behavior (a/d >2.5). Arch action controls the behavior for deeper beams, while beam action is the dominant mechanism for slender beams (Park and Paulay, 1975). One can see that no shear reduction is observed for a/d values smaller than 1 and no shear effects will occur beyond a certain a/d ratio as a function of the reinforcement ratio ρ. For example, if ρ is 0.028, no shear reduction will be noted for a/d >7. However, if ρ is 0.0188, no shear reduction will be noted for a/d >6.5. Despite this observation, the current design procedures (ACI 318-05, 2005) provide a fictitious separation between the shear resistance and the flexural capacity of a member subjected to the combination of flexure and shear.

Several analytical models were proposed to evaluate the reduction of the flexural capacity due to the effect of shear (Krauthammer et al., 1987; Russo et al., 1991 and 1997). The model proposed by Krauthammer et al. (1987) was based on a simplified numerical evaluation of the experimental data shown in Figure 8.15 with modifications to include the contributions of shear reinforcement. The models by Russo et al. (1991 and 1997) were based on theoretical representations of the beam and arch actions, as discussed by Park and Paulay (1975). These approaches were evaluated recently by Kyung and Krauthammer (2004) by analyses of previously tested reinforced concrete beams under both static and dynamic loading effects. They concluded that although both approaches provided similar results, the model by Krauthammer et al. (1987) showed better results in the dynamic domain. Both approaches were incorporated in the structural analysis code DSAS (Krauthammer et al., September 2004).

5.13.4 Size Effects and Combined Size and Rate Effects

Previous tests and theoretical investigations showed that structural concrete behavior loaded in tension, compression, shear, or torsion is influenced by specimen size (Bazant and Planas, 1998). Studies showed that larger compression specimens had steeper softening paths and larger beams were weaker in bending, shear and torsion. This phenomenon is termed *size effect*. Various explanations were proposed for these observations, for example, related to boundary layer effects, differences in rates of diffusive phenomena, heat of hydration, other phenomena related to chemical reactions, statistical facts based on the number of defects in the volume, or related energy dissipation during the evolution of fracture and damage. Accordingly, Bazant proposed the following size effect law in the static domain (Figure 5.27):

$$\sigma_{Nc} = (Bf_t)/[1 + (D/D_0)]^f \qquad (5.39)$$

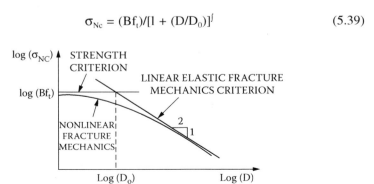

FIGURE 5.27 Size effect law in static domain (Bazant and Planas, 1998).

Behaviors of Structural Elements

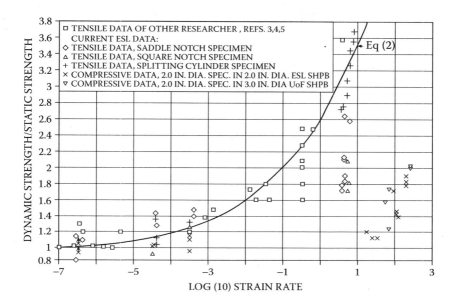

FIGURE 5.28 Strain rate effects in concrete under tension and compression (Ross et al., 1989).

in which B is a constant, f_t is the size-independent tensile strength, D is the specimen size, and D_0 is the reference specimen size.

Concrete, like various other materials, is sensitive to loading rate effects, as noted in Section 5.4.2. Actually, available test data show that the strain rate effects shown in Figure 5.28 are more pronounced than those shown in Figure 5.6. The concept of size effects has not been studied extensively in the dynamic domain, and its form under short duration dynamic (blast or impact) loading conditions is not known. Nevertheless, more recent studies showed that the size and rate effects are coupled phenomena (Krauthammer et al., October 2003; Elfahal and Krauthammer, 2005 and March–April 2005). Additional studies are required to clearly define how these phenomena are related, and what physical laws govern them.

5.13.5 Connections and Support Conditions

It has been shown that the effect of the lateral restraint of a support is very significant to the development of membrane enhancement in slabs. Although even small magnitudes of support stiffness can introduce some membrane enhancement, well designed supports would allow more. Attention to connection details, therefore, becomes an important factor. Reinforced concrete structures cannot develop their full capacities if their reinforcement details are inadequately designed.

In connections, for example, joint size is often limited by the size of the elements framing into it. This restriction, along with poor reinforcement detailing, may create connections without sufficient capacity to develop the required strength of adjoining elements. Knee joints are often the most difficult to design when continuity of adjoining members is required. Although connections will be addressed in more detail in Chapter 7, a brief discussion is presented here.

The work of Nilsson (1973) provided extensive insight into the behavior of joints under static loads and into the relationships between performance and internal details. He showed that even slight changes in a connection detail had significant effects on the strength and behavior of joints. Park and Paulay (1975) proposed the diagonal strut and truss mechanisms to describe a joint's internal resistance to applied loads. The diagonal compression strut is obtained from the resultant of the vertical and horizontal compression stresses and shear stresses. When yielding in the flexural reinforcement occurs, the shear forces in the adjoining members are transferred to the joint core through the concrete compression zones in the beams and columns. The truss mechanism is formed with uniformly distributed diagonal compression, tensile stresses in the vertical and horizontal reinforcement, and the bond stresses acting along the beam and column exterior bars. These two mechanisms can transmit shearing forces from one face of a joint to the other, and their contributions are assumed to be additive.

The design approach for reinforced concrete connections by Park and Paulay (1975) was adopted by the Department of the Army (1990). Results from recent investigations of a blast containment structure by Otani and Krauthammer (March–April 1977), Krauthammer (May 1998), and Ku and Krauthammer (March–April 1999) can be used to illustrate the importance of structural concrete detailing in connection regions.

The findings showed that the location of the diagonal bar across the inner corner of the connection affected the joint's strength. The relationships between the maximum stress and the location of the diagonal bar and/or its cross-sectional area could be examined to produce design recommendations that would ensure a desired level of performance. It was observed that the radial reinforcing bars across the connection (in the direction normal to the diagonal bar) affected the tensile stress in the diagonal bar at the inner corner of the connection. Furthermore, the diagonal compressive strut to resist the applied loads could be mobilized effectively by a proper combination of radial and diagonal bars.

The strengthening of joint regions by diagonal bars caused the formation of plastic hinge regions in the walls near the ends of the diagonal bars. The relocation of the largest rotations from the support faces to the plastic hinge regions showed a shift of maximum moment and shear along the slabs. Examination of the stresses in the flexural bars along the slabs' planes of symmetry revealed that yielding and the maximum stresses in the interior flexural and tension bars were in the plastic hinge regions. Besides the damage to the flexural reinforcement, maximum shear stresses in the concrete and large tensile shock wave stresses also occurred in the plastic hinge regions. These stress patterns in both concrete and steel were in addition to the presence of direct in-plane tension in the slab (because of the

Behaviors of Structural Elements

expansion of the walls and roof caused by the interior explosion). They showed that excessive damage could occur at these regions if not properly designed.

Current design procedures in TM 5-1300 (Department of the Army, 1990) do not address the issue of plastic hinge regions nor do they anticipate that the location of maximum negative moment (negative yield line location) may occur near the ends of the diagonal bars. Furthermore, the computation of structural capacity in current design procedures is based on the assumption of yield line formation at the supports. Clearly, a shift in hinge location must be included to derive a more realistic structural capacity. Also, the classical yield line theory treats yield lines that have zero thicknesses. In reality, the plastic hinge regions have finite thicknesses (about the slab's thickness), and this issue should be addressed in design procedures.

Using support rotations as a parameter that can indicate the extent of damage is customary, as defined in TM 5-1300. For the containment structure under consideration, it was required that such rotations would not exceed 2 degrees. A support rotation (sometimes defined as global rotation) is estimated by the slab's peak midspan deflection divided by half the span length. Although both the localized and global rotations computed in the hinge regions varied, they were much larger than 2 degrees and the local rotations were much larger than the global rotations. This observation confirmed that the magnitudes of the local and global rotations provide good indications of damage to the slab. Furthermore, these findings emphasize the need to address both local and global rotations in design approaches.

Steel connections have not been studied extensively for blast-resistant design. Nevertheless, it was noted recently (Krauthammer, May 1998; Krauthammer et al., June, August, and November 2002) that steel connections exhibited deficient behavior under blast loads, even if they were designed to resist seismic loads. Such deficiencies should be addressed by modifying steel connections for blast resistance.

5.14 APPLICATION TO STRUCTURAL DESIGN

With any structural design process, a designer is faced with a given or selected loading condition and performance objectives for a protective system. Once the material selections are made (structural concrete, structural steel, masonry, or combinations), one must develop preliminary designs of the various components. The preliminary designs are prepared as for static loads using the previously summarized design equations and/or those for typical design manuals (Department of the Army 1986, and 1990; ASCE, 1985). However, one must employ enhanced material properties to address loading rate effects. Practical values of strain rate enhancements for such materials (Department of the Army, 1990) are provided in Table 5.3 and illustrated in Figure 5.28. One may apply the average strain rate to calculate enhancement factors for the stress parameters of the material models, then employ the modified stress-strain relationships for deriving the resistance functions used in the dynamic analysis.

TABLE 5.3
Reinforced Concrete Design Dynamic Increase Factors

	Far Design Range			Close-In Design Range		
	Reinforcing Bars		Concrete	Reinforcing Bars		Concrete
Type of Stress	f_{dy}/f_y	f_{du}/f_u	f'_{dc}/f'_c	f_{dy}/f_y	f_{du}/f_u	f'_{dc}/f'_c
Bending	1.17	1.05	1.19	1.23	1.05	1.25
Diagonal tension	1.00	—	1.00	1.10	1.00	1.00
Direct shear	1.10	1.00	1.10	1.10	1.00	1.10
Bond	1.17	1.05	1.00	1.23	1.05	1.00
Compression	1.10	—	1.12	1.13	—	1.16

Source: Department of the Army, 1990.

One should note in Table 5.3 that the far design range represents detonations at larger scaled distances than the close-in design range. Consequently, since the rise times are much shorter for close-in explosions, the loading rates are much higher. Therefore, the recommended dynamic increase factors are higher for the close-in design range. These dynamic increase factors are intentionally conservative to prevent one from over-estimating structural capacity. Furthermore, the level of such conservatism is directly related to the type of structural behavior under consideration. Since a flexural response is more ductile than a shear response, one may use higher dynamic increase factors for flexural design.

It was also shown experimentally that steel yield strength is more sensitive to rate effects than the ultimate strength, and one notes that the dynamic increase factors for the steel yield strength are much higher than for the ultimate strength. This observation is also illustrated in Figure 5.29. For concrete, the entire static stress–strain curve is scaled by the appropriate dynamic increase factor. However, for steel, the yield strength is scaled by one factor, while the ultimate is scaled by another. Additional figures in Chapter 4 of TM 5-1300 (Department of the Army, 1990) provide further refinements to the dynamic increase factor values cited in Table 5.3.

5.15 PRACTICAL DAMAGE AND RESPONSE LIMITS

Once a preliminary static design is obtained, it can be analyzed under the anticipated dynamic loads, as will be discussed in Chapter 6. One can use the calculated peak central deflection for estimating the damage to be induced by such loads as follows. Assuming symmetric load and deflection distributions, the support rotation is defined by the ratio of the calculated peak deflection to half the span length:

$$\tan \theta = \Delta_{max}/(0.5 \, L) \qquad (5.40)$$

Behaviors of Structural Elements

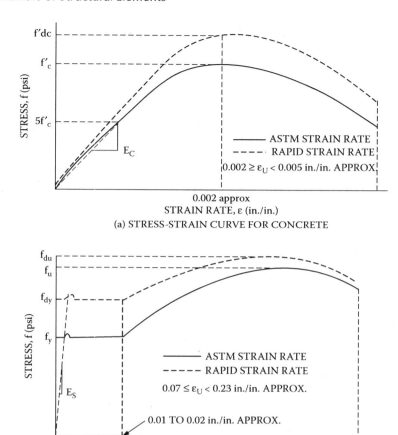

FIGURE 5.29 Strain rate material models (Department of the Army, 1990).

The following damage ranges are suggested in TM 5-1300 (Department of the Army, 1990):

$$0° \leq \theta \leq 2° \quad \text{Light damage}$$

$$2° \leq \theta \leq 5° \quad \text{Moderate damage}$$

$$5° \leq \theta \leq 12° \quad \text{Severe damage}$$

One may use various other computational approaches to estimate the actual damage.

After a preliminary design is obtained, one can estimate the natural periods for the various structural elements and perform dynamic analyses to assess their responses as discussed in Chapter 6. If a response is acceptable, no redesign is necessary. However, if the response is unacceptable, one needs to either redesign the structural elements, modify the loads, or both. The loads can be modified if one can impose stricter controls on both the range R and the charge weight W. Other parameters may involve the structural shape, shielding, etc., as discussed in previous chapters.

6 Dynamic Response and Analysis

6.1 INTRODUCTION

Chapters 2 and 3 showed that the loading environments associated with many relevant threats (such as impact, explosion, and penetration) are extremely energetic and their duration is measured in milliseconds (about one thousand times shorter than typical earthquakes). Accordingly, the structural response under short duration dynamic effects can be significantly different from those under static or slower dynamic loading cases. Therefore, a designer must provide suitable structural detailing to account for these short duration effects. Due to the complexities associated with these severe loading environments, designers must employ appropriate computational tools in support of their designs. Furthermore, one must be able to address the various structural response and failure modes to ensure that the design will be capable of mitigating such conditions.

When subjected to severe short duration loads, structures may fail in a variety of ways. The mode of failure will depend upon the characteristics of the structure and the loading, as well as the proximity and intensity of the blast to a structural member. Current approaches separate structural responses (e.g., flexure, diagonal, or direct shear, etc.) from localized responses. Localized responses are primarily governed by material properties and the associated structural behavior is negligible. A contact or close-in explosion produces a crater on the near side of a reinforced concrete element and spalling on the opposite face. Both cratering and spalling weaken a structural element by reducing the cross-sectional area that can resist the applied loads.

When the spall and crater zones merge, the cross-section is breached. The capacities of different materials to resist cratering, spalling, and breach dictate the thickness required to maintain structural integrity, or the need to provide protective features to mitigate these effects (e.g., front face armor, and/or rear face spall plates). Typically, the material behavior dictates the modes of deformation and the resulting patterns of failure.

Other modes of local failure that may result from direct blast involve shear failures. Shear responses are related either to flexural behavior (diagonal tension or diagonal compression) or direct shear that does not involve a flexural response (ASCE, 1999; Beck et al., 1981; Krauthammer, 1984 and 1986; Krauthammer et al., 1986 and 1990). Typically, brittle materials such as concrete are weak in tension and shear. When subjected to a principal tension that exceeds the tensile capacity of the material, the reinforcing steel crossing the failure plane holds the structural element together. Similarly under shear stresses, the reinforcing steel crossing the failure plane holds the two parts on each side of the failure section together, allowing shear forces to be transferred across the section through aggregate interlock and dowel action.

When the capacity of the reinforcing steel is reached, its ability to hold the parts together across the failure plane is diminished, and the ability of the section to resist either tension or shear is reduced significantly. Similarly, flat plate slabs must transfer the loads to the columns through shear at the column–slab interface. Although the area and resistance of this interface may be increased by means of drop panels, dowel action by reinforcing bars and/or shear heads, the capacity of the slab to transfer the load to the columns is strongly affected by the shear capacity of the concrete.

Extreme loads on short deep members may result in a direct shear failure before the development of a flexural mode of response. This direct shear response is typically associated with a lower ductility or a brittle behavior mode and may occur at areas of structural or load discontinuity, as discussed in Chapter 5. For instance, when a slab is supported on walls, supported areas of the slab are restricted from moving with the severe load, while the unsupported areas will be accelerated in the load direction. This very quick relative motion between the supported and unsupported parts of the slab will induce a sharp stress discontinuity at the slab–wall interfaces that can fail the slab along vertical failure planes.

Furthermore, the shock waves traveling through the supported areas of the slab will continue into the wall and will not be affected by discontinuities. At the same time, the shock waves traveling across the thickness of the unsupported part of the slab will reach the rear wall surface and will reflect as tensile waves. These wave propagation phenomena are discontinuous at the slab–wall interfaces and they can enhance the previously discussed stress discontinuity caused by the motion differences between the supported and unsupported slab regions along vertical shear planes.

Closed form solutions are limited to simple loading, simple geometric and support conditions, and mostly linear materials. Therefore, these methods will not be addressed herein. Approximate simplified methods are based on assumed response modes, and the recommendation is to use such methods with data from computer codes based on current design manuals. Advanced numerical methods require significant resources and are usually not utilized in typical design activities. Nevertheless, these advanced computational approaches can be used in the final design stages for detailed assessment.

Advanced structural analyses can be used for obtaining design guidelines and/or for the detailed evaluation of the anticipated structural response. As advanced modeling capabilities and computational hardware continue to evolve, they are expected to be fully integrated into simulation-based design approaches. Since very severe damage should be expected for scaled ranges $(R/W)^{1/3}$ less than 1 (in ft/lbs$^{1/3}$), structural response calculations may not be valid for such scaled ranges. Therefore, for such severe conditions, structural breaching calculations must be performed separately from structural response analyses. However, breaching calculations require the use of advanced computational tools (e.g., hydro-codes, SPH codes, etc.).

Each of the preceding behavioral modes is associated with a corresponding response frequency range that influences the characteristics of the load–structure interaction and structural resistance. The dynamic response of a structural element to the transient blast loads will determine how much deformation the element undergoes and the possible consequences. Very short duration loads (as will be shown later, these types of loads are defined as impulsive) are no longer acting on the structure by the time the structure reaches its peak response, and the system may be idealized by a spring mass system set into motion by an initial velocity.

Stiffer structural elements or modes of deformation associated with higher frequencies of response may respond to the same blast environment with a greater dynamic amplification factor. However, the characteristics of the higher frequency system will determine the peak deformations associated with the transient loading. Therefore, considering an elastic–plastic SDOF analogy for each of the various modes of failure could provide a reasonable means to determine the dynamic response of an element and its ability to deform within prescribed limits. Each analogy, however, depends on the proper detailing and simulation of the member and its connections to guarantee that the capacity can be achieved and the ultimate deformation can be withstood.

A critical stage in the evaluation of a hardened structure (or any other dynamically loaded system) is dynamic analysis. It has been shown (Biggs, 1964; Clough and Penzien, 1993; Tedesco et al., 1999; Chopra, 2001; Humar, 2002) that the need for a dynamic analysis can be determined by the ratio of load duration to the natural period of the element in question (t_d/T_n). Usually if that ratio is larger than about five, the response will be similar to that in the static domain behavior. However, for much smaller ratios of t_d/T_n the response will be similar to that of a system subjected to a very short loading pulse. These two types of behavior can be defined based on theoretical solutions. A dynamic analysis is required if the ratio t_d/T_n is between the previous limits.

This chapter will discuss the dynamic responses of structural systems to blast, shock, and impact. It will not address the general treatment of structural dynamics and readers should refer to the above-cited references for such information. Additional treatment of the three domains in the structural response is provided in Chapter 8.

In general, a dynamic analysis should be considered for critical structural elements (such as those to be exposed to the load) during the design process and for the entire system upon completion of the design. Different computational approaches can be used to address specific analysis needs. The following discussion will address the various computational approaches, with emphasis on methods that are typically used in this field. The requirements for dynamic analysis are such that various methods can be employed. Some are more efficient than others and some are applicable only for narrow and specific types of computations. Although structural elements are continuous systems, they can be treated

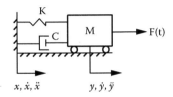

FIGURE 6.1 Single degree of freedom system.

at various levels of sophistication, e.g., as either lumped mass models or as continuous systems. These approaches are presented next.

6.2 SIMPLE SINGLE DEGREE OF FREEDOM (SDOF) ANALYSIS

Consider the single degree of freedom (SDOF) system shown in Figure 6.1 and summarize the treatment of this system based on information in various books (Biggs, 1964; Clough and Penzien, 1993; Tedesco et al., 1999; Chopra, 2001). In this case, one may have relative motions between the mass and its support as follows:

$$u = y - x \tag{6.1a}$$

$$\dot{u} = \dot{y} - \dot{x} \tag{6.1b}$$

$$\ddot{u} = \ddot{y} - \ddot{x} \tag{6.1c}$$

in which $x, \dot{x}, \ddot{x}$, are the displacement, velocity, and acceleration of the supports, respectively, and $y, \dot{y}$, and $\ddot{y}$ are the displacement, velocity, and acceleration of the SDOF mass, respectively. For simplicity, one may consider the SDOF system without support motions and write the corresponding equation of equilibrium, as follows:

$$M\ddot{y} + C\dot{y} + Ky = F(t) \tag{6.2}$$

One may divide the entire equation by M to obtain the following:

$$\ddot{y} + 2\xi\omega\dot{y} + \omega^2 y = \frac{F(t)}{M} \tag{6.3}$$

in which

$$\xi = \frac{C}{C_{cr}} \tag{6.4}$$

Dynamic Response and Analysis

$$C_{cr} = 2\sqrt{KM} \tag{6.5}$$

$$\omega^2 = \frac{K}{M} \tag{6.6}$$

ω is the circular natural frequency and is equal to $2\pi f$, where f is the frequency in Hz. The frequency f is 1/T where T is the natural frequency. In order to reflect damping effects on frequency, one should use the following relationship:

$$\omega' = \omega\sqrt{1 - \xi^2} \tag{6.7}$$

Therefore, the equation of motion should be:

$$\ddot{y} + 2\xi\omega'\dot{y} + (\omega')^2 y = \sum_{i=1}^{N} \left(\frac{F(t)}{M}\right)_i \tag{6.8}$$

The summation on the right side of Equation (6.8) represents all forcing functions that may affect the SDOF system under consideration. Now, consider Equation (6.2) and assume that no damping is involved. The resulting equation of motion will be:

$$M\ddot{y} + Ky = F(t) \tag{6.9}$$

One can treat the following two general cases: free vibrations when no forcing function is applied and forced vibrations when a forcing function is applied. These cases are discussed next. For the case of free vibrations, Equation (6.9) is rewritten, as follows:

$$\ddot{y} + (K/M)y = 0 \tag{6.10}$$

The solution of this equation is the following:

$$y = C_1 \sin \omega t + C_2 \cos \omega t \tag{6.11}$$

in which C_1 and C_2 are constants of integration to be determined based on initial conditions and ω was defined in Equation (6.4). For example, assume that at t = 0 the displacement and velocity are y_0 and v_0, respectively. This will result in the following solution:

$$y = (v_0/\omega) \sin \omega t + y_0 \cos \omega t \tag{6.12}$$

This solution represents a harmonic vibration whose displacement amplitude is y_0, velocity amplitude is v_0/ω, and a natural circular frequency ω. This oscillatory motion will continue without change because no energy loss (damping) is present. When a forcing function is applied, the solution to Equation (6.9) is given by combining Equation (6.12) with a particular solution for the forced vibration. For example, if the forcing function is harmonic, e.g., $F(t) = F_0 \sin \omega_0 t$, the following solution is derived:

$$y = C_1 \sin \omega t + C_2 \cos \omega t + \{(F_0/K)/[1 - (\omega_0/\omega)^2]\} \sin \omega_0 t \quad (6.13)$$

Again, C_1 and C_2 are constants of integration to be determined based on initial conditions. It can be shown that the SDOF system will exhibit a steady state oscillation due to the forcing function on which one can note the transient free vibration. When the frequency of the forcing function approaches the natural frequency of the system $[1 - (\omega_0/\omega)^2] \to 0$ and the forced dynamic displacement $\{(F_0/K)/[1 - (\omega_0/\omega)^2]\}$ approaches ∞, the condition is termed *resonance*. Another useful parameter is the dynamic load factor (DLF) that is the ratio of the dynamic deflection and the static deflection under the peak constant force F_0, as follows:

$$DLF = y/y_{st} = y/(F_0/K) = yK/F_0 \quad (6.14)$$

The introduction of damping into this system will cause both the free and forced vibrations to exhibit gradual decays of the dynamic response as discussed elsewhere (Biggs, 1964; Clough and Penzien, 1993; Tedesco et al., 1999; Chopra, 2001). Here, only brief summaries of relevant issues are provided and the reader is encouraged to review such background information as needed.

6.2.1 Theoretical Solution for SDOF Systems: Duhamel's Integral

Considering Equation (6.3), one may treat the forcing function F(t) as shown in Figure 6.2. The impulse in a time interval $\Delta\tau$ is the area under the load function for that time interval as follows:

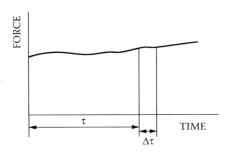

FIGURE 6.2 General forcing function.

Dynamic Response and Analysis

$$I = F(\tau)\Delta\tau = \Delta(mv) \tag{6.15}$$

in which m is the mass and v is the velocity. The change in momentum from time $t = \tau$ to time $t = \tau + \Delta\tau$ is derived as follows:

$$\Delta(mv) = (mv)_{\tau+\Delta\tau} - (mv)_{\tau} = m(\Delta v) \tag{6.16}$$

This is based on the original definition of Newton's second law:

$$F\frac{d}{dt}(mv) = \dot{m}v + m\dot{v} \tag{6.17}$$

in which $\dot{m}$ is dm/dt, $\dot{v} = dv/dt = a$, and a is the acceleration. Since the mass is assumed constant for velocities that are much smaller than the speed of light (v<<c and c is the speed of light), $\dot{m} = 0$. Therefore:

$$F = m\dot{v} = ma \tag{6.18}$$

and for each time interval $\Delta\tau$, one derives the following velocity:

$$v = \frac{F\Delta\tau}{m} \tag{6.19}$$

Since the impulse is the change in momentum, $I = \Delta(mv)$, one can approximate the force pulse by a series of pulses (or steps). The solution is given by the following expression:

$$y = +\frac{1}{M\omega'}\int_0^t F(\tau) e^{-\xi\omega_i(t-\tau)} \sin\left[\omega'(t-\tau)\right] d\tau \tag{6.20}$$

from which one can compute $\dot{y}(t)$. Clearly, these responses are functions of the frequency (or natural period), thus linking the solution to the ratio t_d/T_n, as previously noted.

For design purposes, one would like to use maximum values that are functions of frequency and damping. For such applications, it is convenient to introduce the concept of a response spectrum. This is a plot of the maximum response of a behavioral parameter (e.g., d, v, a, σ, etc.) for a broad range of possible linear SDOF systems (variable magnitudes of the stiffness-to-mass ratio K/M) to a given excitation. Accordingly, one can define the following maximum response parameters:

$$D = |y(t)|_{max} \tag{6.21}$$

$$V = |v(t)|_{max} = \omega D \tag{6.22}$$

$$A = |a(t)|_{max} = \omega V = \omega^2 D \tag{6.23}$$

From the above expressions, one can derive the following relationship of the peak displacement, pseudo velocity, acceleration of the mass, and the natural frequency:

$$\omega D = V = A/\omega \tag{6.24}$$

In seismic engineering (i.e., the excitation is caused by base motion), one can show (Clough and Penzien, 1993; Tedesco et al., 1999; Chopra, 2001) that there are two limiting cases (assuming no damping, $\xi = 0$), as follows:

Case A is for a large mass and a very weak spring. In this case $\omega = K/M$ is very low and the maximum mass displacement, $u_{max} = d \Rightarrow$ base displacement, therefore, the pseudo velocity is:

$$V = \omega D = 2\pi f d \tag{6.25}$$

Case B is for a small mass and a very strong spring. In this case $\omega = K/M$ is very large and the mass will be accelerated by the base at the same acceleration. Therefore,

$$V = \frac{A}{\omega} = \frac{a}{2\pi f} \tag{6.26}$$

These results show that $2\pi f$ will be an asymptote for very low frequencies and $1/(2\pi f)$ will be an asymptote for very high frequencies. Accordingly, a response spectrum is a plot of a system maximum response parameter (e.g., pseudo velocity V), as a function of frequency in the form of log V versus log f. Taking the logarithm of all parts of Equation (6.24) leads to the following relationship:

$$\log \omega + \log D = \log V = \log A - \log \omega \tag{6.27}$$

This expression can be illustrated qualitatively as shown in Figure 6.3. Note on the figure that any horizontal line represents constant values of log V and any vertical line represents constant values of log ω. Returning to Equation (6.27), one can obtain the following two relationships:

$$\log D = \log V - \log \omega \tag{6.28}$$

$$\log A = \log V + \log \omega \tag{6.29}$$

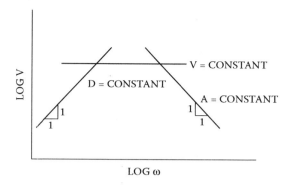

FIGURE 6.3 Four-way logarithmic response spectrum for undamped SDOF.

These relationships show that the lines of peak displacement and peak acceleration have slopes of +1 and −1, respectively (i.e., these lines are inclined at 45° and −45°, respectively).

Generally, one can employ any forcing function and apply it to a broad range of SDOF systems to obtain a response spectrum for that frequency range. One can apply this concept to derive the response spectra for the El Centro, California, earthquake of 1940 as shown in Figures 6.4 and 6.5. Figure 6.4 shows the recorded

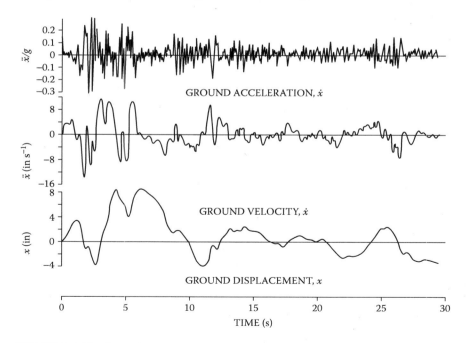

FIGURE 6.4 North–south component of 1940 El Centro, California, earthquake (Krauthammer and Chen, 1988).

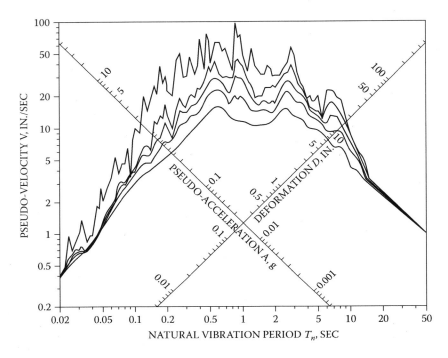

FIGURE 6.5 Response spectra for elastic systems, 1940 El Centro earthquake. $\zeta = 0, 2, 5, 10,$ and 20% (Chopra, 2001).

ground acceleration and its corresponding velocity and displacement time histories. By applying the ground acceleration to a broad range of SDOF systems (i.e., changing K and M to obtain various values of ω), one can derive many values of the pseudo velocity V that can be plotted in the form of a response spectrum as shown in Figure 6.5. One may repeat such computations for various damping values to obtain a family of response spectra.

Similar computations can be done for structures subjected to blast loads. The corresponding graphs are called shock response spectra, as discussed by Crawford et al. (1973). The analytical derivation of such response spectra may not be simple for complicated forcing functions, and one must employ more advanced computational approaches. The analytical methods for dynamic analysis, those that consider specific problems and design aspects, have existed for half a century (the classical methods have existed even longer). Books on structural dynamics provide the necessary tools for analysis of and the response of structural systems (Biggs, 1964; Berg, 1989; Clough and Penzien, 1993; Tedesco et al., 1999; Chopra, 2001). These analytical methods provided solutions to rather simple systems under harmonic loads. Only within the last 40 years has it become possible to perform analyses on more complex systems and loading conditions thanks to the introduction of large scientific computers and the application of numerical methods.

Dynamic Response and Analysis

One may also consider nonlinear systems (in most cases elasto-plastic) by introducing a new concept known as the *ductility factor* as follows:

$$\mu \equiv \frac{\Delta_m}{\Delta_y} \tag{6.30}$$

where μ is the ductility factor, Δ_m is the maximum displacement, and Δ_y is the yield displacement. Accordingly, one may derive response spectra for elasto-plastic systems whose shapes would be similar to the elastic response spectra, but they will account for responses beyond the linear behavioral range.

6.2.2 Graphical Presentations of Solutions for SDOF Systems

As noted above, analytical solutions for SDOF systems may be very cumbersome, even for simple loading functions. Such computations become much harder for nonlinear resistance functions and complicated loading pulses. To simplify such computations in support of design activities, one may use dynamic response charts that enable an analyst to estimate the values of key parameters for assessing the suitability of a tentative structural design. Such charts were derived by employing the general approaches described above, as illustrated in Figure 6.6 and Figure 6.7 for a right angle triangular load and an assumed elasto-plastic resistance function. Figure 6.6 illustrates a family of curves that relate the ductility factor μ to the nondimensional time ratio t_d/T (pulse duration to natural period).

Each curve represents a different yield resistance to peak load ratio R_m/F_1. For a tentative design, one knows the values of t_d/T, R_m/F_1, and the peak elastic deflection y_{el}. These values will define uniquely the ductility factor μ from which one computes the maximum displacement y_m. The derived value of y_m can be used to assess the adequacy of the selected design. For example, if y_m is the peak displacement for a beam, dividing it by half the span length $L/2$ yields the support rotation θ. Comparing the derived support rotation with recommended damage levels (Department of the Army, 1990; ASCE, 1999) will show whether the beam will meet its performance or operational requirements. Similarly, from Figure 6.7, one can derive the time at which the maximum displacement will be reached, and this can be compared with possible performance or operational requirements.

6.2.3 Numerical Solutions of SDOF Systems

The closed form solution of the equation of motion by the approach described earlier may not be possible for highly nonlinear cases. Furthermore, it is necessary to derive an efficient numerical integration procedure that will be valid for a wide range of cases. One such technique is the Newmark β Method (Newmark, 1962), as described next. Consider the equation of motion:

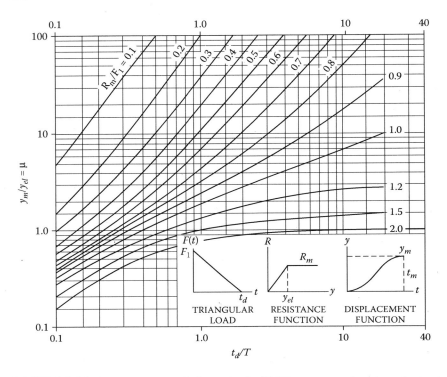

FIGURE 6.6 Maximum response of elasto-plastic SDOF systems under triangular load pulse with zero rise time (U.S. Army, 1957).

$$M\ddot{y} + C\dot{y} + Ky = F(t) \qquad (6.31)$$

in which y is the relative displacement of the mass with respect to the base. Following the procedure discussed above, we can obtain:

$$\ddot{y} + 2\xi\omega' \dot{y} + \omega'^2 y = f(t) \qquad (6.32)$$

All these parameters were defined earlier.

Newmark (1962) proposed the following numerical solution technique for this equation, as discussed also by Newmark and Rosenblueth (1971):

1. Assume that y, $\dot{y}$, and $\ddot{y}$ are known at time $t = t_i$, and assign them the subscript i.
2. $t_{i+1} = t_i + \Delta t$, and assume a value $\ddot{y}_{i+1}$ for the acceleration at time t_{i+1}.
3. Compute a velocity at t_{i+1}: $\dot{y}_{i+1} \cong \dot{y}_i + (\ddot{y} + \ddot{y}_{i+1}) \dfrac{\Delta t}{2}$

Dynamic Response and Analysis

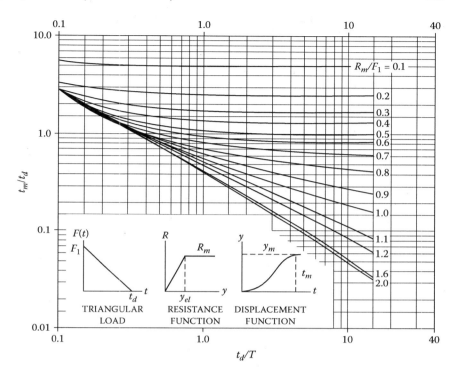

FIGURE 6.7 Time of maximum response of elasto-plastic SDOF systems under triangular load pulse with zero rise time (U.S. Army, 1957).

4. Compute a displacement at t_{i+1}: $y_{i+1} \cong y_i + \dot{y}_i \Delta t + (1/2 - \beta)\ddot{y}_i(\Delta t)^2 + \beta \ddot{y}_{i+1}(\Delta t)^2$
5. Substitute y_{i+1} and $\dot{y}_{i+1}$ into Equation (6.32) and compute a new value for $\ddot{y}_{i+1}$: $\ddot{y}_{i+1} = -2\xi\omega'_l \dot{y}_{i+1} - \omega_l'^2(y_{i+1} - y_o) - \ddot{x}_o$ in which y_o is the static displacement relative to the base and $\ddot{x}_o$ is the base acceleration. Note that y_o and $\ddot{x}_o$ may be zero for certain cases.
6. Check whether $\ddot{y}_{i+1}$ from step 2 is close enough to $\ddot{y}_{i+1}$ from step 5. If they are close, the solution is achieved. If they are not close, repeat steps 2 through 6 until convergence is attained. Since one needs to assume a new value for $\ddot{y}_{i+1}$ in step 2, it is quite efficient to select the value obtained in step 5 for this purpose.

In step 2, it was assumed that $\ddot{y}$ varies linearly over Δt. In step 3, if $\beta = 1/4$, it will correspond to a linear variation of $\dot{y}$ (constant acceleration) over Δt, but if $\beta = 1/6$, it will correspond to a parabolic variation of $\dot{y}$ (linear acceleration) over Δt. β values between 1/4 and 1/6 are recommended for convergence and stability.

If $\beta > 1/8$, the method will converge only if it is stable. The choice of $\Delta t \leq 0.1\, T_n$ and $\beta = 1/6$ will ensure a very fast convergence to the solution.

6.2.4 Advanced SDOF Approaches

It was noted above that several characteristic parameters must be known about a structure under consideration to employ an SDOF model of the structure: the blast load as a function of time F(t), mass M, damping C, and stiffness K. Mass and stiffness parameters for the structural system under consideration are selected based on the load source, type of structure, general conditions for load application to the structure, (for example, localized load on a small part of the structure, distributed load over a larger part of the structure, the load distribution function, i.e., uniform or nonuniform, etc.), and the expected behavioral domain (linear elastic, elastic–perfectly plastic, nonlinear, etc.).

The previous sections provide information on the treatment of either linearly elastic or elastic–plastic structural systems by SDOF simulations. However, such approaches are still simplistic and may not represent accurately the more realistic behaviors of fully nonlinear structural systems.

For example, ideal elastic–plastic systems cannot account for fully nonlinear concrete and steel material models, plastic hinge formation and growth, coupling of several failure modes (flexure, diagonal shear, direct shear, axial loads, etc.), membrane effects (both in compression and tension), consideration of confinement effects in a cross-section, and the contributions of confined and unconfined concrete zones, variable strain rate effects, etc. Nevertheless, these issues were addressed in the literature (Beck et al., 1981; Krauthammer, 1984 and 1986; Krauthammer et al., 1986 and 1990), and one can employ such advanced concepts for analysis and design, as briefly discussed next.

The continuous structural element represented by an equivalent SDOF system as defined in the above-mentioned references is a modified version of Equation (6.32):

$$\ddot{y} + 2\xi\omega_1'\dot{y} + \frac{R(y)}{M} = f(t) \tag{6.33}$$

Here R(y) replaces Ky and it can be any form of load deformation (resistance) function for the structural mechanism or system under consideration. Although the role of damping may not be very significant in typical structural systems (e.g., ζ is usually smaller than 10%), it was found that damping may play a significant role when medium–structure interaction involves large structural deformations (Krauthammer, 1986; Krauthammer et al., 1986). The nonlinear load–deflection relationship serves as the skeleton resistance curve for the numerical computations, as shown in Figure 6.8, based on which the dynamic resistance function and the dissipation of energy are evaluated.

Dynamic Response and Analysis

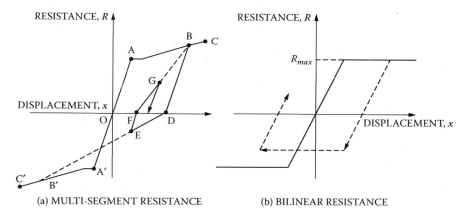

FIGURE 6.8 Dynamic resistance functions for flexural behavior (Krauthammer et al., 1990).

Similarly to previous methods (Department of the Army, 1986 and 1990; Biggs, 1964), the equivalent mass of the SDOF system is derived based on the deflected shape of the structural member, as will be discussed later. The derivation of nonlinear resistance functions for various structural behavior mechanisms and other parameters is accounted for and discussed in other publications (Beck et al., 1981; Krauthammer, 1984 and 1986; Krauthammer et al., 1986 and 1990). Under such an approach, one can address the true flexural response that includes explicit interactions of bending moment, diagonal shear, and axial force by one SDOF system and the direct shear response by another SDOF system. The two SDOF systems are loosely coupled by the support reactions, as shown in Figure 6.9. The support reactions (discussed further in this chapter) are computed from the flexural response and represent the load for computing the direct shear response. Clearly, this method is suitable for nonlinear structural analysis.

We have considered the behavior of SDOF systems, and the effects of loading conditions, damping, and structural properties. However, real structures usually tend to exhibit more complicated types of behaviors and therefore more advanced techniques and models must be employed. One may consider a structure as a multi-degree-of-freedom (MDOF) or continuous system, as briefly discussed in the following sections.

6.3 MULTI-DEGREE-OF-FREEDOM (MDOF) SYSTEMS

6.3.1 Introduction to MDOF

Background information on MDOF systems can be found in books on structural dynamics (Biggs, 1964; Berg, 1989; Clough and Penzien, 1993; Tedesco et al., 1999; Chopra, 2001). Such cases can be addressed both theoretically or numerically as lumped mass systems. It is assumed that a structure's mass can be lumped

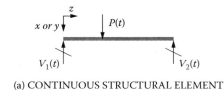
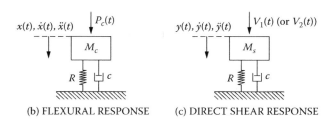

FIGURE 6.9 Coupled equivalent SDOF systems (Krauthammer et al., 1990).

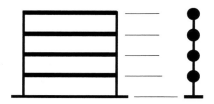

FIGURE 6.10 Lumped mass structural model.

into several representative floor masses and the stiffness can be illustrated by column springs, as shown in Figure 6.10.

Using free body diagrams for each floor, one can derive the following equation of motion for a MDOF system in matrix form:

$$[M]\{\ddot{x}\} + [C]\{\dot{x}\} + [K]\{x\} = \{P(t)\} \tag{6.34}$$

where [M] is the mass matrix, representing all system masses, [C] is the damping matrix representing damping of various masses, [K] is the stiffness matrix representing the stiffness contributions of various levels (i.e., floors), {P(t)} is the force vector representing forces applied to the various masses, and {x} represents the displacement vector (for the displacement of the masses). The velocity and acceleration vectors are noted by the first and second time derivatives of {x}.

Various methods can be employed to obtain solutions for such systems. In order to simplify the discussion, one may ignore damping, and thus:

$$[M]\{\ddot{x}\} + [K]\{x\} = \{P(t)\} \tag{6.35}$$

Dynamic Response and Analysis

One may use a base motion as additional input and employ the relative mass-to-base motion to formulate the equation of motion. Here, however, the treatment will be similar to that for SDOF systems. Only a summary of the basic problem and its solution will be provided, and the reader is urged to review the appropriate background in the literature.

For free oscillations, $\{P(t)\} = 0$. The equation for this case becomes:

$$[M]\{\ddot{x}\} + [K]\{x\} = 0 \qquad (6.36)$$

The structure will vibrate in its *natural modes*, and, using separation of variables, the displacement will be of the form:

$$\{x(t)\} = Z_n \theta_n(t) \qquad (6.37)$$

where Z_n is the time-independent mode shape vector and θ_n is a time-dependent scalar function. As discussed for SDOF systems, such a solution might have the following form:

$$\{x(t)\} = Z_n(A_n \sin \omega_n t + B_n \cos \omega_n t) \qquad (6.38)$$

The system oscillates in one of its natural modes, assuming that the base remains motionless. All the system's masses describe a synchronous motion. The shape of the configuration does not depend on time, although its magnitude varies with t as prescribed by the n-th function θ. Substituting Equation (6.37) into Equation (6.36) and rearranging terms will provide the following expression:

$$(-\omega_n^2[M]Z_n + [K]Z_n)\theta_n(t) = 0 \qquad (6.39)$$

Note that the above expression is a product of a time-independent term by a time-dependent term and that it is equal to zero. This can be achieved if either one of the terms is zero. However, one has a trivial solution if $\theta_n(t) = 0$, which means that the following option should be examined:

$$(-\omega_n^2[M] + [K])Z_n = 0 \qquad (6.40)$$

which leads to solving the following eigenvalue or characteristic value problem:

$$\det(-\omega_n^2[M] + [K]) = 0 \qquad (6.41)$$

Equation (6.41) is a polynomial of order N in ω_n^2 that provides N real and positive roots for ω_n^2 because both the mass and stiffness matrices are symmetric and positive definite. Once the N characteristic values of ω_n are obtained, one can find corresponding N values for the mode shapes Z_n from Equation (6.40). Once all these are obtained, one can formulate the solution for the problem. These mode shapes are orthogonal (independent of each other), which means that the following expressions exist:

$$Z_m^T \, MZ_n = 0 \quad \text{for } m \neq n \qquad (6.42a)$$

and

$$Z_m^T \, KZ_n = 0 \quad \text{for } m \neq n \qquad (6.42b)$$

If a system has more than one mode with the same natural frequency, it may undergo free vibrations in that natural frequency with a configuration that is a linear combination of these modes. Since the natural modes constitute a complete set, any configuration X that satisfies the boundary conditions can be expressed as a linear combination of natural modes:

$$y = \sum_n \alpha_n Z_n \qquad (6.43)$$

in which a_n is a dimensionless factor known as a participation factor (or coefficient). Therefore, by substitution, one can show that the general solution for a free vibration problem is expressed as:

$$y(t) = \sum_n a_n Z_n \sin \omega_n (t - t_n) \qquad (6.44)$$

in which t_n is a time shift for each natural mode. t_n and a_n must be chosen for each problem to satisfy initial conditions.

One can treat the damped free vibration problem similarly, but this will not be presented here. When damped systems are considered, we may not obtain real natural modes. In order to ensure real natural modes, one must have the following values for the damping matrix:

$$[C] = \alpha[M] + \beta[K] \qquad (6.45)$$

where M and K are diagonal matrices. This condition is termed Rayleigh damping. Consider the case when the ground remains motionless while a system of external forces that vary in time as simple harmonic functions acts on a structure. All forces have the same ω and all are in phase. The vector of the forces can be written as $P = b \sin \omega t$. It can be shown (Newmark and Rosenblueth, 1971) that the general solution can be written as:

Dynamic Response and Analysis

$$y(t) = \sum_n (B_d) \frac{1}{\omega_n^2} \frac{Z_n^T b Z_n}{Z_n^T M Z_n} \sin(\omega t - \phi_n) \quad (6.46)$$

where

$$(B_d)_n = \left[\left(1 - \frac{\omega^2}{\omega_n^2}\right)^2 + \left(2\xi_n \frac{\omega}{\omega_n}\right)^2\right]^{-\frac{1}{2}} \quad (6.47)$$

$$\phi_n = \tan^{-1}\left[\frac{2\xi_n \frac{\omega}{\omega_n}}{1 - \left(\frac{\omega}{\omega_n}\right)^2}\right] \quad (6.48)$$

and ξ_n is the damping ratio in the n-th natural mode.

One can address transient effects, similarly (Biggs, 1964; Berg, 1989; Clough and Penzien, 1993; Tedesco et al., 1999; Chopra, 2001), and the reader is encouraged to review such theoretical approaches. Since such treatment may not be practical for supporting design activities, the next section will focus on a brief summary of numerical approaches for this problem.

6.3.2 Numerical Methods for MDOF Transient Responses Analysis

The most direct approach is to expand the Newmark β method (see Section 6.2.3) to MDOF systems as discussed in Newmark and Rosenblueth (1971). Alternatively, one could use various classical approaches cited in the above references and discussed next. Most of these approaches are based on the Rayleigh quotient and its various improvements. The basis for the Rayleigh methods is the fundamental requirement for energy conservation for undamped systems as follows:

$$\text{Maximum strain energy} = \text{maximum kinetic energy} \quad (6.49)$$

For example, if the displacement in a SDOF system is given in the following form:

$$u = u_0 \sin \omega t \quad (6.50)$$

the velocity is obtained from the derivative of Equation (6.50) with respect to time, as follows:

$$v = \omega u_0 \sin \omega t \qquad (6.51)$$

Now the potential energy (entire strain energy) can be computed (work done by the spring when the mass undergoes the noted displacement) as follows:

$$V = 0.5 \, ku^2 = 0.5 \, k \, (u_0 \sin \omega t)^2 \qquad (6.52)$$

Similarly, one can derive the following expression for the kinetic energy:

$$U = 0.5 \, mv^2 = 0.5 \, m(\omega u_0 \sin \omega t)^2 \qquad (6.53)$$

By inspection, one notes that at $t = T/4 = \pi/2\omega$ in which T is the natural period, the kinetic energy will be zero (since $\sin \omega t = 0$ at that time). Therefore, the potential energy reaches its maximum value at this time as follows:

$$V_{max} = 0.5 \, ku_0^2 \qquad (6.54)$$

Similarly, the potential energy vanishes at $T/2 = \pi/\omega$ and the kinetic energy reaches its maximum value as follows:

$$U_{max} = 0.5 \, mu_0^2 \omega^2 \qquad (6.55)$$

Imposing the principle of energy conservation, $V_{max} = U_{max}$ leads to the following equation:

$$0.5 \, ku_0^2 = 0.5 \, mu_0^2 \omega^2 \qquad (6.56)$$

from which one obtains the previously derived expression for the circular natural frequency:

$$\omega^2 = k/m, \text{ and } \omega = 2\pi f \qquad (6.57)$$

where f is the natural frequency.

One can employ this approach for solving the general MDOF problem as follows:

$$u(x, t) = \psi(x) Z_0 \sin(\omega t) \qquad (6.58)$$

where: $u(x, t)$ is the displacement as a function of the coordinate x (geometry) and time, $\psi(x)$ is the shape function ratio of the displacement to any point x to the reference displacement or to z, and $z(t)$ is a generalized coordinate. One can

Dynamic Response and Analysis

compute the corresponding potential and kinetic energies, e.g., assuming a beam. To determine potential energy:

$$V = \frac{1}{2} \int_0^L EI(x) \left(\frac{\partial^2 u}{\partial x^2}\right)^2 dx \qquad (6.59)$$

and

$$V_{max} = \frac{1}{2} Z_0^2 \int_0^L EI(x)[\psi''(x)]^2 dx \qquad (6.60)$$

in which ψ'' indicates a second derivative with respect to x. The kinetic energy is obtained from the following expression:

$$U = \frac{1}{2} \int_0^L m(x)(\dot{u})^2 dx \qquad (6.61)$$

from which one can find the maximum value as follows:

$$U_{max} = \frac{1}{2} Z_0^2 \omega^2 \int_0^L m(x)[\psi(x)]^2 dx \qquad (6.62)$$

Imposing the conservation of energy, $V_{max} = U_{max}$ leads to the following:

$$\omega^2 = \frac{\int_0^L EI(x)[\psi''(x)]^2 dx}{\int_0^L m(x)[\psi(x)]^2 dx} = \frac{k^*}{m^*} \qquad (6.63)$$

where k^* is the generalized stiffness and m^* is the generalized mass. The main problem arises if the shape function $\psi(x)$ is not known beforehand. As a result, one may assume a shape and thus obtain an approximate solution, and the static deflection under the inertia forces might be such an approximate shape. Using this approach will lead to the basic Rayleigh quotient solution as shown below:

$$\omega^2 = \frac{\int_0^L EI(x)\left(\psi''^{(0)}\right)^2 dx}{\int_0^L m(x)\left(\psi''^{(0)}\right)^2 dx} \tag{6.64}$$

This approach can be improved by using the computed frequency to compute new inertia forces and the corresponding deflection and potential energy. Again, imposing the conservation of energy will lead to a modified value of ω^2. A further modification is obtained by computing a new kinetic energy and repeating the process. This method can be extended to the MDOF cases discussed previously. The same concept can be employed to derive a matrix iterative operation method, e.g., the Stodola method based on the flexibility formulation for vibration analysis whose basis is summarized below.

1. Compute inertia forces corresponding to an assumed shape.
2. Compute the deflections resulting from these forces.
3. Compute inertia forces corresponding to the computed deflections.
4. Iterate further until convergence.

The Stodola method will enable one to derive the lowest modal frequency (i.e., the first mode). Once the first mode has been derived, one can obtain the second mode, provided that any trace of the first mode is removed from the trial solution. Such elimination of the first mode is obtained through the orthogonality condition. The same process is employed for obtaining sequentially higher modes. This method is practical for computing manually up to about five modes; it becomes too cumbersome for deriving more modes. Also, it is possible to use the same approach for computing the highest mode and derive sequentially the lower modes.

Another computational procedure is the Holzer method that adopted an approach like an inverse of the Stodola method. In the Stodola method, one assumes a mode shape, and continuously adjusts it until the true mode shape is obtained, then computes the corresponding modal frequency. In the Holzer method, one assumes a modal frequency that is adjusted until the true modal frequency is obtained, then computes the true mode shape. The Holzer method is best suited for structures arranged along a base axis, e.g., multi-story buildings, but it can be modified for other structural configurations. Further information about these procedures is found in the literature.

6.4 CONTINUOUS SYSTEMS

Continuous systems can be analyzed without transforming them to equivalent lumped mass systems, as discussed next.

Dynamic Response and Analysis

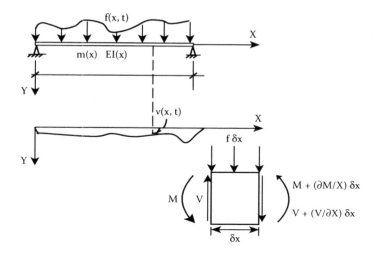

FIGURE 6.11 Beam vibration problem.

6.4.1 GENERAL CONTINUOUS SYSTEMS

Consider a continuous beam vibrating transversely without damping and supported in such a manner that the end conditions (i.e., supports) do not contribute to the system's strain energy (Figure 6.11). This requires that the supports be simple, fixed, guided, and free. One may neglect shear deformations and rotary inertia effects and solve this problem as discussed in various books on structural dynamics and noted earlier. The following parameters are defined:

- $m(x)$ = mass/unit length
- $EI(x)$ = flexural stiffness
- $v(x,t)$ = transverse displacement
- $f(x,t)$ = dynamic force/unit length
- $V(x,t)$ = transverse shear force
- $M(x,t)$ = bending moment

The equation of motion is obtained from equilibrium considerations for the beam element as follows:

$$f\delta x + \left(\frac{\partial V}{\partial x}\right)\delta x = m\frac{\partial^2 v}{\partial x^2}\delta x \qquad (6.65)$$

Define the following moment–curvature relationship (neglecting shear deformation):

$$M = -EI\frac{\partial^2 v}{\partial x^2} \qquad (6.66)$$

For rotational equilibrium (ignoring rotary inertia) one obtains:

$$V = \frac{\partial M}{\partial x} \tag{6.67}$$

Introducing Equation (6.66) and Equation (6.67) into Equation (6.65) will result in the following:

$$m\frac{\partial^2 v}{\partial t^2} + EI\frac{\partial^4 v}{\partial x^4} = f(x,t) \tag{6.68}$$

It was assumed that E, I, and m were constant along the beam. Consider the free vibration problem, i.e., f(x,t) = 0. The solution is obtained by separation of variables as done for MDOF systems:

$$v(x,t) = \psi(x)q(t) \tag{6.69}$$

One can show that this would lead to the following:

$$m\psi\ddot{q} + qEI\psi^{iv} = 0 \tag{6.70}$$

in which $\ddot{q}$ denotes a second derivative with respect to time and ψ^{iv} denotes a fourth derivative with respect to x. Dividing the equation by $m\psi q$ leads to:

$$\frac{\ddot{q}}{q} = \frac{EI\psi^{iv}}{m\psi} = 0 \tag{6.71}$$

Since we have two terms that are functions of separate variables, the following relationship must exist:

$$-\frac{\ddot{q}}{q} = \frac{EI\psi^{iv}}{m\psi} = \text{constant} = \omega^2 \tag{6.72}$$

Thus we transformed the partial differential equation of motion to the following two ordinary differential equations:

$$\ddot{q} + \omega^2 q = 0 \tag{6.73}$$

and

$$EI\psi^{iv} = \omega^2\psi \tag{6.74}$$

Dynamic Response and Analysis

If E, I, and m vary along the beam, one must modify the above procedure so the appropriate derivatives are included. For any given combination of m(x) and EI(x), there is an infinite set of frequencies ω_n^2 and mode shapes $\psi_n(x)$ that satisfies the deformation and support conditions of the beam. One can show that for a simply supported uniform beam of length 1, the mode shapes and their corresponding circular natural frequencies will be defined as shown below. For a complete derivation see Biggs (1964), Clough and Penzien (1993), and Tedesco et al. (1999).

$$\psi_n(x) = \sin(n\pi x/1) \tag{6.75a}$$

and

$$\omega_n = (n\pi/l)^2 (EI/m)^{1/2} \tag{6.75b}$$

Clearly, the first mode shape is a half sine wave with the following frequency ω_1:

$$\omega_1 = \pi^2 \sqrt{\frac{EI}{m\ell^4}} \tag{6.76}$$

The second mode shape is a complete sine wave at four times the first frequency ($\omega_2 = 4\omega_1$); the third mode shape is 1.5 sine waves at nine times the first frequency ($\omega_3 = 9\omega_1$), etc. The arbitrary beam displacement can be obtained by orthogonality as a Fourier series in $\Delta < x < \ell$, as follows:

$$v(x,t) = \sum_{n=1}^{\infty} q_n(t) \sin\left(\frac{n\pi x}{\ell}\right) \tag{6.77}$$

in which $q_n(t)$ is determined by the loading conditions. One can show the following relationship if two different mode shapes and frequencies are involved such that $\psi_r \neq \psi_s$, $\omega_r \neq \omega_s$ (Biggs, 1964; Clough and Penzien, 1993; Tedesco et al., 1999):

$$\int_0^t m\psi_r \psi_s \, dx = 0 \tag{6.78}$$

and

$$\int_0^t EI\psi_r'' \psi_s'' = 0 \tag{6.79}$$

Equations (6.78) and (6.79) are then orthogonality relations for the modes, similarly to the earlier discussion for MDOF systems. If there are repeated frequencies, functions ψ still exist such that the orthogonality relation should hold for all $r \neq s$ including those for which $\omega_r = \omega_s$. Also, as discussed for MDOF systems, one can derive a modal equation of motion that shows the contribution of all the modes to the total behavior. Interested readers can find such information in books on structural dynamics (Biggs, 1964; Clough and Penzien, 1993; Tedesco et al., 1999).

This brief summary of closed form solutions highlights the fact that these solutions are limited to simple loading, simple geometrics and support conditions, and mostly linear materials and systems. Therefore, these methods will not be addressed further. Various approaches that lead to approximate simplified methods are based on assumed behavioral models, and it is recommended to use such methods with data from computer codes based on current design manuals. The following discussions are aimed at providing information on such approximate methods and on how they can be applied for the analysis and design of structural systems under severe short duration loads.

6.5 INTERMEDIATE AND ADVANCED COMPUTATIONAL APPROACHES

As noted previously, typical SDOF methods can provide information for only one assumed mode of deformation and one location in a structure. Although such information is valuable for supporting design activities, it may be a serious limitation when the assumed deformation mode does not accurately represent the behavior of the structure under consideration or one requires information for larger areas in a structure.

Although advanced SDOF approaches can address several response modes and locations in a structure, they may not be sufficiently detailed for various advanced design applications. Therefore, cases that require more advanced design and analysis may be treated with more advanced structural theories. Unfortunately, close form solutions that can provide a complete description of the structural behavior are limited to linear materials and simple structural systems. Consequently, analysts and designers have looked for intermediate computational approaches that can provide more information than typical SDOF approaches but would not require extensive resources.

For example, for the analysis of structural elements subjected to blast, shock, and impact, one could employ improved Timoshenko beam or Mindlin plate formulations that account for shear deformations, rotatory inertia, and more involved material models that include multi-axial stress–strain relationships and strain rate effects. Such procedures could be defined as intermediate approximate computational methods. One may also adopt more advanced approximate computational approaches (such as finite element or finite difference algorithms) that involve detailed representation of structural systems and require extensive computational resources. These capabilities are briefly discussed next.

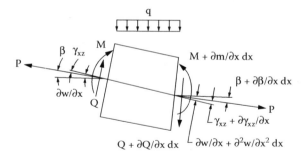

FIGURE 6.12 Timoshenko beam element.

6.5.1 INTERMEDIATE APPROXIMATE METHODS

The Timoshenko beam theory (Timoshenko, 1921 and 1922) accounts for shear deformation and rotary inertia — essential factors for a realistic analysis of the problem at hand, as discussed earlier. The general equations of motion of the Timoshenko beam can be derived from the equilibrium of a Timoshenko beam differential element, as shown in Figure 6.12, as follows (Krauthammer et al., 1993a, b and 1994):

$$\frac{\partial M}{\partial x} - Q = -\rho I \frac{\partial^2 \beta}{\partial t^2} \qquad (6.80a)$$

$$\frac{\partial Q}{\partial x} + q(x,t) + P \frac{\partial \beta}{\partial x} = \rho A \frac{\partial^2 w}{\partial t^2} \qquad (6.80b)$$

where M and Q are the bending moment and shear force, respectively, I is the moment of inertia, A is the cross-sectional area, ρ is the material mass density, β is the rotation of the cross-section due to bending, w is the transverse displacement of the midplane of the beam, P is the axial force on the member acting at the plastic centroid of the cross section, q is the distributed dynamic load transverse to beam length, x is the distance along the beam, and t is the time. Equation (6.80a) and Equation (6.80b) are used directly for obtaining the solution in the nonlinear material domain. Fully nonlinear material models for concrete (both confined and unconfined addressing the axial and shear behaviors), reinforcement, and bond can be used to derive the moment–curvature relationships. In addition to the dynamic equilibrium equations above, the Timoshenko beam theory imposes the following compatibility relationships among the deformations, as shown in Figure 6.13.

$$\gamma_{xz} = \partial w/\partial x - \beta \qquad (6.81a)$$

and

$$\varphi = \partial \beta/\partial x \qquad (6.81b)$$

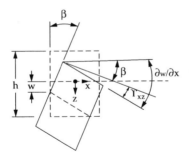

FIGURE 6.13 Timoshenko beam element deformation.

where γ_{xz} is the shear strain, φ is the curvature of the cross-section, and w, x, and β are as defined earlier.

Recent developments in applying this approach to studying structural responses to blasts showed that these methods could be successfully employed to obtain detailed behavioral information considering that the improved theories include the combination of flexure, diagonal shear, direct shear, and membrane/axial modes of deformation. These methods require one to use either finite difference or finite element approaches. Finite difference computational methods were utilized successfully to solve the above noted differential equations (Krauthammer et al., 1994a, b, and December 2002). Once the stability criterion for the convergence is determined, the required computational resources are very modest. Although such approaches are much more advanced than SDOF methods, they are limited to the consideration of single structural elements or simple structural systems. Despite their enhanced capabilities as compared with SDOF approaches, the intermediate numerical procedures have been used infrequently in support of structural design, primarily because very few reliable computer codes are available. However, they could become much more popular in the future when such computer codes may become available to help designers.

6.5.2 Advanced Approximate Computational Methods

Advanced computational methods include various finite element, finite difference, mesh-free, and hybrid codes but they require large computational resources, cannot be used efficiently for typical design activities, and should be employed selectively. Nevertheless, they can be very useful for the assessment of structural behavior under expected loading conditions for design purposes or for forensic investigations following explosive loading incidents. During the last couple of decades, finite element codes evolved very significantly and have become the tools of choice for comprehensive structural analysis, including the ability to address structural response to short duration loading effects. Therefore, the following discussion is dedicated to an overview of finite element approaches and their application to the transient loading regime.

Dynamic Response and Analysis

The finite element approach is based on replacing a continuum (in either two or three dimensions) with an assemblage of discrete elements (Bathe, 1996; Cook et al., 2002; Reddy, 2006). Different element types are available (line elements, triangular or quadrilateral area or surface elements, tetrahedral or hexahedral volume elements, etc.) and they are derived based on assumed displacement fields or extrapolation algorithms (constant or linear strain, regular or isoparametric, etc.). Furthermore, such elements are derived while addressing the requirements of equilibrium and compatibility and an assumed material model. Generally, the transformation from a continuum to the finite element model will result in the following equation of equilibrium:

$$[M]\{\ddot{x}\} + [C]\{\dot{x}\} + [K]\{x\} = \{F(t)\} \tag{6.82}$$

in which $\{F(t)\}$ is the vector describing the force–time histories applied to specific nodes in the model, $[K]$ is the structural stiffness matrix, $[M]$ is the mass matrix, and $[C]$ is the damping matrix. The vectors $\{\ddot{x}\}$, $\{\dot{x}\}$, and $\{x\}$ represent the nodal accelerations, velocities, and displacements of the structure at any time (t), respectively. The resulting systems of simultaneous equations are solved numerically and one may use either implicit (iterative) or explicit (forward marching in time) algorithms for such purpose.

Several recent studies of such applications can be used to illustrate these capabilities. A hybrid coupled finite element and finite difference (FE/FD) approach was used for the analysis of shallow-buried arch structures subjected to surface blast loads. The analyses showed the evolution of damage for the dynamic medium–structure interaction and structural response phenomena (Steven and Krauthammer, January 1991). A similar approach was used to study the behavior of structural concrete beam–column connections and assess the influence of detailing parameters on their performance (Krauthammer and Ku, 1996).

An advanced finite element code was used to study the behavior of three-dimensional structural concrete connections and the specific contributions of reinforcement details on behavior (Otani and Krauthammer, 1997). Although such examples demonstrate the effectiveness of advanced numerical analysis approaches, they can be used also to highlight various possible difficulties and complications that could be associated with their application.

6.5.3 Material Models

One of the most difficult problems in finite element analysis is the availability and selection of appropriate material models. In the early days, analyses were limited to linear elastic models. However, the expansion of these approaches into the nonlinear domain enabled one to use a wide range of plasticity based models (Owen and Hinton, 1980). Such models are based on the theory of plasticity and the experience gained by applying it for the analysis of metals. For example, the

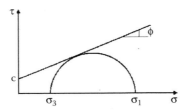

FIGURE 6.14 Mohr–Coulomb criterion.

Tresca yield criterion was modified for application to soils and other geologic materials by using a combination of the Mohr circle and the Coulomb friction model to derive the Mohr–Coulomb criterion, as follows:

$$\tau = c + \sigma \tan\varphi \tag{6.83}$$

in which τ is the shear stress on the failure plane, c is the material cohesion, σ is the normal effective stress on the failure surface, and φ is the angle of internal friction, as shown in Figure 6.14. This approach leads to the formation of an irregular hexagonal pyramid in the principal stress space (σ_1, σ_2, σ_3). According to this criterion, the yield strength in compression is higher than the tensile strength, indicating its dependence on the third invariant of the stress tensor. Since the Mohr–Coulomb criterion is expressed in terms of maximum and minimum stresses, it is inconvenient to express it in terms of a general three-dimensional stress state and its six stress vector components. This criterion was generalized by using the invariants of the stress tensor (Drucker and Prager, 1952) as follows:

$$f = (J_{2D})^{0.5} - \alpha J_1 - k \tag{6.84}$$

in which α and k are positive material parameters, J_1 is the first invariant of the stress tensor, and J_{2D} is the second invariant of the deviatoric stress tensor. This criterion (also known as the Drucker–Prager failure criterion) is represented by a right circular cone in the three-dimensional stress space (σ_1, σ_2, σ_3), while the Mohr–Coulomb criterion forms an irregular hexagonal pyramid inscribed in that cone. According to this criterion, any state of stress outside this surface is unstable, and stress points on the surface represent plastic deformations of the material.

Concrete and many geologic materials exhibit irreversible deformations even at early loading stages, and researchers have proposed to use plasticity-based material models to represent such behavior. Models with series of yield surfaces that led to the failure surface were proposed. In metals, successive yielding caused material hardening, while in geologic materials a distinct softening trend was observed. This is represented by a series of hardening caps on the Drucker–Prager model (Figure 6.15), and the shape of such caps is determined based on test data.

Dynamic Response and Analysis

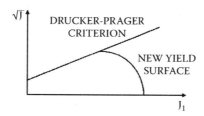

FIGURE 6.15 Drucker–Prager model with cap.

Variations of this approach have been proposed and applied successfully in advanced numerical analysis procedures (Meyer and Okamura, 1986; Stevens and Krauthammer, September 1989; Malvar et al., 1997). Such applications included the consideration of concrete–steel bond and slip effects, rate and size effect assessments, and issues related to code validation by comparison with precision test data (Krauthammer, June 1993; Krauthammer et al., May 1996 and May 2001; Meley and Krauthammer, 2003). Other material models represent the shear behavior of concrete (Collins and Mitchell, 1991; Hsu, 1993), and were employed successfully to simulate the behavior of structural concrete systems under both static and dynamic loading conditions (Krauthammer et al., 1994 and December 2002).

Nevertheless, researchers noted the artificial adaptation of plasticity-based material models to describe the brittle behavior of rock and concrete and explored the application of fracture mechanics-based approaches to address the observed phenomena that include the evolution of cracking, size, and rate effects (Chandra and Krauthammer, May 1996; Krauthammer et al., October 2003; Elfahal et al., 2005; Elfahal and Krauthammer, 2005). Some models have been also implemented successfully in advanced numerical codes, but considerable issues need to be addressed further.

6.6 VALIDATION REQUIREMENTS FOR COMPUTATIONAL CAPABILITIES

As with requirements in other scientific areas, modern protective technologies are founded on a combination of precision tests and various numerical simulations. The linkage of these two essential components has increased gradually over the last half century and the driving force behind both the capabilities and requirements has been the rapid evolution in computer power. Nevertheless, because of current enhanced capabilities in both experimental and numerical analysis, the need for stronger interaction and collaboration among researchers in these areas is greater than ever. This point of view has been expressed, discussed, and accepted at several technical meetings and workshops on this subject (Krauthammer et al., May 1996 and 1998; Namburu and Mastin, 1998; DTRA 2001).

It is clearly recognized that numerical simulations will play an increasingly more important role, eventually replacing many experimental studies. This shift is expected to translate into very significant cost reductions in future fortification-related research, analysis, and design. However, to enable this transition and ensure that it will be effective, one requirement is to treat explicitly the validation of numerical capabilities and the precision testing data to support it. The following general conclusions and recommendations were developed, based on the workshops mentioned above:

- Successful precision testing and computer code validation have been achieved by the transportation equipment industry. A similar degree of success could be expected in the fortifications area, but much more work is required to reach that goal.
- The guidelines and requirements for precision testing should be adopted and followed. This will ensure that the test data are both valid and useful for computer code validation.
- Validation activities should include parameters whose characteristics are as similar as possible to those expected to be related to the real problem. However, these parameters must be selected to meet the precision requirements of such activities.
- Precision reporting of specific details in both tests and numerical simulations should be adopted to ensure that all involved parties have all the required information for effective computer code validation and verification efforts.
- Innovative precision test methods are needed for studying both materials and structures in the impulsive loading domain. Shock tube and explosive testing approaches introduce various difficulties and limitations and they cannot be considered precision tests.
- Although code validation activities do not have to be directly related to weapon effects, it would be very helpful if the parameters under consideration had similar characteristics.
- The emphasis should be on simple and well-defined tests.
- Impact testing can be used effectively for studying both material behavior and structural response. Such approaches, however, should be developed and used according to the recommendations presented in the workshop reports.
- It has been recognized that successful computer code validation activities cannot be achieved with inadequate material models. This is especially true in the high pressure and high strain rate domains. Careful attention must be given to the materials' initial conditions, scaling and size effects, and proper use of physics for material models derivations. Considering the structural behavior of material specimens is important, and the role of various parameters on their behavior. Unlike for metals, the behavior of brittle materials may not be adequately represented by plasticity-based models. One should consider

Dynamic Response and Analysis

- the application of damage and/or fracture-based approaches, possibly combined with statistical methods, for such cases.
- Future activities should address both the consideration of material constitutive formulations and the behaviors of simple structural components. These activities should be performed in parallel and the efforts should be collaborative to ensure efficient information exchange among the parties.
- Future activities should address simultaneously theoretical, numerical, and experimental issues. These activities should be performed in parallel and the efforts should be collaborative to ensure efficient information exchange among the parties.
- Code validation activities should take place within a collaborative framework to enable quality control, wise use of resources, and synergy of research capabilities.
- The scientific and technical communities should develop ideas and proposals on how to pursue the required developments. The sponsors should work closely with these communities to implement their recommendations and develop quality assurance procedures for these activities.

Various commercial structural analysis computer codes are available, and their accuracies for simulating structural behavior under impulsive loading must be determined. It is required that a numerical solution be within an acceptable range from the experimental data. This is defined as code validation, and various studies were conducted according to the workshop recommendations (Krauthammer et al., May 2001; O'Daniel et al., 2002; Meley and Krauthammer, 2003).

These activities included the precision testing and numerical simulation of structural concrete, aluminum, and steel systems under impact loads. The mesh geometry and the material properties were developed based on experimental data, and parametric studies were performed to examine their effects on the analysis. Several mesh sizes were employed to investigate the effects of mesh geometry on the analysis. Since the impact strikers could have had complicated geometries, various striker models were used in the investigations. Impact loading generates large local plastic deformation near an impacted area. Therefore, the effects of the key material properties on the solution were also investigated.

This approach is illustrated next. A simple test case for which precision test data were available was selected and used for the validation of the various numerical approaches, according to the procedures outlined by Krauthammer et al. (May 1996). Accordingly, as noted above, test data for reinforced concrete beams tested by impact were used for this purpose (Feldman and Siess, 1958). One such case (beam 1-c) was used to illustrate this approach. The experiment setup and material properties of the tested specimens are illustrated in Figure 6.16 and the material properties are defined in Table 6.1.

Beam 1-c was analyzed with the following numerical approaches. The computer code DSAS (Krauthammer et al., August 2003 and September 2004; Kyung

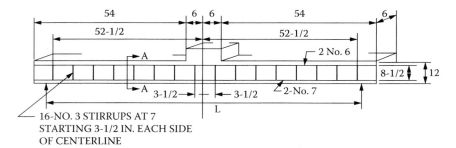

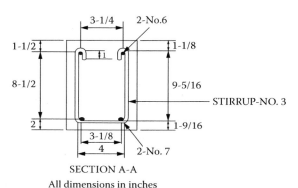

FIGURE 6.16 Test beam 1-c (Feldman and Siess, 1958).

TABLE 6.1
Material Properties for Beam 1-c

Beam	Concrete		Tension Steel		Compression Steel		Shear Steel	
	f'_c (ksi)	E_c (ksi)	f'_y (ksi)	E_s (ksi)	f'_y (ksi)	E_s (ksi)	f'_y (ksi)	E_s (ksi)
1-c	6.000	4234	46.08	29520	46.70	29520	46.08	29520

and Krauthammer, June 2004) was used for the application of the advanced (fully nonlinear dynamic) SDOF and Timoshenko beam. The finite element code ABAQUS/Explicit (Hibbitt et al., 2005) was used for the fully nonlinear dynamic three-dimensional computation.

The material models of reinforcing steel bars in all three codes were based on using an effective stress–strain relationship that included considerations of steel–concrete bonds (Hsu, 1993). The concrete model for ABAQUS/Explicit was a modified Drucker–Prager model with the following parameters: d = 0.85 psi, $\beta = 60°$, R = 0.5, a = 0.01, and $e^{pl} = 0.003$. The concrete material model in the Timoshenko beam code was based on the modified compression field theory discussed elsewhere (Krauthammer et al. 1993, 1994, and December 2002). In the SDOF code, it was based on a nonlinear approach that included considerations

Dynamic Response and Analysis 271

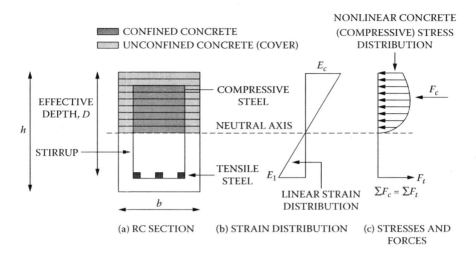

FIGURE 6.17 Cross-sectional equilibrium.

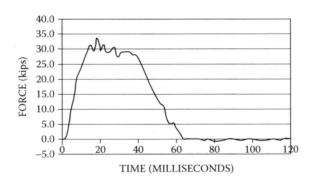

FIGURE 6.18 Measured impact force–time history (Feldman and Siess, 1958).

of flexure, diagonal shear, and axial forces. The beam resistance functions for the SDOF and Timoshenko beam were derived based on cross-sectional equilibrium considerations, as shown in Figure 6.17 and coupled with the appropriate boundary conditions.

Both the SDOF and Timoshenko beam computations addressed the issue of direct shear behavior, as discussed in Chapter 5, and the direct shear resistance models for beam 1-c were based on those parameters. However, direct shear was not noticed in the tested beams, and the DSAS-based analyses did not indicate such response. Beam 1-c was loaded by localized impact on the center stub, and the loading function for this case is described in Figure 6.18.

The finite element model for this case is shown in Figure 6.19; both the concrete and reinforcement meshes are presented. The steel was represented by

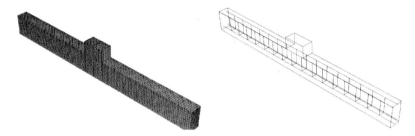

FIGURE 6.19 Finite element models for concrete beam and reinforcement.

discrete space linear two-node beam elements whose nodes were attached to nodes in the concrete mesh at the right location. The effective stress–strain material model for the steel accounted for bond effects, as noted above. The concrete was modeled using eight-node three-dimensional continuum elements with reduced integration, hourglass control. The concrete elements had an aspect ratio of approximately 1:1:1 based on about a 1-in. dimension in each direction. These dimensions were chosen to ensure that the concrete node lines coincided with reinforcement locations, resulting in a total of approximately 13,500 elements for the beam.

Figure 6.20 and Figure 6.21 illustrate comparisons of experimental and analytical results. These graphs show good agreement in deflection, reaction time histories, and times for peak values. The numerical results from all approaches show very good agreement with the reported test data. They all captured quite well the peak and residual responses as well as the times to peak. Similar outcomes are noted for the support reactions, although the post-load free vibration is less accurate than the computed responses during the load application. This inconsistency, however, is probably a result of the unloading paths in the various resistance functions.

Clearly, using the SDOF approach is very attractive in support of design activities because it requires the fewest human and computational resources. One can assess very quickly a tentative design, modify it, and reanalyze it within a very short time. Once the preliminary design is done, one may use the Timoshenko beam approach to obtain more information about different locations along the beam. The finite element approach is very detailed, but it requires far more human and computational resources and it should not be undertaken instead of a SDOF analysis.

6.7 PRACTICAL COMPUTATION SUPPORT FOR PROTECTIVE ANALYSIS AND DESIGN ACTIVITIES

The previous sections of this chapter were devoted to reviews of various theoretical and numerical structural analysis approaches. It was noted that closed-form solutions are limited to simple loading, simple geometric and support conditions,

Dynamic Response and Analysis

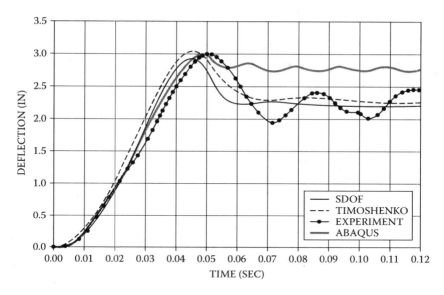

FIGURE 6.20 Comparisons of computed deflections.

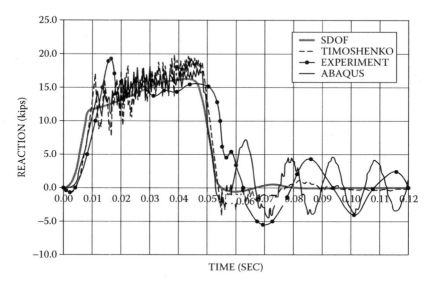

FIGURE 6.21 Comparisons of computed support reactions.

and mostly linear materials. Approximate simplified methods are based on assumed response modes and are implemented in computer codes based on design manuals.

Advanced numerical methods require significant resources, must be validated against precision tests, and are usually not utilized in typical design activities. Nevertheless, these advanced computational approaches can be used in the final design stages for detailed performance assessment. Clearly, one needs an efficient and expedient computational support capability for design activities, and either an SDOF-based approach or one of the intermediate analysis approaches should be considered. SDOF-based structural analysis approaches have been used extensively in support of protective design, and they are discussed in more detail next.

6.7.1 Equivalent SDOF System Approach

We noted that SDOF-type analyses are very efficient computationally. However, the structural systems that need to be analyzed are continuous (e.g., beams, slabs, frames, and their combinations), and analyzing them as continuous systems is less suitable for expedient assessment and design activities. Therefore, one needs to devise an approach that could apply the SDOF analysis approach for continuous systems. For this purpose, one needs to transform the continuous system to an equivalent SDOF system whose analysis would provide results that could represent the behavior of the continuous system. To achieve this goal, one needs to understand well the structural behavior, so that the relevant behavioral parameters can be represented accurately. Consider the behavior of a beam with fixed supports, subjected to a uniformly distributed load, as shown in Figure 6.22.

Two beam response modes are shown in the figure. The first mode flexural response is depicted in the three drawings on the left. The direct shear response is shown on the right. In the flexural response sequence, one notes a gradual transition from the elastic to the plastic domains. Initially the beam deflects in the elastic domain and the deflected shape maintains a rotation of 0° at the supports. When the moments at the support reach the yield moment, one develops plastic hinges and the beam will deflect like a simply supported beam with finite

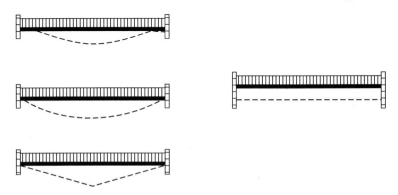

FIGURE 6.22 Continuous beam response mechanisms.

Dynamic Response and Analysis

rotations at the plastic hinge locations. Finally, a third plastic hinge develops at the center, and a three-hinge mechanism is formed. Beyond this stage, the beam can resist additional loads in the tension membrane mode, as discussed in Chapter 5. In the direct shear mode, the beam does not exhibit flexural behavior and the entire beam will slip at the supports if its shear capacity is exceeded, as discussed in Chapter 5. Clearly, one needs to capture these behavioral mechanisms when transforming the continuous beam to an equivalent SDOF system.

Note that different points along the beam have different values of transverse displacement at the different behavioral stages. Since the acceleration distribution along the beam is proportional to the deflected shape, the inertia forces (mass times acceleration) will vary accordingly. These lead to the concept of an equivalent mass that can be used to transform the continuous mass of the beam to an equivalent concentrated mass in the SDOF system (Biggs, 1964; Clough and Penzien, 1993; Tedesco, 1999; Chopra, 2001). The equivalent mass is computed based on the kinetic energy of the moving parts of the beam. One must sum up the product $0.5mv^2$ along the beam that is composed of the mass per unit length and the local velocity of that mass. The effective mass will depend on the deflected shape, and the mode shape $\varphi(x)$ for the appropriate behavioral stage is used for this purpose. Accordingly, the equivalent mass is derived as follows:

$$M_e = \int_0^L m\varphi^2(x)dx \tag{6.85}$$

Therefore, equivalent mass factor is defined as follows:

$$K_M = M_e/M_{Total} \text{ or } K_M = M_e/M \tag{6.86}$$

Similarly, one can derive the following equivalent load factor:

$$K_L = F_e/F_{total} \tag{6.87}$$

in which

$$F_e = \int_0^L w(x)\varphi(x)dx \tag{6.88}$$

Since the structural resistance represents a distributed force system that pushes against the applied forces, the resistance factor will be the same as the load factor:

$$K_R = K_L \tag{6.89}$$

Having derived these factors enables one to write the following equivalent SDOF equation of motion (damping is ignored here to simplify the discussion):

$$K_M M\ddot{y} + K_L Ky = K_L F(t) \qquad (6.90)$$

Also, one can compute the equivalent circular natural frequency as follows:

$$\omega_e = (K_e/M_e)^{1/2} = (K_L K/K_M M)^{1/2} = (K/K_{LM} M)^{1/2} \qquad (6.91)$$

in which $K_{LM} = K_M/K_L$. Similar considerations can be applied for beams with different boundary and loading conditions and for slabs with different boundary conditions, as shown in several references (U.S. Army, 1957; Biggs, 1964; Department of the Army, 1986).

6.7.2 Applications to Support Analysis and Design

Some of the available cases are provided in Table 6.2 through Table 6.8; others can be obtained from the cited references. These tables also provide expressions for the dynamic support reactions that can be used to assess the forces at or along the supports. Such forces are used to analyze the direct shear response of the structural system under consideration. The dynamic reactions are derived based on a force equilibrium consideration for the structural members that includes the inertia forces acting on it (similar to the calculations of reactions under static loads). This general approach enables one to convert typical continuous structural systems to equivalent SDOF systems and derive their dynamic response parameters. These analyses can be performed either numerically or graphically, as described in Chapter 5. The graphical approach shown in Chapter 5 is illustrated again in Figure 6.23a and Figure 6.23b.

If the ratio of t_d/T is shorter than the range shown in these figures, the system may be considered to respond in the impulsive domain. However, if that ratio is greater than the range shown, one may estimate the system to behave as if it is loaded instantaneously by a constant force (pseudo static domain). Although various sources provide approximate solutions for this response domain, this general area will be discussed again in much more detail in Chapter 8. Meanwhile, one can employ the following empirical relationship for the required maximum structural resistance (Department of the Army, 1986):

$$(R_m)_{required} = I\omega/(2\mu - 1)^{1/2} \qquad (6.92)$$

in which I is the impulse for the applied forcing function, ω is the circular natural frequency, and μ is the ductility (all parameters were defined in Chapter 5). For a peak triangular load F_1 with a duration t_d (Figure 6.23), $I = 0.5(F_1 t_d)$ and $\omega = 2\pi f = 2\pi/T$. If both R_m and F_1 are known, one can rewrite Equation (6.92) to estimate the resulting ductility. In the pseudo static domain, one assumes an instantaneously applied constant force of magnitude F_1 and the required peak resistance will be:

$$(R_m)_{required} = F_1/(I - 0.5\mu) \qquad (6.93)$$

TABLE 6.2
SDOF Parameters for Simply Supported Beams

Loading Diagram	Strain Range	Load Factor K_L	Mass Factor K_M		Load-Mass Factor K_{LM}		Maximum Resistance R_m	Spring Constant k	Dynamic Reaction V
			Concentrated Mass*	Uniform Mass	Concentrated Mass*	Uniform Mass			
	Elastic	0.64		0.50		0.78	$\dfrac{8M_p}{L}$	$\dfrac{384EI}{5L^3}$	$0.39R + 0.11P$
	Plastic	0.50		0.33		0.66	$\dfrac{8M_p}{L}$	0	$0.38R_m + 0.12P$
	Elastic	1	1.0	0.49	1.0	0.49	$\dfrac{4M_p}{L}$	$\dfrac{48EI}{L^3}$	$0.78R - 0.28P$
	Plastic	1	1.0	0.49	1.0	0.33	$\dfrac{4M_p}{L}$	0	$0.75R_m - 0.25P$
	Elastic	0.87	0.76	0.52	0.87	0.60	$\dfrac{8M_p}{L}$	$\dfrac{56.4EI}{L^3}$	$0.525R - 0.25F$
	Plastic	1	1.0	0.56	1.0	0.56	$\dfrac{6M_p}{L}$	0	$0.52R_m - 0.02F$

* Equal parts of concentrated mass are lumped at each concentrated load.

Source: Department of the Army, 1986.

TABLE 6.3
SDOF Parameters for Fixed Beams

Loading Diagram	Strain Range	Load Factor K_L	Mass Factor K_M Concentrated Mass*	Mass Factor K_M Uniform Mass	Load-Mass Factor K_{LM} Concentrated Mass*	Load-Mass Factor K_{LM} Uniform Mass	Maximum Resistance R_m	Spring Constant k	Effective Spring Constant k_E	Dynamic Reaction V
	Elastic	0.53		0.41		0.77	$\dfrac{12M_{Ps}}{L}$	$\dfrac{384EI}{L^3}$		$0.36R + 0.14P$
	Elasto-Plastic	0.64		0.50		0.78	$\dfrac{8}{L}(M_{Ps}+M_{Pm})$	$\dfrac{384EI}{5L^3}$	$\dfrac{307EI}{L^3}$	$0.39R + 0.11P$
	Plastic	0.50		0.33		0.68	$\dfrac{8}{L}(M_{Ps}+M_{Pm})$	0		$0.38R_m + 0.12P$
	Elastic	1	1.0	0.37	1.0	0.37	$\dfrac{4}{L}(M_{Ps}+M_{Pm})$	$\dfrac{192EI}{L^3}$		$0.71R - 0.21P$
	Plastic	1	1.0	0.38	1.0	0.38	$\dfrac{4}{L}(M_{Ps}+M_{Pm})$	0		$0.75R_m - 0.25P$

* Concentrated mass is lumped at the concentrated load.

M_{Ps} = Ultimate moment capacity at support.
M_{Pm} = Ultimate moment capacity at midspan.

Source: Department of the Army, 1986.

TABLE 6.4
SDOF Parameters for Fixed Boundary Beams

Loading Diagram	Strain Range	Load Factor K_L	Mass Factor K_M Concentrated Mass*	Mass Factor K_M Uniform Mass	Load-Mass Factor K_{LM} Concentrated Mass*	Load-Mass Factor K_{LM} Uniform Mass	Maximum Resistance R_m	Spring Constant k	Effective Spring Constant k_E	Dynamic Reaction V
	Elastic	0.58		0.45		0.78	$8M_{Ps}/L$	$185EI/L^3$		$V_1 = 0.25R + 0.12P$ $V_2 = 0.43R + 0.19P$
	Elasto-Plastic	0.64		0.50		0.78	$\frac{4}{L}(M_{Ps} + 2M_{Pm})$	$\frac{384EI/L^3}{5}$	$\frac{160EI}{L^3}$	$V = 0.39R + 0.11F \pm M_{Ps}/L$
	Plastic	0.50		0.33		0.66	$\frac{4}{L}(M_{Ps} + 2M_{Pm})$	0		$V = 0.38R_m + 0.12F \pm M_{Ps}/L$
	Elastic	1.0	1.0	0.43	1.0	0.43	$16M_{Ps}/3L$	$107EI/L^3$		$V_1 = 0.25R + 0.07F$ $V_2 = 0.54R + 0.14F$
	Elasto-Plastic	1.0	1.0	0.49	1.0	0.49	$\frac{2}{L}(M_{Ps} + 2M_{Pm})$	$48EI/L^3$	$\frac{180}{L^3}$	$V = 0.78R - 0.28F \pm M_{Ps}/L$
	Plastic	1.0	1.0	0.33	1.0	0.33	$\frac{2}{L}(M_{Ps} + 2M_{Pm})$	0		$V = 0.75R_m - 0.25F \pm M_{Ps}/L$
	Elastic	0.81	0.67	0.45	0.83	0.55	$6M_{Ps}/L$	$132EI/L^3$		$V_1 = 0.17R + 0.17P$ $V_2 = 0.33R + 0.33P$

TABLE 6.4 (CONTINUED)
SDOF Parameters for Fixed Boundary Beams

Loading Diagram	Strain Range	Load Factor K_L	Mass Factor K_M		Load-Mass Factor K_{LM}		Maximum Resistance R_m	Spring Constant k	Effective Spring Constant k_E	Dynamic Reaction V
			Concentrated Mass*	Uniform Mass	Concentrated Mass*	Uniform Mass				
	Elasto-Plastic	0.87	0.76	0.52	0.87	0.60	$\frac{2}{L}(M_{Ps} + 3M_{Pm})$	$\frac{56EI}{L^3}$	$\frac{122EI}{L^3}$	$V = 0.52R - 0.025F \pm M_{Ps}/L$
	Plastic	1.0	1.0	0.56	1.0	0.56	$\frac{2}{L}(M_{Ps} + 3M_{Pm})$			$V = 0.52R_m - 0.02F \pm M_{Ps}/L$

* Equal parts of the concentrated mass are lumped at each concentrated load.

M_{Ps} = Ultimate moment capacity at support.
M_{Pm} = Ultimate moment capacity at midspan.

Source: Department of the Army, 1986.

Dynamic Response and Analysis

TABLE 6.5
SDOF Parameters for Simply Supported Slabs

Strain Range	a/b	Load Factor K_L	Mass Factor K_M	Load–Mass Factor K_{LM}	Maximum Resistance	Spring Constant k	Dynamic Reactions V_A	Dynamic Reactions V_B
Elastic	1.0	0.45	0.31	0.68	$\frac{12}{a}(M_{Pfa} + M_{Pfb})$	$252EI_a/a^2$	$0.07P + 0.18R$	$0.07P + 0.18R$
	0.9	0.47	0.33	0.70	$\frac{1}{a}(12M_{Pfa} + 11M_{Pfb})$	$230EI_a/a^2$	$0.06P + 0.16R$	$0.08P + 0.20R$
	0.8	0.49	0.35	0.71	$\frac{1}{a}(12M_{Pfa} + 10.3M_{Pfb})$	$212EI_a/a^2$	$0.06P + 0.14R$	$0.08P + 0.22R$
	0.7	0.51	0.37	0.73	$\frac{1}{a}(12M_{Pfa} + 9.8M_{Pfb})$	$201EI_a/a^2$	$0.05P + 0.13R$	$0.08P + 0.24R$
	0.6	0.53	0.39	0.74	$\frac{1}{a}(12M_{Pfa} + 9.3M_{Pfb})$	$197EI_a/a^2$	$0.04P + 0.11R$	$0.09P + 0.26R$
	0.5	0.55	0.41	0.75	$\frac{1}{a}(12M_{Pfa} + 9.0M_{Pfb})$	$201EI_a/a^2$	$0.04P + 0.09R$	$0.09P + 0.28R$
Plastic	1.0	0.33	0.17	0.51	$\frac{12}{a}(M_{Pfa} + M_{Pfb})$	0	$0.09P + 0.16R_m$	$0.09P + 0.16R_m$
	0.9	0.35	0.18	0.51	$\frac{1}{a}(12M_{Pfa} + 11M_{Pfb})$	0	$0.08P + 0.15R_m$	$0.09P + 0.18R_m$
	0.8	0.37	0.20	0.54	$\frac{1}{a}(12M_{Pfa} + 10.3M_{Pfb})$	0	$0.07P + 0.13R_m$	$0.01P + 0.20R_m$
	0.7	0.38	0.22	0.58	$\frac{1}{a}(12M_{Pfa} + 9.8M_{Pfb})$	0	$0.06P + 0.12R_m$	$0.10P + 0.22R_m$
	0.6	0.40	0.23	0.58	$\frac{1}{a}(12M_{Pfa} + 9.3M_{Pfb})$	0	$0.05P + 0.10R_m$	$0.01P + 0.25R_m$
	0.5	0.42	0.25	0.59	$\frac{1}{a}(12M_{Pfa} + 9.0M_{Pfb})$	0	$0.04P + 0.08R_m$	$0.11P + 0.27R_m$

M_{Pfa} and M_{Pfb} = Total positive ultimate moment capacity along midspan section parallel to edges a and b, respectively.

Source: Department of the Army, 1986.

If both R_m and F_1 are known, one can rewrite Equation (6.93) to estimate the resulting ductility.

6.7.3 Approximate Procedure for Multi-Segmental Forcing Functions

The loading function considered in the SDOF analysis presented in the previous section was a right angle triangle with a peak value F and duration t_d, as shown in Figure 6.23. However, blast loads are known to have an exponentially decaying

TABLE 6.6
SDOF Parameters for Fixed Slabs

Strain Range	a/b	Load Factor K_L	Mass Factor K_M	Load–Mass Factor K_{LM}	Maximum Resistance	Spring Constant k	Dynamic Reactions V_A	Dynamic Reactions V_B
Elastic	1.0	0.33	0.21	0.63	$29.2 M^o_{Psb}$	$810 EI_a/a^2$	$0.10P + 0.15R$	$0.10P + 0.15R$
	0.9	0.34	0.23	0.68	$27.4 M^o_{Psb}$	$743 EI_a/a^2$	$0.09P + 0.14R$	$0.10P + 0.17R$
	0.8	0.36	0.25	0.69	$26.4 M^o_{Psb}$	$705 EI_a/a^2$	$0.08P + 0.12R$	$0.11P + 0.19R$
	0.7	0.38	0.27	0.71	$26.2 M^o_{Psb}$	$692 EI_a/a^2$	$0.07P + 0.11R$	$0.11P + 0.21R$
	0.6	0.41	0.29	0.71	$27.3 M^o_{Psb}$	$724 EI_a/a^2$	$0.06P + 0.09R$	$0.12P + 0.23R$
	0.5	0.43	0.31	0.72	$30.2 M^o_{Psb}$	$806 EI_a/a^2$	$0.05P + 0.08R$	$0.12P + 0.25R$
Elasto-Plastic	1.0	0.46	0.31	0.67	$\frac{1}{a}[12(M_{Pra} + M_{Psa}) + 12(M_{Prb} + M_{Psb})]$	$252 EI_a/a^2$	$0.07P + 0.18R$	$0.07P + 0.18R$
	0.9	0.47	0.33	0.70	$\frac{1}{a}[12(M_{Pra} + M_{Psa}) + 11(M_{Prb} + M_{Psb})]$	$230 EI_a/a^2$	$0.06P + 0.16R$	$0.08P + 0.20R$
	0.8	0.49	0.35	0.71	$\frac{1}{a}[12(M_{Pra} + M_{Psa}) + 10.3(M_{Prb} + M_{Psb})]$	$212 EI_a/a^2$	$0.06P + 0.14R$	$0.08P + 0.22R$
	0.7	0.51	0.37	0.73	$\frac{1}{a}[12(M_{Pra} + M_{Psa}) + 9.8(M_{Prb} + M_{Psb})]$	$201 EI_a/a^2$	$0.5P + 0.13R$	$0.08P + 0.24R$

	0.6	0.53	0.74	$\frac{1}{a}[12(M_{Pfa}+M_{Psa})+9.3(M_{Pfb}+M_{Psb})]$	$197EI_a/a^2$	$0.04P+0.11R$	$0.09P+0.26R$
	0.5	0.55	0.75	$\frac{1}{a}[12(M_{Pfa}+M_{Psa})+9.0(M_{Pfb}+M_{Psb})]$	$201EI_a/a^2$	$0.04P+0.09R$	$0.09P+0.28R$
Plastic	1.0	0.33	0.51	$\frac{1}{a}[12(M_{Pfa}+M_{Psa})+12(M_{Pfb}+M_{Psb})]$	0	$0.09P+0.16R_m$	$0.09P+0.16R_m$
	0.9	0.35	0.51	$\frac{1}{a}[12(M_{Pfa}+M_{Psa})+11(M_{Pfb}+M_{Psb})]$	0	$0.08P+0.15R_m$	$0.09P+0.18R_m$
	0.8	0.37	0.54	$\frac{1}{a}[12(M_{Pfa}+M_{Psa})+10.3(M_{Pfb}+M_{Psb})]$	0	$0.07P+0.13R_m$	$0.10P+0.20R_m$
	0.7	0.38	0.58	$\frac{1}{a}[12(M_{Pfa}+M_{Psa})+9.8(M_{Pfb}+M_{Psb})]$	0	$0.06P+0.12R_m$	$0.10P+0.22R_m$
	0.6	0.40	0.58	$\frac{1}{a}[12(M_{Pfa}+M_{Psa})+9.3(M_{Pfb}+M_{Psb})]$	0	$0.05P+0.10R_m$	$0.10P+0.25R_m$
	0.5	0.42	0.59	$\frac{1}{a}[12(M_{Pfa}+M_{Psa})+9.0(M_{Pfb}+M_{Psb})]$	0	$0.04P+0.08R_m$	$0.11P+0.27R_m$

M_{Psa} and M_{Psb} = Total negative ultimate moment capacity along edges a and b, respectively.

M^o_{Psa} and M^o_{Psb} = Negative ultimate moment capacities per unit width at edges a and b, respectively.

Source: Department of the Army, 1986.

TABLE 6.7
SDOF Parameters for Fixed One-Way Slabs: Short Edges Fixed and Long Edges Simply Supported

Strain Range	a/b	Load Factor K_L	Mass Factor K_M	Load-Mass Factor K_{LM}	Maximum Resistance	Spring Constant k	Dynamic Reactions V_A	V_B
Elastic	1.0	0.39	0.26	0.67	$20.4 M^o_{Psa}$	$575 EI_a/a^2$	$0.09P + 0.16R$	$0.07P + 0.18R$
	0.9	0.41	0.28	0.68	$10.2 M^o_{Psa} + \dfrac{11}{a} M_{Pfb}$	$476 EI_a/a^2$	$0.08P + 0.14R$	$0.08P + 0.20R$
	0.8	0.44	0.30	0.68	$10.2 M^o_{Psa} + \dfrac{10.3}{a} M_{Pfb}$	$396 EI_a/a^2$	$0.08P + 0.12R$	$0.08P + 0.22R$
	0.7	0.46	0.33	0.72	$9.3 M^o_{Psb} + \dfrac{9.7}{a} M_{Pfb}$	$328 EI_a/a^2$	$0.07P + 0.11R$	$0.08P + 0.24R$
	0.6	0.48	0.35	0.73	$8.5 M^o_{Psb} + \dfrac{9.3}{a} M_{Pfb}$	$283 EI_a/a^2$	$0.06P + 0.09R$	$0.09P + 0.26R$
	0.5	0.51	0.37	0.73	$7.4 M^o_{Psb} + \dfrac{9.0}{a} M_{Pfb}$	$243 EI_a/a^2$	$0.05P + 0.08R$	$0.09P + 0.28R$
Elasto-Plastic	1.0	0.46	0.31	0.67	$\dfrac{1}{a}[12(M_{Pfa} + M_{Psa}) + 12(M_{Pfb})]$	$271 EI_a/a^2$	$0.07P + 0.18R$	$0.07P + 0.18R$
	0.9	0.47	0.33	0.70	$\dfrac{1}{a}[12(M_{Pfa} + M_{Psa}) + 12(M_{Pfb})]$	$248 EI_a/a^2$	$0.06P + 0.16R$	$0.08P + 0.20R$
	0.8	0.49	0.35	0.71	$\dfrac{1}{a}[12(M_{Pfa} + M_{Psa}) + 10.3(M_{Pfb})]$	$228 EI_a/a^2$	$0.06P + 0.14R$	$0.08P + 0.22R$
	0.7	0.51	0.37	0.72	$\dfrac{1}{a}[12(M_{Pfa} + M_{Psa}) + 9.7(M_{Pfb})]$	$216 EI_a/a^2$	$0.05P + 0.13R$	$0.08P + 0.24R$

Dynamic Response and Analysis

0.6	0.53	0.37	0.70	$\frac{1}{a}[12(M_{Pfa} + M_{Psa}) + 9.3(M_{Pfb})]$	$212EI_a/a^2$	$0.04P + 0.11R$	$0.09P + 0.26R$
0.5	0.55	0.41	0.74	$\frac{1}{a}[12(M_{Pfa} + M_{Psa}) + 9.0(M_{Pfb})]$	$216EI_a/a^2$	$0.04P + .09R$	$0.09P + 0.28R$
Plastic							
1.0	0.33	0.17	0.51	$\frac{1}{a}[12(M_{Pfa} + M_{Psa}) + 12(M_{Pfb})]$	0	$0.09P + 0.16R_m$	$0.09P + 0.16R_m$
0.9	0.35	0.18	0.51	$\frac{1}{a}[12(M_{Pfa} + M_{Psa}) + 11(M_{Pfb})]$	0	$0.08P + 0.15R_m$	$0.09P + 0.18R_m$
0.8	0.37	0.20	0.54	$\frac{1}{a}[12(M_{Pfa} + M_{Psa}) + 10.3(M_{Pfb})]$	0	$0.07P + 0.13R_m$	$0.10P + 0.20R_m$
0.7	0.38	0.22	0.58	$\frac{1}{a}[12(M_{Pfa} + M_{Psa}) + 9.7(M_{Pfb})]$	0	$0.06P + 0.12R_m$	$0.10P + 0.22R_m$
0.6	0.40	0.23	0.58	$\frac{1}{a}[12(M_{Pfa} + M_{Psa}) + 9.3(M_{Pfb})]$	0	$0.05P + 0.10R_m$	$0.10P + 0.25R_m$
0.5	0.42	0.25	0.59	$\frac{1}{a}[12(M_{Pfa} + M_{Psa}) + 9.0(M_{Pfb})]$	0	$0.04P + 0.08R_m$	$0.11P + 0.27R_m$

Source: Department of the Army, 1986.

TABLE 6.8
SDOF Parameters for Fixed One-Way Slabs: Long Edges Fixed and Short Edges Simply Supported

Strain Range	a/b	Load Factor K_L	Mass Factor K_M	Load-Mass Factor K_{LM}	Maximum Resistance	Spring Constant k	Dynamic Reactions V_A	V_B
Elastic	1.0	0.39	0.26	0.67	$20.4 M_{Psb}^o$	$575 EI_a/a^2$	$0.07P + 0.187R$	$0.09P + 0.16R$
	0.9	0.40	0.28	0.70	$19.5 M_{Psb}^o$	$600 EI_a/a^2$	$0.06P + 0.16R$	$0.10P + 0.18R$
	0.8	0.42	0.29	0.69	$19.5 M_{Psb}^o$	$610 EI_a/a^2$	$0.06P + 0.14R$	$0.11P + 0.19R$
	0.7	0.43	0.31	0.71	$20.2 M_{Psb}^o$	$662 EI_a/a^2$	$0.05P + 0.13R$	$0.11P + 0.21R$
	0.6	0.45	0.33	0.73	$21.2 M_{Psb}^o$	$731 EI_a/a^2$	$0.04P + 0.11R$	$0.12P + 0.23R$
	0.5	0.47	0.34	0.72	$22.2 M_{Psb}^o$	$850 EI_a/a^2$	$0.04P + 0.09R$	$0.12P + 0.25R$
Elasto-Plastic	1.0	0.46	0.31	0.67	$\frac{1}{a}[12 M_{Psb} + 12(M_{Psb} + M_{Pfb})]$	$271 EI_a/a^2$	$0.07P + 0.18R$	$0.07P + 0.18R$
	0.9	0.47	0.33	0.70	$\frac{1}{a}[12 M_{Psb} + 11(M_{Psb} + M_{Pfb})]$	$248 EI_a/a^2$	$0.06P + 0.16R$	$0.08P + 0.20R$
	0.8	0.49	0.35	0.71	$\frac{1}{a}[12 M_{Pfa} + 10.3(M_{Psb} + M_{Pfb})]$	$228 EI_a/a^2$	$0.06P + 0.14R$	$0.08P + 0.22R$
	0.7	0.51	0.37	0.73	$\frac{1}{a}[12 M_{Pfa} + 9.8(M_{Psb} + M_{Pfb})]$	$216 EI_a/a^2$	$0.05P + 0.13R$	$0.06P + 0.24R$
	0.6	0.53	0.39	0.74	$\frac{1}{a}[12 M_{Pfa} + 9.3(M_{Psb} + M_{Pfb})]$	$212 EI_a/a^2$	$0.04P + 0.11R$	$0.09P + 0.26R$

	0.5	0.55	0.41	0.74	$216EI_a/a^2$	$0.04P + 0.09R$	$0.09P + 0.26R$
Plastic	1.0	0.33	0.17	0.51	$\frac{1}{a}[12M_{Pfa} + 12(M_{Psb} + M_{Pfb})]$	$0.09P + 0.16R_m$	$0.09P + 0.16R_m$
	0.9	0.35	0.18	0.51	$\frac{1}{a}[12M_{Pfa} + 11(M_{Psb} + M_{Pfb})]$	$0.08P + 0.15R_m$	$0.09P + 0.18R_m$
	0.8	0.37	0.20	0.54	$\frac{1}{a}[12M_{Pfa} + 10.3(M_{Psb} + M_{Pfb})]$	$0.07P + 0.13R_m$	$0.10P + 0.20R_m$
	0.7	0.36	0.22	0.58	$\frac{1}{a}[12M_{Pfa} + 9.8(M_{Psb} + M_{Pfb})]$	$0.06P + 0.12R_m$	$0.10P + 0.22R_m$
	0.6	0.40	0.23	0.58	$\frac{1}{a}[12M_{Pfa} + 9.3(M_{Psb} + M_{Pfb})]$	$0.05P + 0.10R_m$	$0.10P + 0.25R_m$
	0.5	0.42	0.25	0.59	$\frac{1}{a}[12M_{Pfa} + 9.0(M_{Psb} + M_{Pfb})]$	$0.04P + 0.08R_m$	$0.11P + 0.27R_m$

Source: Department of the Army, 1986.

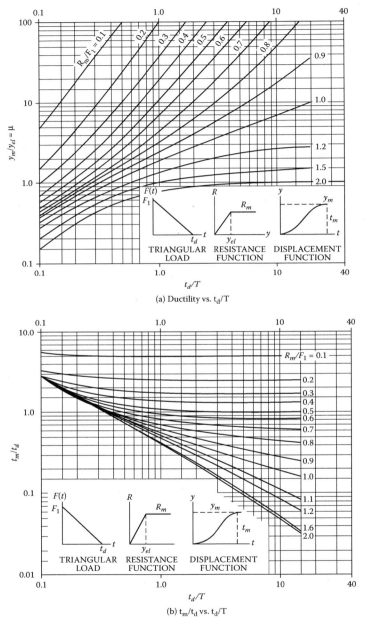

FIGURE 6.23 SDOF analysis charts.

Dynamic Response and Analysis

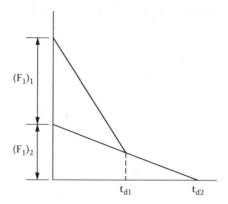

FIGURE 6.24 Bilinear triangular load pulse.

part after they reach their peak pressure (either the incident pressure P_{so} or the reflected pressure P_r), as discussed in Chapter 2. Although a triangular load pulse may be a reasonable assumption for preliminary design considerations, one may wish to employ a blast pulse that resembles a blast pressure time history more closely. This can be achieved with a multi-linear signal, e.g., as illustrated in Figure 6.24 for a two-segment decaying pulse.

The following expression is attributed to Newmark (Newmark et al., 1961; Newmark and Haltiwanger, 1962; Biggs, 1964; Department of the Army, 1986):

$$[(F_1)_1/R_m]C_1(\mu) + [(F_1)_2/R_m]C_2(\mu) = 1 \qquad (6.94)$$

$(F_1)_1$, $(F_1)_2$, t_{d1}, and t_{d2} are defined in Figure 6.24. $C_1(\mu)$, and $C_2(\mu)$ are the values of Rm/F that correspond to a specific ductility ratio (selected from Figure 6.23, as discussed in the previous section) and the ratios t_{d1}/T and t_{d2}/T, respectively. One employs a trial-and-error procedure to obtain the appropriate ductility ratio, while T and R_m are known for the structural system under consideration. One may employ more than two line segments for the load pulse, as discussed elsewhere (Newmark et al., 1961; Newmark and Haltiwanger, 1962).

7 Connections, Openings, Interfaces, and Internal Shock

This chapter is focused on the behavior of specific structural connections and other structural and/or mechanical systems, on the corresponding implementation of design recommendations, and on the issue of internal shock conditions due to explosion effects.

7.1 CONNECTIONS

This section focuses on the behavior of structural connections and on the corresponding implementations in design and construction. Improved design approaches for both structural concrete and structural steel connections are needed. Such design approaches should be derived on the basis of additional studies that must be supported by combined theoretical, numerical, and experimental efforts.

Connections (including various types of supports, joints, etc.) are structural elements that tend to affect overall structural performance. They must be designed and constructed with extreme care in order to allow the adjoining structural elements (beams, columns, slabs, walls, etc.) to be utilized fully. It is vital to remember that *bad connections prevent a structure from mobilizing its resistance*. It does not make sense to have strong structural elements and weak connections. In order to reach the full capacity of the structure, a *balanced design* is needed, in which all structural elements including connections can provide their full resistance and in which all resistances are compatible. To illustrate the influences of connections on structural performance, we shall discuss various types of joints and their capabilities.

7.1.1 INTRODUCTION

We have discussed the behavior of beams and columns: the effects of thrust, flexure, and shear. In all cases, certain assumptions had to be made regarding boundary conditions or support conditions for the structural elements. The support conditions were defined as simply supported, fixed, hinged, free, etc. Now we must look into the issue of support conditions and try to understand better how structural elements are supported. Since many typical blast-resistant structures are made of structural concrete, this chapter focuses primarily on structural concrete joints. Recent studies on the behavior of blast-resistant structural steel connections will be addressed also, and they will form the basis for future design recommendations.

Consider real structures made of beams, columns, slabs, and other types of structural or nonstructural elements. In order to construct a system, these elements must be joined so that the structural system may function as required by the specified performance criteria, design and building codes, etc. Furthermore, the connections of structural elements serve as their supports and it must be recognized that connections involve complicated details, as shown in Figure 7.1, and Figure 7.2. These details tend to introduce certain types of effects on the adjoining structural elements — effects that extend beyond the simple definition of support condition. The designs of supports, connections, and joints must receive as much attention (and sometimes more) as the designs of other structural components in order to ensure required structural performance.

7.1.2 BACKGROUND

The preceding discussions on the design of individual structural elements (beams, slabs, columns, etc.) were based on simple assumptions concerning their behavior. For example, beams are typically assumed to behave according to the Euler–Bernoulli models (named after the two 18th century mathematicians who developed closed-form solutions for these structural elements). This approach, however, is a first-order approximation for real behavior and it is based only on the bending behavioral aspects of such structural elements. These regions are defined as *beam regions* or *B regions*. As noted by Park and Pauley (1975) and MacGregor (1997), this approach is good as long as one is not close to regions of discontinuity. Such discontinuities can be associated with geometric and/or load characteristics. This issue is illustrated in Figure 7.3 in which several structural elements are presented.

Regions where discontinuities exist are those near openings, corners, severe cracking, or localized loads (supports, concentrated forces, ends of distributed loads). In those regions, the stress distribution is much more complicated or disturbed than it is in a B region. Stress distribution is strongly affected by the discontinuities and the behavior cannot be defined with the behavioral models used for B regions. Such regions are defined as *discontinuous* or *disturbed* and they are termed *D regions*. Accordingly, one cannot use a classical approach (beam theory) to describe the state of internal stresses or to design the detailing in a D region. For such regions, the recommendation is to use an alternative approach such as the strut-and-tie method discussed below.

Extensive studies during the last five decades have shown that short duration, high magnitude loading conditions significantly influence structural responses. Explosive loads are typically applied to structures at rates approximately 1,000 times faster than earthquake-induced loads. The corresponding structural response frequencies can be much higher than those induced by conventional loads. Furthermore, short duration dynamic loads often exhibit strong spatial and time variations, resulting in sharp stress gradients in structures. High strain rates also

Connections, Openings, Interfaces, and Internal Shock

(a) DETAILS

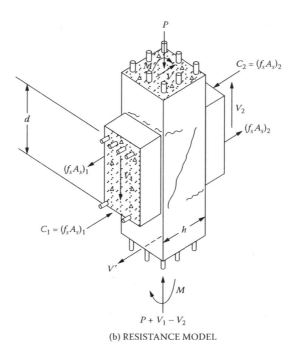

(b) RESISTANCE MODEL

FIGURE 7.1 Typical reinforced concrete connection (Park and Paulay, 1975).

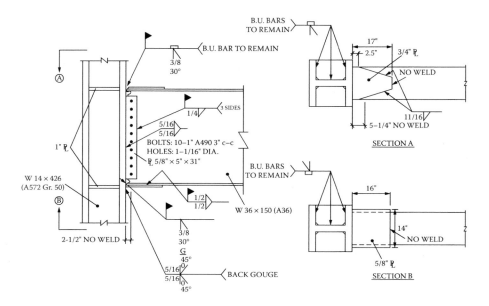

FIGURE 7.2 Typical steel connection (Engelhardt et al., 1995; FEMA, 2000).

affect the strength and ductility of structural materials, bond relationships for reinforcement, failure modes, and structural energy absorption capabilities. The definition of the loading function often involves many assumptions and strongly random variables. This is particularly true for blast and impact loadings that are typically defined in the form of pressure–time or force–time histories. Such load functions are used also in various structural analyses, from single-degree-of-freedom (SDOF) to advanced numerical methods. Structural detailing in blast-resistant construction is being gradually transformed from an art to a well-defined technical area. Although differences exist between structural concrete and structural steel construction, the underlying theories are similar.

Structures cannot develop their full resistance if their details are inadequate. This is also true for structures designed to resist blast loads. Since many buildings could be subjected to blast loads, building behavior and its relationship with the detailing performance under blast loads are of great interest.

Blast-resistant structures must be robust (i.e., have well-defined redundancies) to ensure alternative load paths in case of localized failures. Ensuring robustness may not be possible if the structural details cannot perform as expected.

Connections, Openings, Interfaces, and Internal Shock

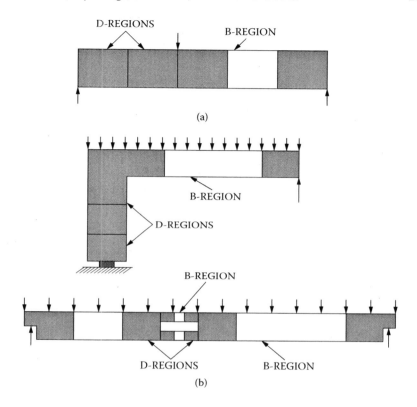

FIGURE 7.3 B and D regions in structural elements (MacGregor and Wight, 2005).

Recent earthquakes in the United States and Japan have highlighted troublesome weaknesses in design and construction technologies of structural connections. Both steel and concrete connections performed surprisingly poorly because of brittle failures. It has been shown for facilities subjected to blast loads that structural details played significant behavioral roles. Therefore, a better understanding of the behavior of structural details under blast loads is important, and a strong interest exists in developing better approaches to ensure improved structural behavior.

Although considerable information on this subject can be found in various references (ASCE, 1985; Baker et al., 1983; Department of the Army, 1986 and 1990; Drake et al., 1989; Department of Energy 1992), current design procedures usually do not explicitly address these issues. TM 5-1300 (Department of the Army, 1990) contains guidelines for the safe design of blast-resistant structural concrete and steel connections. However, the adequacy of these design procedures may not be well defined because of insufficient information about their behaviors under blast loads.

7.1.3 Studies on Behavior of Structural Concrete Detailing

In both structural concrete and structural steel, joint size is often limited by the size of the elements framing into it. This restriction along with poor reinforcement detailing may create connections without sufficient capacities to develop the required strengths of adjoining elements. Knee joints are often the most difficult to design when continuity between beam and column is required. The work of Nilsson (1973) provided extensive insight into the behavior of joints under static loads and into the relationships between performance and internal details. He showed that even slight changes in a connection detail produced significant effects on strength and behavior as presented in Table 7.1 and discussed in more detail later.

Park and Paulay (1975) proposed diagonal strut and truss mechanisms to describe a joint's internal resistance to applied loads as shown in Figure 7.4. The resultant compression and shear stresses form the diagonal compression strut. These two mechanisms, capable of transmitting shear forces from one face of a joint to the other, were assumed to be additive.

More recently, Paulay (1989) and Paulay and Priestly (1992) used equilibrium criteria to address the joint reinforcement necessary for sustaining a diagonal compression field in a joint core. Their approach addressed the strength requirements for beam column joints under severe quasi-static loads. They showed how external loads are resisted by internal forces and how these forces vary as the damage in a joint accumulates. This resistance mechanism and the corresponding forces conform to the state of internal stresses as shown in Figure 7.5.

By understanding these stresses and the corresponding damages, one may derive an idealized truss model that represents a joint's internal resistance mechanism. The opening moment can cause the formation of a diagonal crack propagating from the inside corner and continuing toward the outside corner. Reinforcing the inside corner with steel perpendicular to the crack can prevent crack growth. However, the consequence will be the formation of an interior crack perpendicular to the direction of the first crack. This crack too must be prevented by placing steel perpendicular to it.

The most common modes of failure in conventionally reinforced knee joints subjected to either closing or opening static loads were addressed in the literature cited above. According to Nilsson (1973), the detail that provided the most strength with the least amount of congestion used a diagonal bar across the interior corner. That bar resisted the development of the initial tensile crack at the inner corner (i.e., first cracking mode), enabling the stresses to flow around the corner

TABLE 7.1
Behavior of Statically Loaded Knee Joints

Specimen Number	Haunch Size (cm)	Concrete Strength σ_c σ_{sp} kgf/cm²	Failure Moment M_{ut} kgfm	Calculated Ultimate Mom. M_{uc} kgfm	$\dfrac{M_{ut}}{M_{uc}}$ %	Inside Corner Crack Width at $M_{uc}/1.8$ mm
U 21		339 27.3	990	3135	32	Failed
U 27		277 21.1	1840	2990	61	0.60
U 15		289 25.6	2227	3290	68	0.70
U 12		335 31.2	2474	3220	77	0.27
U 28		272 20.5	2540	3185	79	0.26
UV 1	15	305 23.7	3160			
UV 2	5	345 25.2	3120			
U 24		398 25.9	2804	3240	87	0.34
UV 3	10	318 24.1	3712			0.08
UV 4	5	277 22.3	3505			0.06
UV 5		335 26.2	3629	3180	114	0.11
UV 6		292 24.3	3505	3040	115	0.13
UV 7		339 18.4	3773	3070	123	0.13

Source: Nilsson, 1973.

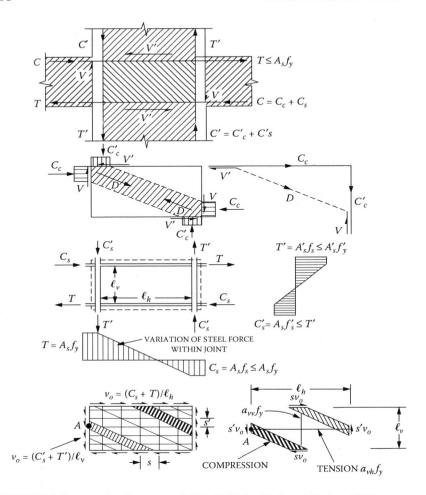

FIGURE 7.4 Structural resistance model for RC joint (Park and Paulay, 1975).

and preventing corner separation and diagonal tension failure. Nilsson (1973) showed that this feature enabled the moment capacities to exceed the design moments for flexural reinforcement amounts of up to 0.76%. For larger steel ratios, the addition of stirrups placed perpendicular to that diagonal bar was recommended for intercepting secondary tensile cracks.

Park and Paulay (1975) recommended such reinforcement for beam–column connections with tensile reinforcement amounts larger than 0.5%. Tests by Balint and Taylor (1972) and Nilsson (1973) showed that full design moment capacities are easily attainable for flexural steel amounts of 1.0% when a concrete haunch was used at the connection's inner corner with 45° diagonal reinforcement. In addition to increasing the moment arm for the diagonal reinforcement, the available space for placing other reinforcements was also increased.

Connections, Openings, Interfaces, and Internal Shock

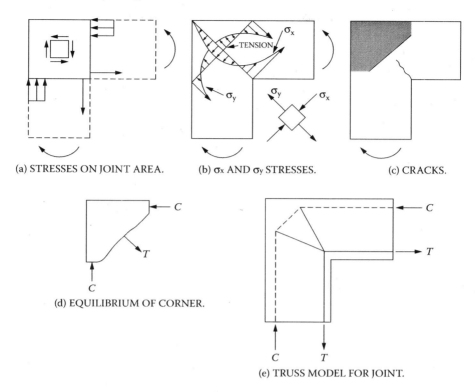

FIGURE 7.5 Internal stresses, damage, and idealized truss model (MacGregor and Wight, 2005).

This brief discussion shows that until the late 1980s, recommendations for the design of connections were empirical, as noted by Jirsa (1991). Nevertheless, when faced with difficult problems such as selecting shear reinforcement, structural engineers introduced alternative approaches for the analysis of reinforced concrete structures. For example, the truss analogy used since the late 1800s (Collins and Mitchell, 1991) is based on representing a structural concrete detail by a virtual truss model that can provide equivalent behavior. In such a model, the compressive forces are carried by concrete struts, while tension is resisted by steel ties, as shown in Figure 7.4 and Figure 7.5. The forces in the truss system are computed by the same procedures used in structural analysis, and engineers must select appropriate cross-sectional areas for steel and concrete components to ensure required performance.

Developments in the application of advanced truss analogies contributed to the development of a rational approach for the design of structural concrete connections. Schlaich et al. (1987) presented an extensive discussion on an innovative approach for designing structural regions in which the strain distributions are significantly nonlinear. The corresponding strut-and-tie design approaches were developed for a wide range of structural details including connections for which traditional design approaches were not applicable as shown in Figure 7.6.

Factors known to affect the behaviors of connections including the effect of lateral restraint and the reduction of concrete strength associated with diagonal tensile strains were addressed by Pantazopoulou and Bonacci (1992). Later, Bonacci and Pantazopoulou (1993) used available data on interior beam–column frame connection assemblies to evaluate the influences of several key variables

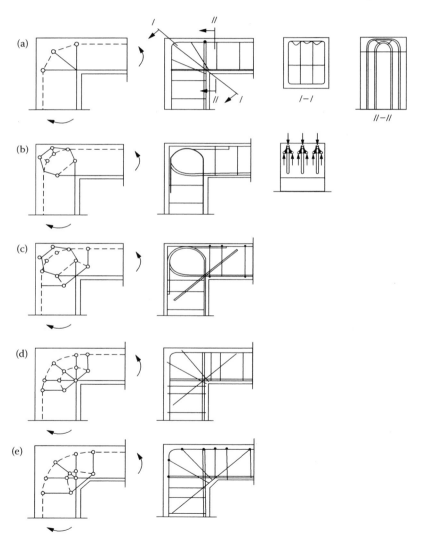

FIGURE 7.6 Strut-and-tie models for connections (Schlaich et al., 1897).

Connections, Openings, Interfaces, and Internal Shock 301

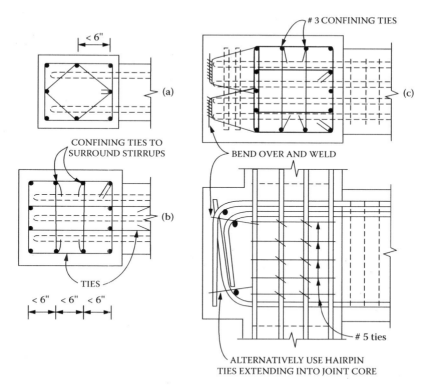

FIGURE 7.7 Optional bar anchorages in knee joints (Park and Paulay, 1975).

on connection behavior. Interestingly, they found that diversity in experimental techniques is largely responsible for the differences in the empirical interpretations of observed joint behavior.

One of the key issues in structural concrete design, particularly the design of effective connections, is reinforcement anchorage. One must ensure that the main reinforcing bars will not be pulled out by strong tensile forces. In T- or cross-joints, where reinforcing bars can be continued into the adjacent structural members, this problem is not as difficult as it is with knee joints. One approach proposed by Park and Paulay (1975) is anchorage of the main bars in a short structural stub on the other side of the column, as shown in Figure 7.7.

To understand the nature of the difficulties in developing effective structural concrete connections, one must understand their behavior. This can be achieved by testing various reinforcement configurations, as done by Nilsson (1973) and discussed next.

If detailing is not important, the connection in Figure 7.8 could be considered. Each connection must be considered for both opening and closing conditions. Under the conditions shown, the reinforcement is placed so that it is continuous from the beam into the column.

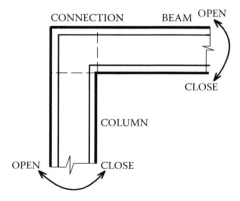

FIGURE 7.8 Simple connection model.

Considering the case of a closing joint, the steel bars at the inner corner will be in compression and those at the outer corner will be in tension. The resultants of those forces will apply compression on the interior of the connection and help confine the concrete there. The outcome seems to be that this connection detail could function well under these conditions. If, however, the joint must operate under opening conditions, the situation would be different. The steel at the inner corner will be in tension and tend to straighten out. This will induce tensile forces on the inner part of the connection and cause tensile failure in that region. Obviously, if the steel bars are continuous at the inner corner, the joint cannot provide adequate support and resistance against opening. This may be prevented by not having continuous steel at the inner corner region as shown in Figure 7.9.

Since we understand the reason for failure, it is simple to propose a solution that may prevent this type of problem and hopefully also improve joint behavior. We must now assess the effects of this detail on behavior. The performance of the joint can be analyzed as before for the opening and closing joints.

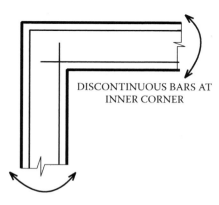

FIGURE 7.9 Modified inner corner reinforcement detail.

Connections, Openings, Interfaces, and Internal Shock

Note that there is no major change for the closing joint, while an improvement can be expected for the opening joint. However, since the reinforcement is not continuous at the inner corner and concrete cannot resist tensile stresses meaningfully, one expects the formation of a diagonal crack from the inner corner toward the interior of the connection causing failure of the inner corner. Furthermore, when the joint is subjected to opening loads, the inner corner bars will be in tension. These tensile forces may exceed the anchorage strength of the bars (i.e., fail the bond between steel and concrete) that may pull out and cause failure of the inner corner. The modified design and detailing eliminated the splitting problem that existed previously, but introduced a possibility of both a bond failure and a diagonal tensile crack failure.

The bond failure can be solved by providing better anchorage (i.e., using mechanical methods) that can prevent this type of problem, but we must investigate the state of forces and the expected behavior in the modified joint, using a free body diagram (Figure 7.10). As mentioned earlier, a free body diagram can be drawn for the case of an opening joint. The types of problems usually associated with this joint include:

- At the inner corner, a diagonal crack may be expected, as noted by Crack 1. This is a result of the tensile state of stress at that location. The existing reinforcement may not be effective in preventing such an opening of the corner and the resulting crack.
- If Crack 1 is prevented by adequate reinforcement, the resultants R will induce tension on the inner corner and form Crack 2. This may represent the boundary between the compressive and tensile regions inside the joint (see Figure 7.6). One may "visualize" this type of crack as the geometric location that separates two regions in the joint that tend to move in opposite directions as a function of the force resultants.

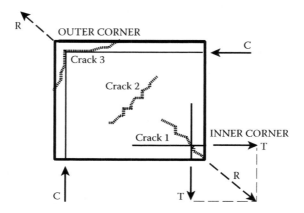

FIGURE 7.10 Free body diagram for opening joint conditions.

The tensile forces at the inner corner pull the concrete core along their resultant, while the compressive forces at the outer corner push the concrete along their resultant. Since two forces are acting in opposite directions, a tensile force or stress will be generated on the concrete core to form Crack 2. Naturally, if Crack 1 forms, then the tensile forces at the inner corner will act to tear the corner apart, thus preventing the formation of Crack 2.
- The opening corner or joint condition may also weaken the bond between steel and concrete at the other corner. The state of compression in that corner will cause the concrete to split along the reinforcement and a concrete wedge to spall off, as noted by Crack 3.

Balint and Taylor (1972) and Somerville and Taylor (1972) provided detailed discussions of these cases. The previous joint was considered by Balint and Taylor who found that for reinforcement ratios of 2% ($\rho = 0.02$), the failure in the joint prevented development of full theoretical capacity. Failure occurred at 44% of the ultimate moment capacity for the system (M'_u, or M_n) similarly to the findings by Nilsson (1973).

At this stage, an intuitive "feel" for joint performance should have been developed, along with an understanding of the causes for premature failure. It would then have been possible to propose further improvements in the joint detailing for added resistance and performance. One such improvement was proposed and tested as reported by Balint and Taylor (1972), and confirmed by Nilsson's findings (1973). The improvements introduced in this joint design (Figure 7.11) include the following:

1. The inner corner is redesigned with a haunch for added resistance and support against closing loads; for opening loads, the next feature is added.
2. The inner corner bars are not continuous as noted earlier, and they are extended and anchored into the compressive zone.
3. A diagonal bar with an area of at least half the beam's tensile steel area is provided to support the inner corner against opening loads. This bar is anchored well into the compressive regions for opening joints (i.e., the top steel). This detail also adds resistance against closing loads.
4. For more than 0.5% longitudinal reinforcement ($\rho > 0.005$), a set of radial bars is introduced to improve the splitting resistance of the joint (i.e., preventing the development of Crack 2 in Figure 7.10) and to improve the internal shear capacity. These are tied to the diagonal bar and possibly also to the outer corner reinforcement. One may use a closed loop where $A_{s1} = A_{s2}$ or an open loop where $A_{s1} \neq A_{s2}$.

This configuration ensures that the adjoining members can provide their full flexural capacities as confirmed by Nilsson (1973). The radial hoops provide

Connections, Openings, Interfaces, and Internal Shock

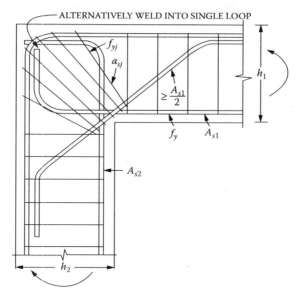

FIGURE 7.11 Improved connection detail (Park and Paulay, 1975).

support to the reinforcement, confinement to the concrete, and shear resistance. Park and Paulay (1975) proposed the following area for each radial hoop:

$$a_{sj} = \left[\frac{f_y}{f_{yj}}\sqrt{1+\left(\frac{h_1}{h_2}\right)^2}\right]\frac{A_{sl}}{n} \qquad (7.1)$$

where n is the number of radial hoops and $\rho = \dfrac{A_{sl}}{bd_1}$. Since radial bars were recommended for reinforcement ratios larger than 0.5%, one can use the following relationship:

$$a_{sy} = \left[\frac{\rho - 0.005}{\rho}\right]\left\{\left[\frac{f_y}{f_{yj}}\sqrt{1+\left(\frac{h_1}{h_2}\right)^2}\right]\frac{A_{sl}}{n}\right\} \qquad (7.2)$$

The introduction of a haunch at the end of the inner corner is recommended, along with placement of the diagonal bar farther away, toward the haunch. Park and Paulay (1975) showed how this type of detail can be used for connections with angles different from 90°.

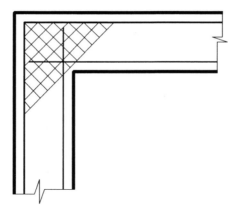

FIGURE 7.12 Alternative connection reinforcing detail.

Another recommended approach based on Balint and Taylor (1972) is shown in Figure 7.12. The detailing for the main reinforcement is simple, but a cage of stirrups is placed over the joint that forms a system of diagonal bars and radial hoops with similar effects as for the previous case. This configuration provides considerable support to the reinforcement, confinement to the concrete core, and shear resistance. Experimental evidence suggests that this approach is quite effective.

Designing blast-resistant reinforced concrete connections is considerably more difficult. Coltharp et al. (1985) performed explosive tests on elements without shear ties and showed a definite diagonal shear failure located immediately near the joint region. Krauthammer et al. (1989) discussed experimental observations on the behavior of buried arches under explosively induced loads in which the behavior of floor-arch joints was unsatisfactory. Intensive inertia forces appear under such conditions in addition to the system of forces described by Paulay (1989).

The magnitudes, intensities, and distributions (both in time and space) of such inertia forces depend on the dynamic behavior of the structure. These parameters can be evaluated either experimentally or numerically. Both approaches pose serious difficulties because of the severe loading environments and their effects on material properties and structural components.

Krauthammer and DeSutter (1989) studied connection details under the effects of localized explosions by simulating numerically physical tests. They recommended preliminary joint details based on their findings. Similar consideration of structural concrete knee joints under internal explosions (Krauthammer and Marx, 1994; Otani and Krauthammer, 1997; Krauthammer, 1998; Ku and Krauthammer, 1999) provided additional information on joint behavior. Further studies of blast-loaded structural concrete and structural steel joints are needed to determine more clearly the contributions of joint detailing.

The findings from these investigations showed that structural details are vital for ensuring the safety of blast-resisting structures. Preliminary studies showed that behavior was closely related to reinforcing details and highlighted the need to study such relationships more closely. The design approach for reinforced concrete connections proposed by Park and Pauley (1975) has been adopted in TM 5-1300 (Department of the Army, 1990). These recommendations were used for a proposed blast containment structure; the three-dimensional behavior of this structure was studied and the findings were published (Krauthammer, 1998).

In structural concrete connections, the location of the diagonal bar across the inner corner significantly affected joint strength. The observed relationships between the maximum stress and the location of the diagonal bar and its cross-sectional area could be examined further to produce design recommendations that can ensure a desired level of performance. It was observed that the radial reinforcing bars affected the tensile stress in the diagonal bar at the inner corner of the connection. Furthermore, the diagonal compressive strut to resist the applied loads can be mobilized effectively by proper combinations of radial and diagonal bars.

Strengthening of the joint regions by diagonal bars was characterized by formation of plastic hinge regions near the ends of the diagonal bars. Relocation of the largest rotations from the support faces to the plastic hinge regions showed a shift of maximum moment and shear along the slabs. Examination of the stresses in the flexural bars along the planes of symmetry revealed that yielding and maximum stresses in the interior flexural and tensile bars occurred in the plastic hinge regions. Besides damage to the flexural reinforcement, maximum shear stresses in the concrete and large tensile shock wave stresses also occurred in the plastic hinge regions. In addition to indicating the presence of direct in-plane tension in the slab (due to the expansion of the walls and roof caused by the interior explosion), these concrete and steel stress patterns showed that excessive damage could occur at these regions if not properly designed.

Current design procedures (e.g., TM 5-1300) do not address the issue of plastic hinge regions, nor do they anticipate that the maximum negative moment (negative yield line location) may occur near the ends of the diagonal bars. Computation of structural capacity in current design procedures is based on the assumption of yield line formation at the supports. Clearly, the shift in hinge location must be included to derive a more realistic structural capacity. Furthermore, the classical yield line theory treats yield lines that have zero thickness. In reality, the plastic hinge regions have finite thicknesses (roughly the member's thickness), and this issue should be addressed in design procedures. As noted previously, the local rotations of all the structural concrete connections exceeded 12° and the global rotations exceeded 2°. The magnitudes of local and global rotations are good indications of the extent of damage to a structure. This finding emphasizes the need to address both local and global rotations in design approaches.

7.1.4 Studies on Behavior of Structural Steel Detailing

The design guidelines for steel connections (AISC, 1992, 1993, and 1997) were developed on the basis of experimental and theoretical investigations. However, recent data from reports on damages to steel structures due to the events associated with the Northridge (California) earthquake in 1994 (Engelhardt et al., 1993 and 1995; Kaufmann, 1996; Richart et al., 1995; Tsai et al., 1995) showed surprisingly poor performance of the steel connections. After an extensive study, recommendations were provided for modification (AISC, 1997; FEMA, 2000) and significant improvement over pre-Northridge connections, as illustrated in Figure 7.13.

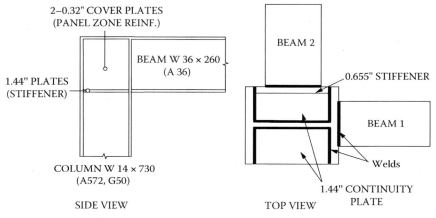

(a) PRE-NORTHRIDGE CONNECTION MODEL

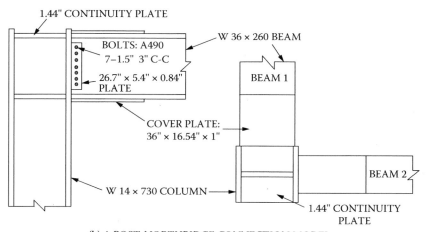

(b) A POST-NORTHRIDGE CONNECTION MODEL

FIGURE 7.13 Steel connection details (Krauthammer et al., 2002).

Connections, Openings, Interfaces, and Internal Shock

After extensive studies of structural concrete connections, similar attention was given to the behavior of structural steel connections. Here too a close relationship was observed between the connection details and the behavior under blast loads. The weaknesses observed under seismic loads may exist also in the current blast design guidelines (TM 5-1300), and they were assessed in recent studies (Krauthammer and Oh, 2000; Krauthammer et al., 2002) using empirical and numerical approaches to explain the behavior of structural steel corner connections subjected to high explosive loads. The studies produced findings that raised concerns about the blast resistance of structural steel connections and the safety of using TM 5-1300 for the design of structural steel connections.

It was shown that structural steel welded connections subjected to "safe" explosive loads may fail because of weld fracture. Furthermore, the corresponding deformations of the structural elements in those cases whose connections did not fail may have exceeded the limits set in TM 5-1300. These deformations also exceeded the values predicted based on the analysis procedures in TM 5-1300. It was shown further that dead loads had adverse effects on behavior because of the added bending and twisting of the beams after they were deformed by the blast effects. Since TM 5-1300 does not address such effects, current design procedures should be modified to reflect the structural damage caused by weak axis deformations.

The issues of material models and strain rate effects were also addressed. All numerical simulations were based on validated material models that incorporated strain rate effects. These were used with computer codes validated with precision test data. Since the concrete material models were developed for impulsive loading conditions, one may not speculate as to how much of an effect the strain rates have on specific structural behavior. However, the approach for analyzing structural steel connections was based on combining dynamic increase factors (DIF) values based on TM 5-1300 (Department of the Army, 1990) and validation against precision test data.

Although steel is not expected to be very sensitive to strain rate effects as compared to concrete, serious strain rate effects were noted in the structural steel connections. The current DIFs must be modified to address more detailed pressure levels and the differences between two- and three-dimensional behaviors. It is recommended that more detailed DIF applications associated with pressure levels may be necessary to avoid possible overestimation of strain rate effects. This could be very important when two-dimensional analyses are used to design three-dimensional structures.

7.1.5 SUMMARY

For the design of joints in frame structures, a combination of procedures acceptable for seismic design should be employed (see ACI-ASCE Committee 352 recommendations 1985, 1988, 1991, ACI 318-02, etc.), and modifications should be based on the material discussed above that has been adopted into TM 5-1300 (Department of the Army, 1990). Additional information on the behaviors of connections under explosive loads based on recent research as discussed above also can be used.

7.2 OPENINGS AND INTERFACES

These elements (such as blast doves, blast valves, cable and conduit penetrations, and emergency exits) are critical for hardened structures. An excellent discussion of this topic appears in ASCE (1985) and a summary of that material is presented here.

7.2.1 ENTRANCE TUNNELS

Many structures, especially underground structures, include entrance tunnels. Such tunnels must be designed to survive dynamic loads and to prevent the accumulation of undesired materials (debris, radioactive fallout, etc.). Some facilities were built with blast doors at their front ends and second doors into the facility. Special attention must be given to the possible differential motions between the tunnel and the facility.

7.2.2 BLAST DOORS

TM 5-1300 (Department of the Army, 1990) addresses designing of steel blast doors. Nevertheless, various configurations for blast doors are presented in Table 7.2 and a typical composite closure is illustrated in Figure 7.14. The responses of such composite systems are illustrated in Figure 7.15 and Figure 7.16 is a recommended design chart for static cases.

The responses of circular closures in the dynamic domain were studied by Gamble et al. (November 1967), and were found to be similar to the responses in static cases. All cases failed in shear and the compression membranes exerted significant influence on behavior, as discussed in Chapter 5.

Little work has been done on door–structure interactions. Tests have shown that blast doors may not survive the applied loads and better designs are required. In some cases, the doors survived, but the door-to-frame connections failed (Coltharp et al., 1985). A door may be blown into a protected space under such conditions. Openings and door supports must be designed very carefully, along the guidelines for connections.

Recommendations by Krauthammer and DeSutter (1989) who studied the issue of how to connect and secure doors into a structure may be useful in overcoming such difficulties. Their recommendations highlighted the need to ensure sufficient seating area between the door and its frame, the internal frame and reinforcement arrangements, and the sealing requirements. The 1989 study by Krauthammer and DeSutter also addressed a generic rectangular protected structure with a blast door in the protected opening, as shown in Figure 7.17. They investigated several reinforcing details for integrating the door frame into the structural system and Figure 7.18 illustrates two such options. Clearly, one

TABLE 7.2
Blast Door Types

Concept	Advantages	Disadvantages
Simple Slab		
Concrete slab with bar reinforcing	Good radiation and thermal protection	Low ductility; low EMP protection; less efficient than composite slab
Steel slab/plate	High ductility; simple construction; good EMP protection	Low thermal protection; inefficient against blast and radiation
Composite Slab		
Steel plate with shear reinforcing STEEL PLATE WITH SHEAR REINFORCING	Most efficient for both blast and radiation; good EMP and thermal protection; high ductility; standard low cost construction	Less efficient against blast than steel bottom and side plate
Steel bottom and side plate STEEL BOTTOM AND SIDE PLATE		Exterior metal strips or compression rebar (metal grills) required for rebound
Metal Grill		
One-way ONE - WAY	Efficient against blast; high ductility; good rebound strength	Complex and very costly construction; inefficient against radiation
Two-way		

Source: ASCE, 1985.

must ensure that a door frame is adequately anchored into the structural reinforcement system, and that the forces transferred from the door into the frame will not exceed the structural capacity.

Another important issue is to treat the door within a comprehensive electromagnetic pulse (EMP) protection for nuclear events. Although EMP protection is not addressed in this book, one should ensure that internal spall plates and/or other types of EMP protection are integrated with the closures. One must provide metallic contacts between all closures, their frames, and the EMP protective system. For CB protection, the blast door, as for other closures, must remain airtight.

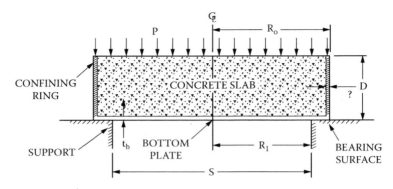

FIGURE 7.14 Composite closure (ASCE, 1985).

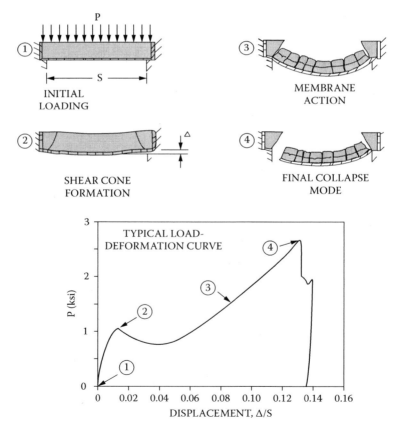

FIGURE 7.15 Subscale closure static response behavior, S/D between 5 and 7 (ASCE, 1985).

Connections, Openings, Interfaces, and Internal Shock

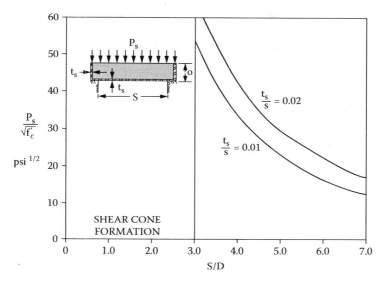

FIGURE 7.16 Closure design chart (ASCE, 1985).

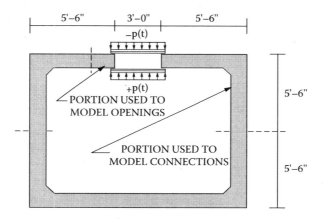

FIGURE 7.17 Simulated protected structure (Krauthammer and DeSutter, 1989).

7.2.3 BLAST VALVES

As illustrated in Figure 7.19, blast valves will close upon the impact of a shock wave. Blast valves are important for nuclear environments and close-in conventional explosions (where relatively large gas volumes may enter a confined space to cause a sharp jump in pressure). Valves must be analyzed for the expected conditions and structural motions in order to verify suitability. Generally, one may consider two types of blast valves: static, and dynamic. A static valve has

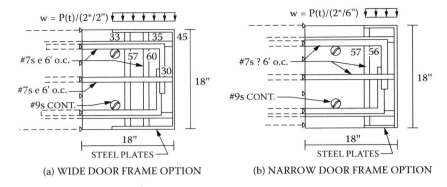

FIGURE 7.18 Effective blast door frame configurations (Krauthammer and DeSutter, 1989).

no moving parts, and it attenuates the pressure pulse by preventing the pressurized air from flowing directly through the opening. This can be achieved by forcing the pressurized air flow to go through a series of small geometrically offset openings or through an opening maze. A dynamic valve, as shown in Figure 7.19, has a pressure sensitive diaphragm that is forced to close by the over-pressure pulse. Once the pressure returns to lower levels, a spring will open the valve to normal air intake.

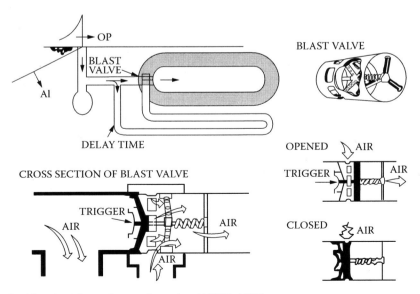

FIGURE 7.19 Blast valve configuration (ASCE, 1985).

7.2.4 Cable and Conduit Penetrations

Various utilities must enter a protective facility. Under a dynamic load, relative structure–medium motions will induce strains on these utilities and may cause severe damage. A designer should use a reliable approach for the analysis of such motions and then provide a design that can accommodate the anticipated displacements. Such utilities may require automatic or remote valves that will close them under predetermined conditions. Some examples for such designs are presented in Figure 7.20 through Figure 7.22.

Similar to using blast valves for preventing airblast pressures from entering a protected space, one can use devices that can prevent pressure pulses from traveling along piping systems. One such device is shown in Figure 7.23. The pressurized fluid is forced into an accumulator rubber bladder through small openings in the center pipe and the pressure in the pipe is reduced accordingly.

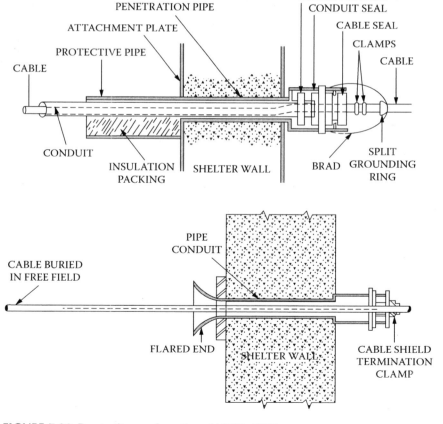

FIGURE 7.20 Penetration configurations (ASCE, 1985).

316 Modern Protective Structures

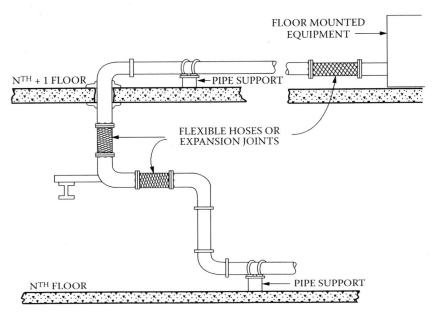

FIGURE 7.21 Exterior wall pipe penetration (TM 5-855-1, 1986).

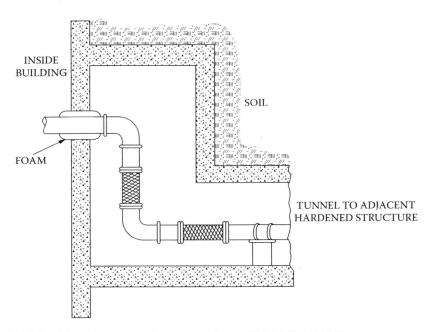

FIGURE 7.22 Piping penetration between floors (TM 5-855-1, 1986).

Connections, Openings, Interfaces, and Internal Shock   317

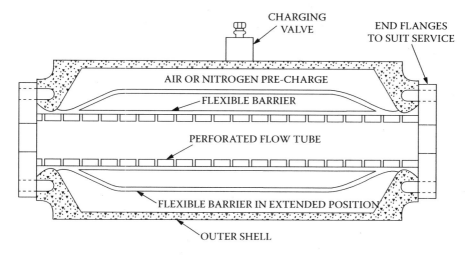

FIGURE 7.23 In-line accumulator with perforated flow tube (TM 5-855-1, 1986).

Once the pressure attenuates, the bladder will contract and push the fluid back into the center pipe.

7.2.5 EMERGENCY EXITS

A critical requirement for a protected facility following an explosive loading incident is ensuring the ability of access into the facility by rescue personnel and the ability of people in the facility to exit safely. In the event that a main entrance is blocked, the use of alternative exits would be required as illustrated in Table 7.3, and Figure 7.24. These exits must be designed using the same guidelines as for the main closures.

7.3 INTERNAL SHOCK AND ITS ISOLATION

Internal shock can be hazardous for both people and equipment. TM 5-1300 (Department of the Army, 1990) provides some shock tolerance levels for people. Other sources provide such information for equipment (ASCE, 1985; Department of the Army, 1986; Harris and Piersol, 2002). One needs to compute or estimate the structural responses to applied blast loads for above-ground and underground systems, as discussed in previous chapters.

The structural motions are then used to analyze the dynamic structure–content interactions from which one obtains the responses of the protected contents. These values should be compared to the available tolerance levels for assessing their survivability. If the computed shock levels exceed allowable limits, one must incorporate shock isolation features into the design. As the explosions approach a structure, the internal shock environment intensifies and the requirements imposed on the shock isolation system increase. Various studies were conducted

TABLE 7.3
Emergency Exit Concepts

Concept	Advantages	Disadvantages
Structurally hardened tunnel	Fast exit; easy maintenance	Hardened tunnel; heavy blast door; difficult blast door operation; security problem
Sand-filled tunnel	Softer blast door; soft tunnel; less expensive; sand is safe and easy to clean	Longer exit time

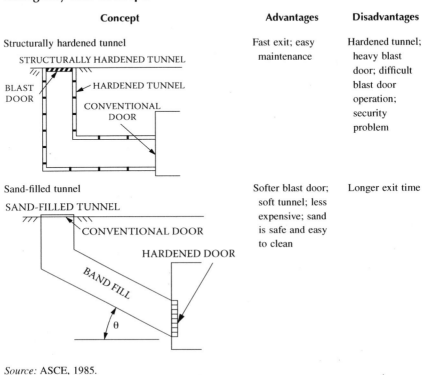

Source: ASCE, 1985.

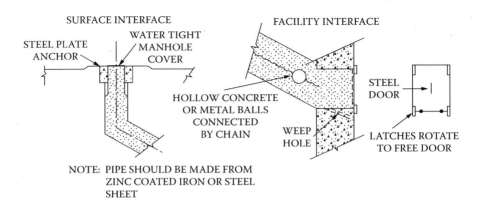

FIGURE 7.24 Emergency exit details (ASCE, 1985).

Connections, Openings, Interfaces, and Internal Shock

in this area and some of the results remain relevant (Crawford et al., October 1974; Harris and Tucker, August 1986; Marquis and Berglund, July 1990 and August 1991; Mett, April 1991; Ball et al., March and April 1991; Dove, 1992). Unfortunately, much of the tested equipment and many of the devices may no longer be used and a designer must obtain updated information from equipment vendors.

Shock isolation is a specialized area of technical activity, and one should employ the procedures discussed for both the analysis and design of MDOF systems. Only general comments are given here; however, additional information is found in the above-cited references. One must address the known limits on human and equipment responses to ensure their safety. Some typical configurations are shown in Table 7.4.

7.4 INTERNAL PRESSURE

7.4.1 Internal Pressure Increases

No relevant data are available on this subject for external pressures over 150 psi. For structures with relatively small opening:volume ratios, the average pressure rise can be estimated from Equation 7.3:

$$\Delta P_i = C_L \left(\frac{A_o}{V_o} \right) \Delta t \tag{7.3}$$

in which ΔP_i is the internal pressure increase (psi), C_L is the factor based on $P - P_i$ (see Figure 7.25), A_o is the opening area (square feet), V_o is the internal volume (cubic feet), and Δt is the time interval for pressure increase (milliseconds). The calculation is performed for 10 to 20 time increments during the pulse duration. For each time step, compute $P_i = P_i + \Delta P_i$, $P - P_i$, C_L, and the new ΔP_i. Repeat to the end of pulse. For negative $P - P_i$, use negative C_L values.

Advanced computer codes may be required for computing accurately internal explosions. The environment could be severe and require sufficient venting. There are empirical methods for this purpose but they may not be very accurate. For example, one could use the procedures in TM 5-1300 (Department of the Army, 1990), and/or the computer codes SHOCK (NCEL, 1988) and FRANG (Wager and Connett, 1989) for quick estimates of internal pressure conditions. Another computer code that would be helpful for more accurate calculations is BlastX (Britt and Ranta, 2001).

7.4.2 Airblast Transmission through Tunnels and Ducts

This is another area where empirical or numerical data exist. Shock tube tests showed that each 90° turn in a tunnel will reduce the pressure by 6%. Some other information is summarized in Figure 7.26. The effect of friction on duct walls is

320 Modern Protective Structures

TABLE 7.4
Shock Isolation Options

Characteristics	Ceiling-Supported			Base-Mounted		
	Pendulous		Horizontal Isolators	Vertical Isolators	Inclined Isolators	Horizontally Stabilized
	Inclined pendulum	Pendulum, low CG	Pendulum, high CG	Horizontal and vertical springs			
Schematic	(inclined pendulum diagram)	(pendulum low CG diagram)	(pendulum high CG diagram)	(horizontal isolators diagram)	(vertical isolators diagram)	(inclined isolators diagram)	(horizontally stabilized diagram)
Feasibility range	For total structure isolation at low frequency and large relative displacements	For equipment and personnel platforms at low frequency and large relative displacements	For missile mounts at low frequency and moderate relative displacements	For equipment isolation at low frequency and moderate to large relative displacements	For equipment and total structure isolation at mid-range frequency and small to moderate relative displacements	For equipment isolation at mid-range frequency and small to moderate relative displacements	For equipment isolation at mid-range frequency and low to moderate relative displacements
Dynamic coupling	System characteristics can be selected to minimize coupling	System characteristics can be selected to minimize coupling	Very significant pitch response induced by horizontal inputs	Negligible	Very significant pitch response induced by horizontal inputs	Negligible	Small pitch response induced

Connections, Openings, Interfaces, and Internal Shock

	SDOF values acceptable when geometry is optimized for minimum coupling	SDOF values amplified due to pitch response	SDOF values amplified due to pitch response	SDOF values (large horizontal space required for hardware)	SDOF values (minimum space required for hardware)	SDOF values (moderate space required for hardware)	SDOF values amplified due to pitch response
Rattlespace requirements							
Dynamic stability	Not critical when optimized	May be critical	Very critical	Not critical	Not critical	Not critical	Not critical
Static stability	Not critical when optimized	May be critical	Very critical	Not critical	May be critical (spring stability must be considered)	Not critical (but spring stability must be considered)	Not critical (but spring stability must be considered)

Low frequency = <1 Hz.

High frequency = 1 to 10 Hz.

Displacement:

	Vertical	Horizontal
Small	< 6 in. (15 cm)	< 2 in. (5 cm)
Moderate	< 15 in. (38 cm)	< 5 in. (13 cm)
Large	< 40 in. (100 cm)	< 14 in. (36 cm)

CG = Center of Gravity.

Source: ASCE, 1985.

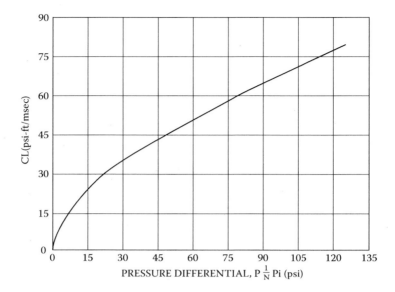

FIGURE 7.25 Leakage pressure coefficient versus pressure differential (TM 5-855-1, 1986).

shown in Figure 7.27 and Figure 7.28. This is another area where more information is badly needed. Some data are available in technical manuals (Department of the Army, 1986; ASCE, 1985; Drake et al., 1989).

Connections, Openings, Interfaces, and Internal Shock

CASE	RATIO OF TRANSMITTED TO PEAK OVERPRESSURE P_T, P_{so}
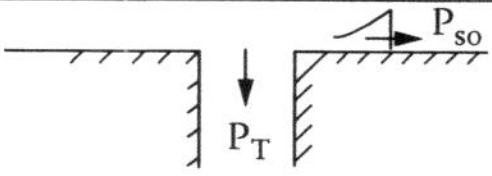 (90° bend, P_{so} horizontal, P_T downward)	0.5
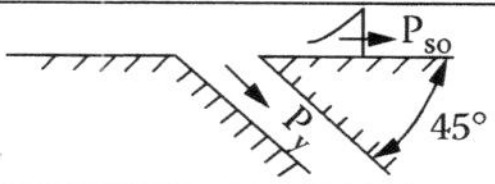 (45° bend)	1.0
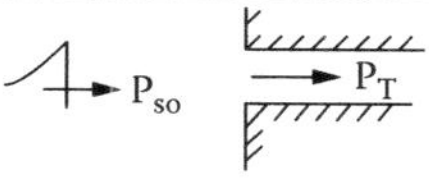 (straight-through, $P_{so} \to P_T$)	1.5
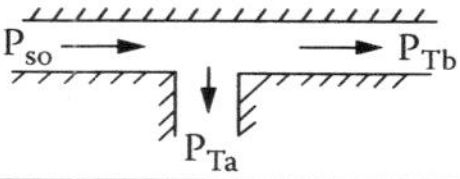 (T-junction, $P_{so} \to P_{Tb}$, P_{Ta} down)	(a) 0.5 (b) 0.8
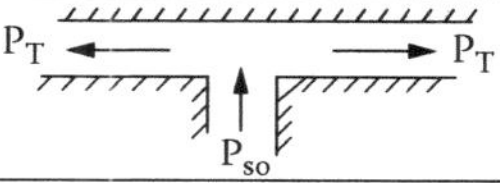 (cross junction, $P_T \leftarrow$, $\to P_T$, P_{so} up)	0.8
90° BENDS	$(0.94)^?$ WHERE ? IS THE NUMBER OF TURNS

FIGURE 7.26 Transmitted over-pressure ($P_{so} \leq 50$ psi) in tunnels (TM 5-855-1, 1986).

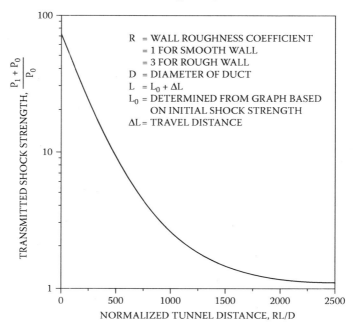

FIGURE 7.27 Shock front attenuation due to duct wall friction (ASCE, 1985).

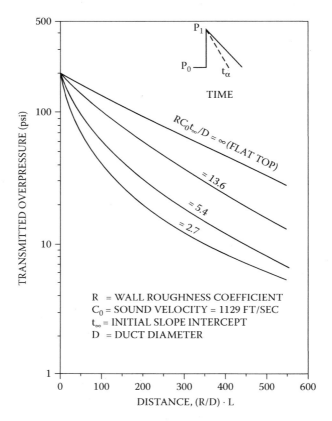

FIGURE 7.28 Attenuation of transmitted over-pressure from initial value of 200 psi (1400 kpa) (ASCE, 1985).

8 Pressure–Impulse Diagrams and Their Applications

8.1 INTRODUCTION

Protective structures are typically made of materials with inherent high mass densities and considerable energy absorption abilities to be suitable and economical to withstand severe impulsive loading arising from blast or impact events. Such designs require some form of dynamic analysis to determine the response characteristics (natural frequency, maximum displacement, reactions, etc.) of elements under consideration. Besides the procedures presented in Chapter 6, the results from the dynamic analysis can be represented as pressure–impulse (P-I) diagrams. The P-I diagram is a useful design tool that permits easy assessment of response to a specified load. With a maximum displacement or damage level defined, the diagram indicates the combinations of load and impulse that will cause failure or a specific damage level.

Traditionally, P-I diagrams for structural elements have been based on single-degree-of-freedom (SDOF) formulations, assuming a flexural mode of response without consideration of any damage due to shear failure (Oswald and Skerhut, 1993). However, experimental results (Kiger, 1980–1984; Slawson, 1984) have shown that reinforced concrete beams and one-way slabs can exhibit a unique form of shearing failure at the supports under certain severe loading conditions. Moreover, these analytical P-I models were often developed using idealized perfectly elastic or elastic–perfectly plastic material models that do not account for the influence of concrete confinement, diagonal shear, or axial compression.

To compensate for these inadequacies, these analytical P-I curves are often "shifted" to match observed experimental results. While these models may be convenient to derive, they do not describe accurately the dynamic response characteristics of a structural element under consideration. This chapter will present the background for the traditional concept of P-I diagrams and also show how they can be modified to represent various behavioral modes other than flexure. The material presented is based on recent studies by Soh and Krauthammer (2004), Ng, Krauthammer (2004) and Blasko and Krauthammer (2007).

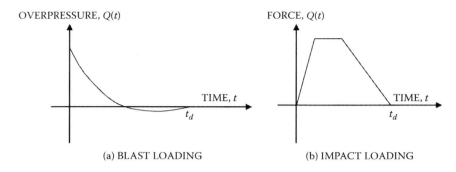

FIGURE 8.1 Idealized transient loading profiles.

8.2 BACKGROUND

Nonperiodic or nonharmonic loads (Figure 8.1) that act for a finite duration are often encountered in the design of protective structures. The transient loads are commonly generated in blast or impact events and these loads are usually defined in terms of peak load (in terms of pressure or force) and impulse (instead of duration). The impulse I of the load is defined as the area enclosed by the load–time curve (defined generically by Q) and the time axis (time integral) and its magnitude are given by the following expression:

$$I = \int_0^{t_d} Q(t)\,dt \qquad (8.1)$$

where Q(t) is the load–time curve and t_d is the duration of the load pulse. When Q(t) is expressed in term of pressure, *specific impulse* becomes an appropriate term. Often for structural dynamic analysis, a designer is mainly concerned with the final states (i.e., maximum displacement and stresses) rather than a detailed knowledge of the response histories of the structure.

Typically, plots of maximum peak response versus the ratio of the load duration and natural period of the system (response spectra) are often used to simplify the design of a dynamic system for a given loading. By defining different sets of axes, the same response spectra for the given dynamic system can be represented in different ways. The P-I diagram is an alternative representation of response spectra and is widely used for structural component damage assessment. Although the various forms of shock spectra may look different, they are basically the same because they describe the relationship between the maximum value of a response parameter and a characteristic of that dynamic system under consideration.

The early application of P-I diagrams saw the British using empirically derived diagrams for brick houses to determine damage criteria for other houses, small office buildings, and light framed industrial buildings (Jarrett, 1968). Currently, the results of such investigations are used as the bases for explosive safe standoff criteria in the United Kingdom (Baker et al., 1983; Mays and Smith, 1995). P-I diagrams were also developed to assess human responses to blast loading and to establish damage criteria to specific organs (eardrum, lungs, etc.) of the human body. This is possible because the body responds to blast loading as a complex mechanical system (Baker et al., 1983).

In protective design, designers have used P-I diagrams extensively to make (preliminary) damage assessments of structural components subjected to blast loading. The Facility and Component Evaluation and Damage Assessment Program (FACEDAP) (Oswald and Skerhut, 1993) is a widely used blast design and analysis program that employs and extends the analytical P-I solutions of Baker et al. (1983) to predict damage to more than 20 structural components based on the assumptions of flexure and buckling modes of failures. It should be pointed out that the P-I diagram should be more correctly referred to as a *load–impulse* diagram because the ordinate can be defined either in terms of pressure or force; in the latter case, the *force–impulse* term becomes appropriate. Traditionally, in specific applications for blast-loaded structures, these load–impulse diagrams appear often with pressure (rather than force) as the ordinate simply because the (blast) load is defined in term of pressure distribution. Moreover, some authors (Baker et al., 1983; Smith and Hetherington, 1994; Mays and Smith, 1995; Krauthammer, 1998) consistently use *pressure–impulse* to describe these diagrams regardless of the nature of the loading. For purposes of this chapter, all load–impulse diagrams (pressure–impulse and force–impulse diagrams) are referred to as P-I diagrams.

8.3 CHARACTERISTICS OF P-I DIAGRAMS

Figure 8.2a shows a typical response spectrum for an undamped, perfectly elastic SDOF system in which x_{max} is the maximum dynamic displacement, K is the spring stiffness, P_0 is the peak force, M is the lumped mass, and T is the natural period of the system. By defining a different set of axes, the same response spectrum can be transformed into what is known as a P-I diagram shown in Figure 8.2b. The primary difference between these two presentations is that the response spectrum emphasizes the influence of scaled time (i.e., t_d/T) on the system response, whereas the P-I diagram emphasizes the combination of peak load and impulse (or equivalent dimensionless quantity) for a given response (or damage level) (Baker et al., 1983).

A P-I diagram, also called an *iso-damage* curve (Mays and Smith, 1995), permits easy assessment of response to a specified load. With a maximum displacement or damage level defined, a P-I curve indicates the combinations of load (or pressure) and impulse that will cause the specified failure (or damage level). In effect, the threshold curve divides the P-I diagram into two distinct regions. Combinations of pressure and impulse that fall to the left of and below the curve will not induce failure while those to the right and above the graph will

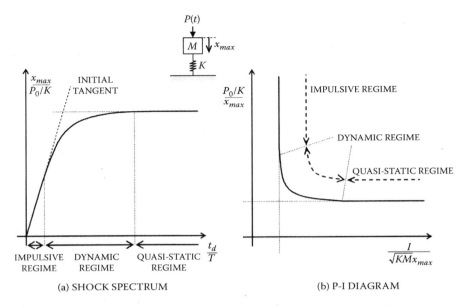

FIGURE 8.2 Typical response spectrum (a) and P-I diagram (b).

produce damage in excess of the allowable limit (i.e., maximum dynamic displacement).

It is well known from structural dynamics that a strong relationship exists between the natural frequency (which directly influences response time) of a structural element and the duration of the forcing or load function (Biggs, 1964; Clough and Penzien, 1993; Humar, 2002). This relationship is normally categorized into three regimes: impulsive, quasi-static, and dynamic (Baker et al., 1983) and they are indicated on the response spectrum and P-I diagram in Figure 8.2. Note that the P-I representation better differentiates the impulsive and quasi-static regimes in the form of vertical and horizontal asymptotes compared to the response spectrum. Because an understanding of these regimes is important in the development of P-I diagrams using the energy balance approach, they are discussed next.

8.3.1 Loading Regimes

In the impulsive loading regime, the load duration is short relative to the response time of the system (which is influenced by the system natural frequency). In effect, the load is applied to the structure and removed before the structure can undergo any significant deformation, as shown in Figure 8.3a. The maximum response (at time t_m) can thus be assumed to be independent of the load time history (or load profile). For the quasi-static regime, the loading duration is significantly longer than the response time. The load dissipates very little before the maximum deformation or resistance is achieved at time t_m (Figure 8.3b).

Pressure–Impulse Diagrams and Their Applications

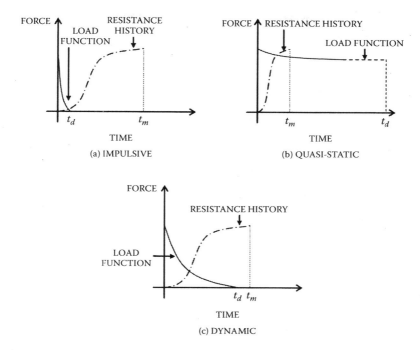

FIGURE 8.3 Comparison of response time loading regimes.

Unlike the impulsive regime, the response in the quasi-static regime depends only upon the peak load (P_0) and structural stiffness (k). However, as with the impulsive regime, the maximum response is not affected by the loading history. The third transition regime, known as the dynamic regime, exists between the impulsive and quasi-static regions and the loading duration and system response time are approximately the same or of the same magnitude as shown in Figure 8.3c. Response in this loading regime is more complex and is significantly influenced by the profile of the load history.

Baker et al. (1983) quantified the three loading regimes for an undamped, perfectly elastic system subjected to an exponentially decaying load (Table 8.1);

TABLE 8.1
Limits for Loading Regimes

Loading Regime	(Approximate) Limits	
Impulsive	$\omega t_d < 0.4$	$t_d/T < 6.37 \times 10^{-2}$
Quasi-static	$\omega t_d > 40$	$t_d/T > 6.37$
Dynamic	$0.4 < \omega t_d < 40$	$6.37 \times 10^{-2} < t_d/T < 6.37$

Source: Baker et al., 1983.

ω is the system natural circular frequency. Humar (2002) suggested that the dynamic response can be assumed to be in the impulsive loading regime for rectangular, triangular, and sinusoidal load pulses with t_d/T ratios less than 0.25.

8.3.2 Influence of System and Loading Parameters

The P-I diagram is affected by a number of parameters including shape of load pulse, load rise time, plasticity, and damping. For a load pulse with an infinitely small or no rise time, the shape of the load pulse has no appreciable effect on the impulsive and quasi-static asymptotes (Baker et al., 1983). However, the pulse shape significantly influences the system response in the dynamic loading regime.

For simple load pulses with zero rise time (rectangular, triangular, and exponential), the dynamic loading regime can be approximated by simple hyperbolic functions. For more complex pulses (Figure 8.4) resulting from nonideal explosions, the response in the dynamic region deviates from the hyperbolic profile and may be more severe as a result of resonance between load rate and structure natural frequency (Baker et al., 1983). Li and Meng (2002) presented a method to eliminate the effects of pulse shape for undamped SDOF system with perfectly elastic, elastic–perfectly plastic, and rigid plastic responses through the use of dimensionless loading parameters. Based on the proposed approach, a unique loading shape-independent P-I diagram was obtained for each type of response.

The effect of finite rise time on the load pulse is to increase the value of the quasi-static asymptote from 0.5 (corresponding to a dynamic load factor of 2.0) to 1.0, which corresponds to static loading. The impulsive asymptote, on the other hand, is not affected by pulse rise time (Baker et al., 1983). However, based on the shock spectra of (isosceles) rectangular and sinusoidal load pulses (i.e., pulses with finite rise time) (Biggs, 1964; Humar, 2002), the P-I curves for pulses with finite rise times are expected to consist of a series of peaks and dips in the dynamic loading regime.

For load pulses with zero rise time and increasing (material) plasticity (or ductility), the P-I curve shifts further away from or nearer to the origin, depending

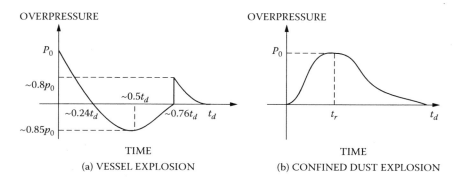

FIGURE 8.4 Non-ideal load pulses (after Baker et al., 1983).

Pressure–Impulse Diagrams and Their Applications   331

on whether the elastic (x_{el}) or maximum (x_{max}) deflection is used in defining the axes. The impulsive and quasi-static asymptotes can be determined from the expressions to be shown in subsequent sections. With increasing ductility, the quasi-static asymptote tends to one corresponding to the static loading case. The shift occurs because more energy can be absorbed whenever the structure deforms plastically.

For a load pulse with a finite rise time, the dynamic overshoot that occurs elastically in the elbow of the curve is damped and the impulsive asymptote is shifted to the right, while no changes are noted for the quasi-static asymptote (Baker et al., 1983). The quasi-static asymptote is not affected since the work applied to the structure and the strain energy absorbed both increase linearly. By far, the discussion on the effect of (viscous) damping appears to absent in the existing literature on P-I diagrams. However, the effects can be deduced from shock spectra. Humar (2002) reported that the maximum response in the quasi-static loading regime reduces by about 13.5% as the damping increases from 0 to 10%. For the impulsive loading regime, it is suggested that damping can be ignored because the amount of energy that can be dissipated in the short duration of motion is quite small (Humar, 2002; Clough and Penzien, 1993).

8.4 ANALYTICAL SOLUTIONS OF P-I DIAGRAMS

8.4.1 CLOSED-FORM SOLUTIONS

For an undamped, perfectly elastic SDOF system, closed-form solutions of P-I curves can be readily obtained from response spectra. The response spectrum for a specific load pulse is derived from the respective response history functions. Table 8.2 summarizes the response history functions for four common load pulses addressed in the literature (Clough and Penzien, 1993; Humar, 2002). These functions were derived assuming zero initial conditions (i.e., zero displacement and velocity). No closed-form solutions of P-I diagrams are documented in existing literature (Baker et al., 1983; Oswald and Skerhut, 1993; Smith and Hetherington, 1994; Mays and Smith, 1995) as the authors relied solely on the balance energy method to obtain approximate P-I solutions, as discussed later. The following section illustrates the derivation of closed-form P-I solutions for an undamped, perfectly elastic SDOF system subjected to rectangular and triangular load pulses.

8.4.1.1 Response to Rectangular Load Pulse

For an undamped elastic SDOF system subjected to a rectangular load pulse, the shock spectrum (Humar, 2002) is given by the following expressions:

$$x_{max}/(P_0/K) = 2 \sin(0.5\omega\, t_d) \quad 0 \leq \omega t_d \leq \pi \tag{8.2a}$$

$$x_{max}/(P_0/K) = 2 \quad \omega t_d > \pi \tag{8.2b}$$

TABLE 8.2
Response Functions of Undamped SDOF Subjected to Transient Loads

Load Type	Loading Function P(t)	Response Functions (assuming $x(0) = \dot{x}(0) = 0$) x(t)
Triangular (with zero rise time) 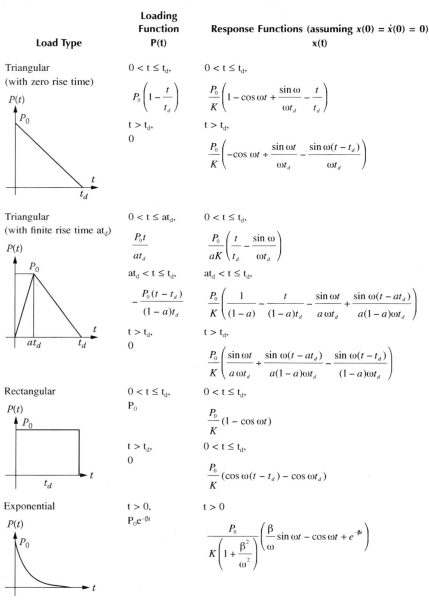	$0 < t \leq t_d$, $P_0\left(1 - \dfrac{t}{t_d}\right)$ $t > t_d$, 0	$0 < t \leq t_d$, $\dfrac{P_0}{K}\left(1 - \cos\omega t + \dfrac{\sin\omega t}{\omega t_d} - \dfrac{t}{t_d}\right)$ $t > t_d$, $\dfrac{P_0}{K}\left(-\cos\omega t + \dfrac{\sin\omega t}{\omega t_d} - \dfrac{\sin\omega(t - t_d)}{\omega t_d}\right)$
Triangular (with finite rise time at$_d$)	$0 < t \leq at_d$, $\dfrac{P_0 t}{at_d}$ $at_d < t \leq t_d$, $-\dfrac{P_0(t - t_d)}{(1 - a)t_d}$ $t > t_d$, 0	$0 < t \leq t_d$, $\dfrac{P_0}{aK}\left(\dfrac{t}{t_d} - \dfrac{\sin\omega t}{\omega t_d}\right)$ $at_d < t \leq t_d$, $\dfrac{P_0}{K}\left(\dfrac{1}{(1-a)} - \dfrac{t}{(1-a)t_d} - \dfrac{\sin\omega t}{a\omega t_d} + \dfrac{\sin\omega(t - at_d)}{a(1-a)\omega t_d}\right)$ $t > t_d$, $\dfrac{P_0}{K}\left(\dfrac{\sin\omega t}{a\omega t_d} + \dfrac{\sin\omega(t - at_d)}{a(1-a)\omega t_d} - \dfrac{\sin\omega(t - t_d)}{(1-a)\omega t_d}\right)$
Rectangular	$0 < t \leq t_d$, P_0 $t > t_d$, 0	$0 < t \leq t_d$, $\dfrac{P_0}{K}(1 - \cos\omega t)$ $0 < t \leq t_d$, $\dfrac{P_0}{K}(\cos\omega(t - t_d) - \cos\omega t_d)$
Exponential	$t > 0$, $P_0 e^{-\beta t}$	$t > 0$ $\dfrac{P_0}{K\left(1 + \dfrac{\beta^2}{\omega^2}\right)}\left(\dfrac{\beta}{\omega}\sin\omega t - \cos\omega t + e^{-\beta t}\right)$

Source: Soh and Krauthammer, 2004.

Pressure–Impulse Diagrams and Their Applications

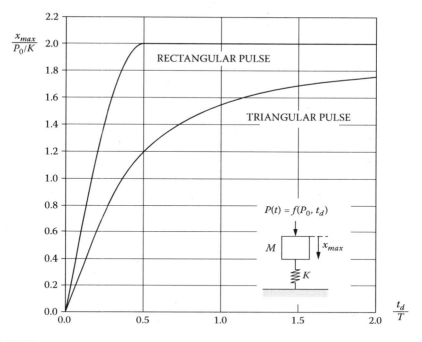

FIGURE 8.5 Shock spectra for rectangular and triangular pulses.

These expressions represent the responses in the free and forced vibration domains, respectively. The shock spectrum is plotted in Figure 8.5. The dimensionless (force and impulse) terms for the vertical and horizontal axes for the P-I plots (of undamped elastic SDOF system) are defined, respectively, as follows:

$$\bar{P} = (P_0/K)/x_{max} \tag{8.3a}$$

$$\bar{I} = I/[(KM)^{0.5} x_{max}] \tag{8.3b}$$

where the load impulse I for a rectangular pulse is given by

$$I = P_0 t_d \tag{8.4}$$

It is obvious that the corresponding P-I curve is defined by two separate functions representing both the free-vibration and the forced-vibration responses. To establish the transition point between the two domains on the P-I plot, one first notes that

$$\bar{I} = \frac{I}{\sqrt{KMx_{max}}} = \frac{P_0 t_d}{\sqrt{KMx_{max}}} = \omega t_d \left(\frac{P_0/K}{x_{max}}\right) \tag{8.5}$$

Combining Equations (8.2b) and (8.5), one derives the following:

$$\bar{P} = \frac{P_0/K}{x_{max}} = \frac{1}{2} \tag{8.6a}$$

$$\bar{I} = \omega t_d \left(\frac{P_0/K}{x_{max}}\right) = \frac{\omega t_d}{2} = \frac{\pi}{2} \tag{8.6b}$$

Thus, the point (0.5π, 0.5) defines the transition point on the P-I curve. To obtain the expression for the free-vibration domain, one combines Equations (8.3a) and (8.5) to obtain:

$$\omega t_d = \frac{\bar{I}}{\bar{P}} \tag{8.7}$$

Substituting the above and Equation (8.5) into Equations (8.2a) and (8.2b), one obtains the following expressions defining the P-I curve for rectangular pulse:

$$\bar{P} \sin\left(\frac{\bar{I}}{2\bar{P}}\right) = \frac{1}{2} \quad 1 \le \bar{I} \frac{\pi}{2} \tag{8.8a}$$

$$\bar{P} = \frac{1}{2} \quad \bar{I} > \frac{\pi}{2} \tag{8.8b}$$

8.4.1.2 Response to Triangular Load Pulse

For a triangular pulse with zero rise time (Table 8.2, first case), the resulting shock spectrum of an undamped elastic SDOF system can be shown (Soh and Krauthammer 2004) to be

$$\frac{x_{max}}{P_0/K} = \frac{\sqrt{2 + (\omega t_d)^2 - 2\omega t_d \sin \omega t_d - 2 \cos \omega t_d}}{\omega t_d} \quad 0 < \omega t_d \le 2.3311223 \tag{8.9a}$$

$$\frac{x_{max}}{P_0/K} = 2\left(1 - \frac{\tan^{-1} \omega t_d}{\omega t_d}\right) \quad \omega t_d > 2.3311223 \tag{8.9b}$$

As before, the above expressions represent the responses in the free and forced vibration domains, respectively. The shock spectrum for the triangular pulse is shown in Figure 8.5. The corresponding P-I curve for the elastic SDOF system can be determined in the same manner as the rectangular pulse. First, it is noted that the impulse for a triangular pulse is given by

$$I = \frac{1}{2} P_0 t_d \tag{8.10}$$

Combining Equations (8.3a), (8.3b), and (8.10) results in:

$$\omega t_d = \frac{2\bar{I}}{\bar{P}} \tag{8.11}$$

The expressions for the P-I curve are obtained by substituting the above and Equation (8.5) into Equation (8.9) for the shock spectrum. The transition point (separating the free and forced-vibration domains) on the P-I curve can be shown to be (1.166, 1.0). The P-I curve for the triangular pulse is thus given by:

$$\left(\frac{2\bar{I}}{\bar{P}^2}\right)^2 = 2 + \left(\frac{2\bar{I}}{\bar{P}}\right)^2 - \frac{4\bar{I}}{\bar{P}} \sin\left(\frac{2\bar{I}}{\bar{P}}\right) - 2\cos\left(\frac{2\bar{I}}{\bar{P}}\right) \quad 1 < \bar{I} \le 1.166 \tag{8.12a}$$

$$\left(\frac{2\bar{I}}{\bar{P}}\right) = \tan\left[\left(\frac{2\bar{I}}{\bar{P}}\right)\left(1 - \frac{1}{2\bar{P}}\right)\right] \quad \bar{I} > 1.166 \tag{8.12b}$$

The analytical P-I curves for rectangular and triangular pulses are plotted in Figure 8.6. It is obvious that the closed-form solutions can be cumbersome, even for simple load pulses. However, unlike the energy balance method (discussed in the next section), the closed-form expressions provide exact descriptions of P-I curves in the dynamic region.

8.4.2 Energy Balance Method

The energy balance method is by far the most common method employed to obtain analytical P-I solutions. The approach, based on principle of conservation of mechanical energy, is convenient to apply because there always exist two distinct energy formulations that separate the impulsive loading regime from the quasi-static loading regime, as discussed earlier.

To obtain the impulsive asymptote, it can be assumed that due to inertia effects the initial total energy imparted to the system is in the form of kinetic energy only. By equating this to the total strain energy stored in the system at its final state (i.e., maximum response) and performing algebraic manipulations, it is possible to obtain expression for the impulsive asymptote. For the quasi-static

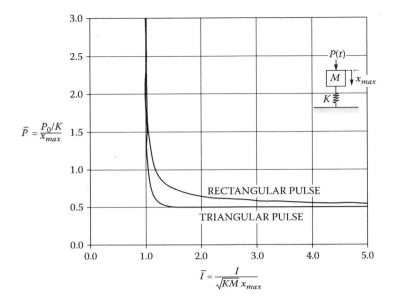

FIGURE 8.6 P-I curves for rectangular and triangular pulses.

loading regime, the load can be assumed to be constant before the maximum deformation is achieved (Figure 8.3b). Now, by equating the work done by load to the total strain energy gained by the system, the expressions for the quasi-static asymptotes are obtained as follows:

$$K.E. = S.E. \quad \text{(impulsive asymptote)} \qquad (8.13)$$

$$W.E. = S.E. \quad \text{(quasi-static asymptote)} \qquad (8.14)$$

where $K.E.$ is the kinetic energy of the system at time zero, $S.E.$ is the strain energy of the system at maximum displacement, and $W.E.$ is the maximum work done by the load to displace the system from rest to the maximum displacement. Based on this approach, the author derived the impulsive and quasi-static asymptotes for three simple SDOF systems illustrated in Table 8.3.

Although the energy balance method greatly reduces computation efforts, its formulation is only applicable to the (two) extremes (impulsive and quasi-static) of the response spectrum and is not valid for the transition zone between impulsive and quasi-static response. The dynamic regime of the P-I curve must be approximated using suitable analytical functions. This is discussed next.

8.4.2.1 Approximating Dynamic Regions

For triangular or exponential load pulses with zero rise time, the common approach is to approximate the dynamic regime of the P-I curve using hyperbolic

TABLE 8.3
Energy Solutions for Simple SDOF Systems[+]

SDOF System[*]	K.E.[#] $\frac{1}{2} M \dot{x}(0)^2$	W.E. $\int P dx$	S.E. $\int R dx$	Impulsive Asymptote K.E. = S.E.	Quasi-static Asymptote W.E. = S.E
Perfect elastic $P_0, I \to M \downarrow x_{max}, K$	$\dfrac{I^2}{2M}$	$P_0 x_{max}$	$\dfrac{1}{2} K x_{max}$	$\dfrac{I}{\sqrt{KM}\, x_{max}} = 1$	$\dfrac{P_0/K}{x_{max}} = \dfrac{1}{2}$
Rigid plastic $P_0, I \to M \downarrow x_{max}, R_p$	$\dfrac{I^2}{2M}$	$P_0 x_{max}$	$R_p x_{max}$	$\dfrac{I}{\sqrt{R_r M\, x_{max}}} = \sqrt{2}$	$\dfrac{P_0}{R_p} = 1$
Elastic–perfect plastic $P_0, I \to M \downarrow x_{max}, R$	$\dfrac{I^2}{2M}$	$P_0 x_{max}$	$K x_{el}^2 \left(\dfrac{x_{max}}{x_{el}} - \dfrac{1}{2} \right)$	$\dfrac{I}{\sqrt{KM}\, x_{max}} = \sqrt{\dfrac{2\mu - 1}{\mu}}$ $\mu = \dfrac{x_{max}}{x_{el}}$	$\dfrac{P_0/K}{x_{max}} = \dfrac{2\mu - 1}{2\mu^2}$

° $\leq x \leq x_{el}\ R = Kx\quad x_{el} < x \leq x_{max}\ R = Kx_{el}$
+ System assumed to be initially at rest (zero displacement and acceleration).
* Idealized resistance functions shown in Figure 8.7.
# It is assumed that the impulse I imparts an instantaneous velocity at time zero, i.e., $\dot{x}(0) = \dfrac{I}{M}$.

approximation. Baker et al. (1983) suggested the use of the following hyperbolic tangent squared relationship:

$$\text{S.E.} = \text{W.E.} \tanh^2 (\text{K.E.}/\text{W.E.})^{1/2} \tag{8.15}$$

For small values of the above expression, the hyperbolic tangent equals its argument, which effectively reduces to Equation (8.13). For large values, the hyperbolic function approaches unity and Equation (8.14) for the quasi-static asymptote is obtained. Baker et al. (1983) reported that less than a 1% error is introduced when Equation (8.15) is used to approximate the transition region for linearly elastic oscillators.

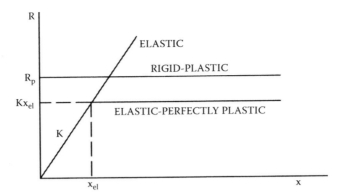

FIGURE 8.7 Idealized resistance functions.

Oswald and Skerhut (1993), based on limited comparison to response curves developed with dynamic SDOF analysis for exponential loading, recommend the following simple hyperbolic function to curve-fit the transition region:

$$(\bar{P} - A)(\bar{I} - B) = 0.4(0.5A + 0.5B)^{1.5} \tag{8.16}$$

where A and B are the values of the vertical (impulsive) asymptote and horizontal (quasi-static) asymptote, respectively. Equation (8.16) can be generalized further as follows:

$$(\bar{P} - A)(\bar{I} - B) = C(A + B)^{D} \tag{8.17}$$

where C and D are constants. For P-I curves of an undamped, perfectly elastic SDOF system subject to rectangular and triangular pulses (Figure 8.7), the approximate values for the constants C and D are computed and shown in Figure 8.8.

8.4.2.2 Continuous Structural Elements

The energy balance method is particularly useful in obtaining approximate P-I solutions for continuous structural elements under both uniaxial and biaxial states of stress. The approach is similar to that for simple mechanical systems discussed earlier except that an appropriate dynamic deformed shape function must be assumed for the structural element under consideration in addition to Equations (8.13) and (8.14). Baker et al. (1983) documented the derivation of analytical P-I solutions for a beam, extensional strip, column, and plate. Table 8.4 summarizes the characteristics (i.e., ordinate, abscissa) of dimensionless P-I diagrams.

Pressure–Impulse Diagrams and Their Applications

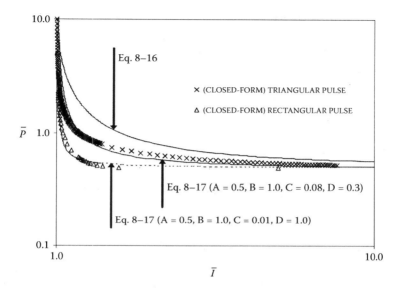

FIGURE 8.8 Approximate solutions for dynamic region.

8.5 NUMERICAL APPROACH TO P-I CURVES

The above discussions were confined to analytical solutions obtained either by closed-form formulation or the approximate energy approach. P-I diagrams can be generated numerically by generating sufficient data points to allow for curve fitting. Each data point on a P-I diagram represents the results from a single dynamic analysis. For each analysis, the maximum response (or damage) may or may not be reached, depending on the combination of pressure and impulse (Figure 8.9).

A search algorithm must be employed to locate the data points that define the transitions from damage to no-damage zones (i.e., the threshold curve). Due to a large number of dynamic analyses that must be performed, the search is best performed with the aid of a computer. This approach gives great flexibility to the type of dynamic model (i.e., equivalent SDOF, MDOF, Timoshenko beam, etc.) to be used so long as it can best describe the experimentally observed dynamic responses of the physical problem. Furthermore, with adequate data points, the numerical approach should be able to accurately describe the nature of the P-I curve in the dynamic loading regime, in particular, for complex load pulses.

Another numerical approach was proposed by Ng and Krauthammer (2004) for RC slabs. An efficient method to find the threshold points was found with the observation that by the definition of a threshold curve, a pressure–impulse (P-I) combination could be classified only as "safe" or "damaged" by the threshold criterion. For a safe combination, increasing the impulse while keeping the pressure constant would eventually find a point that resulted in damaged.

TABLE 8.4
Analytical Expressions for Dynamic Reaction of Beam

No	Element	Material Behavior	Mode of Response	y-axis	x-axis	z-axis
1.	Beam	Perfectly elastic	Bending	$\dfrac{ibh}{\beta_i\sqrt{\rho EI_{xx}A_g}}$	$\dfrac{\rho bhl^2}{\beta_p EI_{xx}}$	$\dfrac{\sigma_{max}}{E}$
2.	Beam	Elastic Perfect-plastic	Bending	$\dfrac{ib\sqrt{EI_{xx}}}{\psi_i\sqrt{\rho A}\sigma_y Z}$	$\dfrac{\rho bl^2}{\psi_p \sigma_y Z}$	$\dfrac{I_{xx}E\varepsilon_{max}}{\psi_e H \sigma_y Z}$
3.	Extension strip	Elastic Perfect-plastic	Extension	$\dfrac{\rho bl\sqrt{E}}{\sigma_y^{3/2} Z}$	$\dfrac{ib\sqrt{E}}{\sqrt{\rho\sigma_y A_g}}$	$\dfrac{E\varepsilon_{max}}{\sigma_y}$
4.	Plate	Elastic Perfect-plastic	Bending	$\dfrac{p_r X^2}{\Phi_p \sigma_y h^2}$	$\dfrac{i_r\sqrt{E}}{\Phi_i \sigma_y\sqrt{\rho h}}$	n.a.
5.	Column	Perfectly elastic	Buckling	$\dfrac{iA_l h\sqrt{E}}{\alpha_i\sqrt{M_{of} l I_{xx}\sigma_y}}$	$\dfrac{\rho A_l l^2}{\alpha_p EI_{xx}}$	n.a.

Definition of Terms: i = Side-on or reflected impulse intensity; p = Side-on or reflected pressusre intensity; i_r = Reflected impulse intensity; p_r = Reflected pressure intensity; l = Total span of member; b = Member width; h = Member depth; X = Shorter of the two half spans of plate; A_g = Cross-sectional area; I_{xx} = Moment of inertia about major axis; Z = Plastic section modulus; E = Elastic modulus; ρ = Uniform mass density; A_l: Loaded area of the roof or floor over the column; M_{of} = Mass of overlaying floor; σ_y: Elastic yield stress; ε_{max} = Maximum strain; σ_{max} = Maximum stress; $\beta_i, \beta_p, \psi_i, \psi_p, \psi_e$ = Beam boundary conditions coefficients; α_i, α_p = Column boundary condition and side-sway coefficients; Φ_i, Φ_p = Plate shapes factors.

Conversely, reducing the impulse for a combination that started with damaged would also find the threshold point. The proposed procedure started with a trial P-I combination. The slab was evaluated for flexure and shear failures under the load. If damaged, the pressure was kept constant and the impulse reduced by a factor of two, until a safe result was found, then the interval between the last two points was divided into equal segments and checked in increasing order until the threshold point was found (switched back to damaged). This procedure is illustrated in Figure 8.10a.

A similar procedure in the reverse direction was used if the first trial P-I combination returned a safe result. After a data point was found, the peak pressure was scaled by a factor of 0.95 for the next data point based on the observation that a curve with the shape illustrated in Figure 8.10(b) has an almost vertical impulsive asymptote, such that large variations in the originate (pressure) would result in small differences in the abscissa (impulse). However, at lower pressure

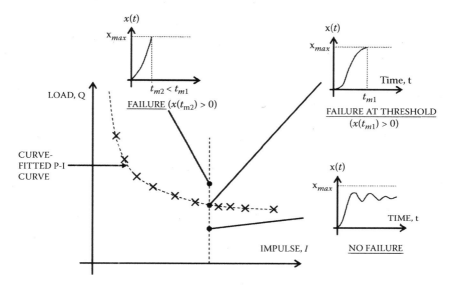

FIGURE 8.9 Numerical derivation of P-I curve.

valves, higher precision of computation was required as impulse varied with pressure in the dynamic regime. At a certain value of pressure near the quasi-static asymptote, impulse varied with small variations of pressure. If a lower pressure value was chosen, there was no solution for the curve.

8.5.1 P-I Curves for Multiple Failure Modes

It should be pointed out that the P-I diagram discussed so far was based on some single mode of failure or damage. In most structures, response and failure can occur in more than one mode. For example, for a reinforced concrete beam, flexure is traditionally assumed to be the dominating failure mode (Baker et al., 1983; Oswald and Skerhut, 1993). However, under certain loading conditions, experimental observations (Kiger et al., 1980–1984; Slawson, 1984) indicate that reinforced concrete element can also fail in direct shear.

If the two modes of failure are to be considered in the overall survivability of an element, the P-I diagram will consist of two threshold curves, each representing a single failure mode as illustrated in Figure 8.11. Thus, the true damage (threshold) curve is represented by the lower bound of the curves (as shown by the dotted line).

8.5.2 Summary

The nature of P-I diagrams has been discussed in this chapter. P-I curves can be derived by closed-form formulation, the approximate energy approach, or numerical method. Closed-form solutions are limited to responses of simple mechanical systems with simple load pulses. The energy balance approach remains a powerful

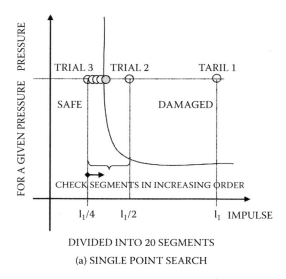

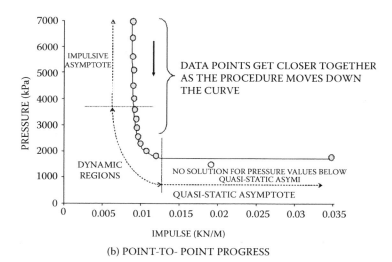

FIGURE 8.10 Numerical search for threshold points.

tool for estimating the asymptotic values of the P-I curves although the response in the dynamic loading regime cannot be determined by this approach and must be approximated by some suitable analytical functions. For a simple SDOF system, the energy balance approach can be extended to continuous structural elements (i.e., beam, plate, and column) by assuming appropriate dynamic deformation shape functions. While a traditional P-I diagram normally assumes a single mode of failure, multimode failures occur in real structures and can be represented

Pressure–Impulse Diagrams and Their Applications

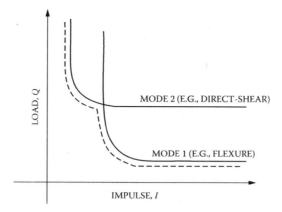

FIGURE 8.11 P-I diagrams with dual failure modes.

by composite P-I diagrams. In particular, one-way spanning structural elements (beams, one-way slabs, and walls) can fail (dynamically) in either flexural or direct-shear and it is expected that the corresponding P-I diagram will comprise of two curves. P-I diagrams can also be generated numerically by performing sufficient dynamic analyses. Using this approach, the type of dynamic model can be independently selected to best describe the dynamic behavior of the physical problem. The next chapter discusses rational structural and material models that can describe the dynamic behavior of reinforced concrete beams subjected to concentrated transient loading.

8.6 DYNAMIC ANALYSIS APPROACH

8.6.1 Introduction

For dynamic analysis of structures, two pieces of information are usually required. First, a dynamic structural model (i.e., SDOF, MDOF, continuous system, etc.) that adequately reflects the mechanical characteristics of the structure under consideration must be assumed. Second, reliable material and constitutive models must be used so that realistic dynamic resistance functions or load deformation characteristics that adequately describe the relationship between the dynamic resistance and the response of the structure can be derived, as discussed in Chapters 5 and 6. An effective approach for P-I analysis is discussed in this section.

8.6.2 Dynamic Material and Constitutive Models

An understanding of the behaviors of materials (concrete and reinforcing steel, steel, etc.) under specific loading conditions and geometrical configurations is essential for deriving reliable load deformation relationships for structural elements. The material stress–strain relationships, interactions of internal forces and moments, influences of external parameters (loading rate, size, etc.) are some important parameters that must be established based on available experimental data.

Traditionally, the dynamic responses of RC beams are assumed to be dominated by flexural modes. However, diagonal shear is present during a flexural response and must be considered, as discussed in Chapter 5 and elsewhere (Krauthammer et al., November–December 1987 and April 1990). Based on experimental observations (Kiger, 1980–1984; Slawson, 1984), RC beams subjected to severe impulsive loading exhibited direct shear modes of failure that were addressed also by numerical studies (Krauthammer et al., 1986, 1993, and 1994). These distinct behaviors are discussed below.

8.6.3 Flexural Behavior

The behavior of RC beams in flexure has been investigated extensively and well documented in literature (Park and Paulay, 1975; MacGregor and Wight, 2005; Nawy, 2003). Typically, flexural behavior is represented by a moment–curvature relationship obtained using internal force equilibrium and compatibility of strain relationships. The load–deformation relationship can be obtained readily from the moment–curvature information and boundary conditions.

In deriving the moment–curvature relationships for RC sections, it is generally assumed that strains have linear distributions over the depth of a beam and that tensile behavior of concrete located below the neutral axis is ignored (Figure 8.12). However, the same approach can be modified to include the behavior of concrete in tension. For design purposes, it is generally acceptable to assume a

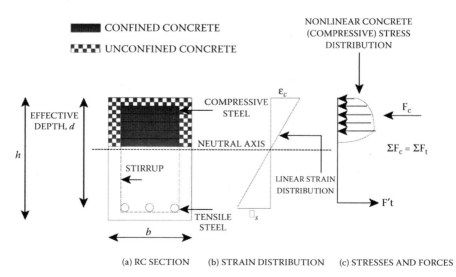

FIGURE 8.12 Reinforced concrete cross-section analysis (Soh and Krauthammer, 2004).

uniform stress distribution (i.e., rectangular stress block) for concrete under compression (ACI 318.05, 2005).

The more general approach, however, is to divide the compression zone into equal layers parallel to the neutral axis and the stresses and forces are determined for all layers of both unconfined and confined concrete within the longitudinal and transverse reinforcing steel, based on appropriate stress–strain relationships (Ghosh and Cohn, 1972; Park and Paulay, 1975; Krauthammer et al., 1982). The stress–strain relationships cited for unconfined and confined concrete were based on the models of Hognestad (1951) and Krauthammer and Hall (1982), respectively, whereas the stress–strain model by Park and Paulay (1975) can be employed for reinforcing steel.

To establish the ultimate capacity of the moment–curvature relationship, the failure criterion must be known. RC sections in flexure may fail in either tension or compression. Tension flexure failure is characterized by the yielding of tensile reinforcement whereas compression flexure failure is associated with crushing of concrete. In the latter mode, concrete may fail either in the unconfined or confined regions. Krauthammer et al. (1987) proposed that the ultimate strain for concrete be modified (for confinement enhancement) and checked at the confined region, instead of the extreme compression fiber, as concrete in the confined zone can sustain higher strains at failure.

It should be pointed out that RC sections can also fail in diagonal shear. Based on experimental results, Krauthammer et al. (1987) applied modification factors (over-reinforcing factor and shear reduction factor) to the moment–curvature relationship to account for the influence of diagonal shear. These factors also included the effects of axial thrust and web reinforcement on flexural capacity. Furthermore, in the post-yielding nonlinear domain, the author modified the length of the plastic hinge to account for the effects of axial load. These considerations are incorporated into the current flexural model.

The preceding discussion is confined to static loading conditions. Studies by Furlong et al. (1968) and Seabold (1970) suggested overall enhancements of 20 to 37% in the strength of RC structural elements subjected to transient loading at high loading rates. Krauthammer (1990) proposed that the average empirical rate effect enhancement factors (Soroushian and Obaseki, 1986; Ross, 1983) be applied to the yield and ultimate stress parameters in the concrete and reinforcing steel stress–strain models, and the modified values be utilized to compute the moment–curvature relationship and resistance functions.

A similar approach can be adopted for addressing the behavior of reinforced concrete slabs beyond the usual design calculations, since an enormous amount of energy may be absorbed beyond the yield state (Krauthammer et al., 1986; Park and Gamble, 2000). The increase in load resistance beyond the serviceability limit of cracking and deflection is important for protective structures because moderate to severe damage is often acceptable if collapse can be avoided. For the flexural response, the load deflection relationship developed in Krauthammer et al. (1986) and Park and Gamble (2000) can be used as illustrated in Figure 8.13 (see Chapter 5).

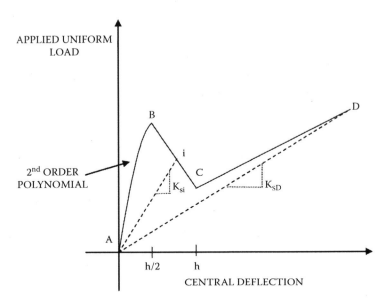

FIGURE 8.13 Resistance model for reinforced concrete slabs (Ng and Krauthammer, 2004).

The AB segment represents the compression membrane behavior that may provide a peak capacity that exceeds considerably the resistance provided by the yield line mechanism. The slab is assumed to snap through at point B and transition into a tensile membrane behavior at point C. The displacement of the central point of the slab at the peak load was approximated to be equivalent to half the slab depth (point B) and to the whole slab depth at the transition from a compression membrane to a tensile membrane when in-plane forces were zero (point C). Point D represents the resistance at the ultimate displacement, when reinforcement fracture disrupts the tensile membrane response. This approach was shown to represent accurately observed test data and was modified recently to address slender, intermediate, and thick slabs (Krauthammer et al., 2003).

8.6.3.1 Dynamic Resistance Function

With the cross-sectional moment–curvature information established for a reinforced concrete beam based on the flexural model discussed above, the corresponding load deflection curve (i.e., resistance function) for a given structural configuration is obtained based on structural mechanics relationships and using a numerical approach similar to that of Krauthammer et al. (1990). The resulting (numerical) resistance function is a piece-wise nonlinear curve (OABC), as shown in Figure 8.14a.

The same approach can be used for reinforced concrete slabs. The modeling of unloading–reloading paths is important for dealing with loading history and dynamic behavior. For bilinear (i.e., elastic–perfectly plastic) resistance functions,

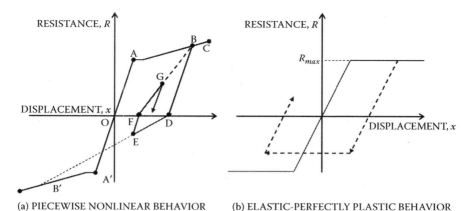

(a) PIECEWISE NONLINEAR BEHAVIOR   (b) ELASTIC-PERFECTLY PLASTIC BEHAVIOR

FIGURE 8.14 Dynamic resistance functions for flexural behavior.

the loading and unloading typically follow the behavior shown in Figure 8.14b. Studies by Newmark and Rosenblueth (1971), Sozen (1974), and Park and Paulay (1975), Krauthammer et al. (1986, 1990) proposed more realistic hysteretic looping as shown in Figure 8.14a.

If the maximum dynamic displacement does not exceed the yield displacement (point A), the beam is assumed to behave elastically and the unloading–reloading occur along line AA about the zero displacement (point O). If the maximum displacement exceeds the yield displacement, the plastic deflection follows the nonlinear portion of the curve (AC). If the flexural failure (point C) is not reached and unloading occurs, the positive unloading path is assumed to follow the straight line BD parallel to the segment OA. If negative unloading occurs beyond point D (i.e., the beam rebounds in the opposite direction), the unloading path is assumed to follow the straight line DB, where B is a point of equivalent damage to that at point B (for symmetric curves in the positive and negative directions, B is a mirror image of point B).

All loading, unloading, or reloading paths are assumed to remain parallel to line OA and subsequent negative unloading will follow the straight line defined by the last point on the displacement axis and the mirror image of the last maximum point on the (negative or positive) resistance curve. It is obvious that the unloading–reloading paths described above determine the amount of internal damping from hysteretic energy dissipation.

8.6.4 Direct Shear Behavior

This phenomenon was discussed in Chapter 5, and only a brief summary is presented here. Kiger (1980–1984) and Slawson (1984) noted that RC slabs exhibited two distinct types of behaviors under severe blast loading. While the familiar flexural response was observed for some structures, one group experienced a special form of shear failure that occurred at an early time — at about

348   Modern Protective Structures

of a millisecond after the load arrival and before any appreciable dynamic flexural response could develop. Furthermore, a flexural response was hardly observed for structures that failed by direct shear. Likewise, shear failures were not observed for structures that failed by flexure at a much later time. The shear failure surface observed in these experiments looked quite similar to the interface shear failure along the vertical shear plane. This type of failure is commonly referred to as *direct shear* and is characterized by sliding (or slipping) or large displacements along the interface shear plane (Krauthammer et al., 1986).

Furthermore, to account for the reversal of loading, the unloading–reloading of the load deflection curve follows the same procedures for defining the unloading–reloading of a flexural resistance curve and is identical to the approach adopted by Krauthammer et al. (1986) and ASCE (1999). One possible load path is shown as dotted lines in Figure 8.15.

8.7 DYNAMIC STRUCTURAL MODEL

While real structures have infinite degrees of freedom, it is often advantageous to model a structure (usually specific structural elements) as a single-degree-of-freedom (SDOF) system. Although SDOF may not adequately describe the detailed response of the structure, SDOF formulation can provide a rapid and easy solution and often gives a designer valuable (but limited) information about dynamic characteristics (fundamental frequency, maximum response, etc.) of the system. The results are often used to aid design activities in preparation of analysis using more advanced methods (Krauthammer, 1998).

In generating a single numerical P-I diagram, a large number of dynamic analyses must be performed to locate and adequately define the threshold curves. While one could employ more advanced formulations like the Timoshenko beam or Mindlin plate for dynamic analysis, a significantly higher computation effort is required by these formulations for a single analysis (Krauthammer et al., 1994), thus making it less attractive for numerical analysis of P-I diagrams.

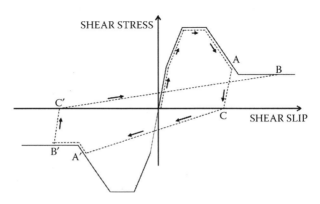

FIGURE 8.15 Dynamic resistance function for direct shear behavior.

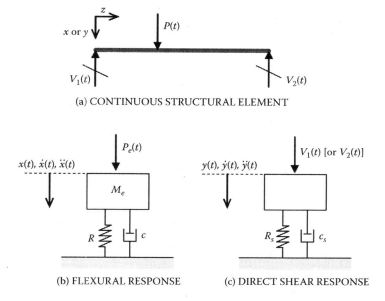

FIGURE 8.16 Equivalent SDOF systems for structural element (Krauthammer et al., 1990).

Krauthammer et al. (1990) presented an analytical technique for obtaining continuous, quantitative descriptions of the dynamic structural responses of impulsively loaded RC beams and one-way slabs. The technique, based on the well-known equivalent SDOF approach as discussed in Chapter 6, utilizes variable response-dependent SDOF parameters (e.g. equivalent mass, equivalent load, etc.) rather than the traditional constant parameters. More importantly, while most equivalent SDOF approaches for one-way spanning RC elements employ only a single mode of response (usually flexure), Krauthammer et al. (1990) proposed two loosely coupled equivalent SDOF systems (Figure 8.16) to better and more accurately capture specific behaviors (i.e., flexure coupled with diagonal shear and direct shear) of the RC element under impulsive loading. The two SDOF systems can be employed, as discussed next.

8.7.1 Flexural Response

8.7.1.1 Equation of Motion

The governing differential equation of motion for the flexural equivalent SDOF system (Figure 8.16b) was presented in Chapter 6 as follows (Krauthammer et al., 1990):

$$\ddot{x}(t) + 2\xi\omega'\dot{x}(t) + \frac{R}{M_e} = \frac{P_e(t)}{M_e} \tag{8.18}$$

where $x(t)$, $\dot{x}(t)$, and $\ddot{x}(t)$ are the flexural displacement, velocity, and acceleration, respectively, M_e is the equivalent mass of the system, R is the flexural dynamic resistance function, ω' is the flexural damped natural circular frequency, ξ is the flexural damping ratio, and $P_e(t)$ is the equivalent forcing function.

It is obvious that the above relationship is nonlinear because the variable system parameters (equivalent mass, flexural resistance, etc.) are dependent on the system response and hence require the use of a numerical integration technique. As discussed in Chapter 6, such analyses are frequently performed using the Newmark β method introduced and explained in that chapter. The variable equivalent mass and loading functions are obtained by use of response-dependent transformation factors, as discussed in Chapter 6 and summarized next.

8.7.1.2 Transformation Factors

To convert a continuous structural element into an equivalent SDOF system, it is necessary to evaluate the equivalent parameters (equivalent mass and equivalent loading function) of a system using transformation (mass, load, and resistance) factors. These constants are evaluated based on an assumed deformation function of the actual structure and often are taken to be the same as those resulting from the static application of dynamic loads. Traditionally, two deformation functions are assumed for responses in both perfectly elastic and rigid plastic ranges and these functions are assumed to remain constant throughout the response range.

Krauthammer et al. (1990) proposed that these factors be evaluated based on actual deformation information at every load step (or time step) of the dynamic response. To do this, the deformed configuration (deflections, slopes, curvatures, etc.) of the structural element at every load step must be computed based on the cross-sectional moment–curvature relationship discussed earlier. The advantage of this method over closed-form solutions is the elimination of the need to explicitly define formulations (or transformation factors) for every type of support condition conceivable. Furthermore, the approach can compute the factors in transition period when behavior changes from elastic to plastic — an aspect normally ignored in the conventional method utilizing constant transformation factors.

At each load step, the normalized deflected shape function $(f(z)_i)$ is computed as:

$$\phi(z)_i = \frac{x(z)_i}{(x_{\max})_i} \tag{8.19}$$

where $x(z)_i$ and $(x_{max})_i$ are the deflected shape function and maximum deflection, respectively, at load step I. From Equation (8.19), the load and mass transformation factors can be calculated, respectively, at every load step (or time step in a dynamic analysis) as follows:

$$(K_L) = \frac{(P_e(t))}{P(t)} i = \frac{1}{L} \int_0^L \phi(z)_i \, dz \tag{8.20}$$

$$(K_M) = \frac{(M_e)}{M} i = \frac{1}{L} \int_0^L (\phi(z)_i)^2 \, dz \tag{8.21}$$

where M is the total lumped mass of the system and $P(t)$ is the actual load history of the forcing function.

8.7.2 Direct Shear Response

8.7.2.1 Equation of Motion

For the direct shear response, as noted earlier, the nonlinear differential equation of motion for the equivalent SDOF system was given by Krauthammer et al. (1990):

$$\ddot{y}(t) + 2\xi_s \, \omega'_s \, \dot{y}(t) + \frac{R_s}{M_s} = \frac{V(t)}{M_s} \tag{8.22}$$

where $y(t)$, $\dot{y}(t)$, and $\ddot{y}(t)$ are the direct shear slip, velocity and acceleration, respectively, M_s is the equivalent shear mass, R_s is the dynamic resistance function for direct shear response, ω_s is the natural circular frequency for direct shear response, x_s is the direct shear damping ratio, and $V(t)$ is the dynamic shear force (or reaction). This nonlinear equation for direct shear is also solved numerically using the Newmark β method. A brief discussion of direct shear mass, damping, and dynamic shear force is presented next.

8.7.2.2 Shear Mass

The equivalent shear mass is computed based on the assumed mode and deformed shape (hence distribution of inertia forces) of a structural element with direct shear failure. If failure is assumed to occur at one end of the support first, the deformed shape of the beam can be assumed to be triangular (Figure 8.17) and

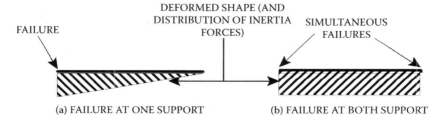

FIGURE 8.17 Deformed shape for direct shear response.

the shear mass can be computed as one-half of the total mass of the beam. On the other hand, for simultaneous shear failures at both supports, the shear mass is the total mass of the structural element. For localized failures under load, the equivalent mass can be computed based on the results of Ross et al. (1981) and Ross and Rosengren (1985).

8.7.2.3 Dynamic Shear Force

The dynamic shear force or reaction can be obtained by considering dynamic equilibrium of a structural element, together with the assumption that the response and the inertia forces along the element span have the same distribution as the elastic deflected shape. Biggs (1964) documented analytical expressions for the dynamic reactions of beams. Ideal boundary conditions are summarized in Table 8.5.

As seen above, the dynamic reactions $V(t)$ are (normally) expressed in terms of the forcing function $P(t)$ and the dynamic resistance function $R(t)$. It is also possible to express the same dynamic reaction in terms of forcing function and inertia force (i.e., product of mass and acceleration). To illustrate this, consider the dynamic equilibrium of a perfectly elastic beam dynamically loaded at mid-span, as shown in Figure 8.18.

The displacement along the length of the beam is assumed to be proportional to the ordinate of the assumed static deflected shape function $f(z)$,

$$x(z,t) = x_{reference}(t)\phi(z) \tag{8.23a}$$

where

$$\phi(z_{reference}) = 1 \tag{8.23b}$$

TABLE 8.5
Analytical Expressions for Dynamic Reaction of Beam

No.	Loading Type	Strain Range	Dynamic Reaction V(t)
1a	Uniformly distributed	Elastic	$0.39R(t) + 0.11P(t)$
1b		Plastic	$Q.38R_m + 0.12P(t)$
2a	Point load at mid-span	Elastic	$0.78R(t) - 0.28P(t)$
2b		Plastic	$0.75R_m - 0.25P(t)$
3a	Point load at one-third	Elastic	$0.525R(t) - 0.025P(t)$
3b		Plastic	$0.52R_m - 0.02P(t)$

R_m = maximum plastic resistance.

Source: Biggs, 1964.

Pressure–Impulse Diagrams and Their Applications

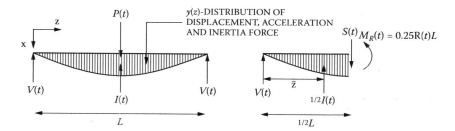

FIGURE 8.18 Dynamic equilibrium of forces (Soh and Krauthammer, 2004).

$z_{reference}$ is the point along the beam where the response ($x_{reference}$) is measured. The acceleration, hence inertia force, along the beam is assumed to follow the same distribution:

$$\ddot{x}(z,t) = \ddot{x}_{reference}(t)\phi(z) \qquad (8.24)$$

The application point of the inertia force for half of the beam with respect to the left support can be seen in Figure 8.18.

$$\bar{z} = \frac{\int_0^{0.5L} z\phi(z)dz}{\int_0^{0.5L} \phi(z)dz} \qquad (8.25)$$

Taking dynamic force and moment equilibrium for half of the beam we have

$$0.5I(t) + V(t) = S(t) \qquad (8.26)$$

$$0.5I(t)\,\bar{z} + 0.25R(t)L = 0.5LS(t) \qquad (8.27)$$

where $I(t)$ is the total inertia force acting on the beam, $S(t)$ is the dynamic shear force at mid-span, and $R(t)$ is the dynamic elastic resistance of the beam as defined in Biggs (1964). Since the distribution of the inertia forces is symmetrical about the mid-span of the beam:

$$S(t) = 0.5P(t) \qquad (8.28)$$

Solving for $V(t)$ using Equations (8.26) through (8.28), we have

$$V(t) = S(t) - \left(\frac{0.5S(t)L - 0.25R(t)L}{\bar{z}}\right) = 0.5P(t) - \frac{L}{\bar{z}}(0.25P(t) - 0.25R(t))$$

or

$$V(t) = \frac{1}{4\bar{z}}\left(2\bar{z} - L\right)P(t) + \frac{L}{4\bar{z}}R(t) \qquad (8.29)$$

In our example, the deflected shape function for a simply supported elastic beam loaded at mid-span (Biggs, 1964) can be shown to be

$$\phi(z) = \frac{z}{L^3}(3L^2 - 4z^2) \qquad \text{for} \qquad z \leq \frac{L}{2} \qquad (8.30a)$$

where

$$\phi\left(z_{reference} = \frac{L}{2}\right) = 1 \qquad (8.30b)$$

Substituting the shape function into Equation (8.25), we have

$$\bar{z} = \frac{\int_0^{0.5L} z\phi(z)dz}{\int_0^{0.5L} \phi(z)dz} = \frac{\int_0^{0.5L} z\frac{z}{L^3}(3L^2 - 4z^2)dz}{\int_0^{0.5L} \frac{z}{L^3}(3L^2 - 4z^2)dz} = \frac{\left(L^2z^3 - \frac{4}{5}z^5\right)\Big|_0^{0.5L}}{\left(\frac{3}{2}L^2z^2 - z^4\right)\Big|_0^{0.5L}} \qquad (8.31)$$

$$= \frac{0.5^3 - (0.2)0.5^5}{(1.5)0.5^2 - 0.5^4} = \frac{8L}{25}$$

The dynamic reaction works out to be

$$V(t) = \frac{1}{4\left(\frac{8L}{25}\right)}\left(2\left(\frac{8L}{25}\right) - L\right)P(t) + \frac{L}{4\left(\frac{8L}{25}\right)}R(t)$$

or

$$V(t) = -0.28125 P(t) + 0.78125 R(t) \qquad (8.32)$$

Pressure–Impulse Diagrams and Their Applications

The above expression for the dynamic shear force agrees with that documented by Biggs (1964) (see Table 8.4, No. 2a). It should be pointed out that the dynamic reactions at the supports are not identical to the reaction (which is equal to the resistance) for the equivalent SDOF system of the same beam element. From Equation (8.26), the inertia force is thus given by

$$I(t) = 1.5625(P(t) - R(t)) \tag{8.33}$$

It should be obvious that the inertia force always has the following form,

$$I(t) = C(P(t) - R(t)) \tag{8.34}$$

where C is a coefficient, depending on loading condition and support conditions. This make sense; for static loading where $P(t)$ equals $R(t)$, $I(t)$ must correspondingly be zero.

Next, the inertia force $I(t)$ is reformulated as a function of mass and acceleration. The beam inertia force is the sum of the inertia forces acting along the beam, i.e.,

$$I(t) = \int_0^L m\,\ddot{x}(z,t)\,dx \tag{8.35}$$

where m is the uniformly distributed mass per unit length (L) of the beam. Since the distribution of acceleration is known from Equation (8.24), we have

$$I(t) = \int_0^L m\,\ddot{x}_{midspan}(t)\phi(z)\,dz$$

$$= \int_0^L \frac{M_t}{L}\,\ddot{x}_{midspan}(t)\phi(z)\,dz$$

or

$$I(t) = K_{IL} M_t\, \ddot{x}_{midspan}(t) \tag{8.36}$$

where M_t is the total mass of the beam, $\ddot{x}_{midspan}(t)$ is the acceleration of the beam at mid-span, and K_{IL} is defined as the inertia load factor (or load factor for the distributed inertia force). Substituting the shape function $f(z)$ from Equation (8.30), the inertia load factor works out to be

$$K_{IL} = \frac{\int_0^{0.5L} \phi(z)dz}{L} = \frac{\int_0^{0.5L} \frac{z}{L^3}(3L^2 - 4z^2)dz}{L}$$

$$K_{IL} = \frac{2}{L^4}\left[\frac{3}{2}L^2z^2 - z^4\right]_0^{0.5L} = 2\left[\left(\frac{3}{2}\right)\left(\frac{1}{4}\right) - \frac{1}{16}\right] = 0.625$$

and therefore,

$$I(t) = 0.625 M_t \ddot{x}_{midspan}(t) \tag{8.37}$$

Now, substituting Equation (8.32) into Equation (8.26), the same dynamic reaction can be expressed as a function of mass and acceleration:

$$V(t) = 0.5P(t) - 0.5(0.625M_t)\ddot{x}_{midspan}(t) \tag{8.38}$$

It can be further shown that for load applied at any points along the beam (Figure 8.18), the general expressions for the dynamic reactions at the supports (ignoring any external mass attached to the beam) are

$$V(t)_1 = \gamma_1 P(t) - \gamma_1' K_{IL} M_t \ddot{x}(t) \tag{8.39a}$$

$$V(t)_2 = \gamma_2 P(t) - \gamma_2' K_{IL} M_t \ddot{x}(t) \tag{8.39b}$$

where γ_1, γ_2 are defined as the load proportionality factors and γ_1' and γ_2' are the inertia proportionality factors (the 1 and 2 subscripts denote left and right supports, respectively). These proportionality factors are dependent on the location of the concentrated load, the geometric properties of the beam, and the support (boundary) conditions. It is obvious that these factors are all equal to half for the case of a simply supported elastic beam loaded at mid-span. Furthermore, unless the beam is assumed to exhibit perfectly elastic behavior, these factors are not constant and must be computed numerically. It should be pointed out that the system responses (i.e., acceleration, displacement, etc.) are measured at the point of load application. Krauthammer et al. (1990) proposed a slightly different formulation to compute dynamic reactions:

$$V(t)_{1i} = \gamma_{1i} P(t) + \gamma_{1i}'(K_{IL})_i M_t \ddot{x}(t) \tag{8.40a}$$

$$V(t)_{2i} = \gamma_{2i} P(t) + \gamma_{2i}'(K_{IL})_i M_t \ddot{x}(t) \tag{8.40b}$$

where the subscript i denotes the load step number.

Pressure–Impulse Diagrams and Their Applications

8.7.3 Summary

Traditionally, the dynamic response of an RC beam has been assumed to be dominated by the flexural mode. Based on experimental observation, RC beams subjected to severe impulsive loading have been shown to also exhibit direct shear modes of failure. The present study employs two loosely coupled equivalent SDOF systems, together with reliable material and constitutive models, to incorporate both flexural and direct shear modes of (dynamic) response. The dynamic flexural model is developed with nonlinear stress distribution of both unconfined and confined concrete and accounts for the effects of diagonal shear, axial compression, web reinforcement, and load rate.

The dynamic direct shear model is based on a static model that describes the interface shear transfer in RC members having well anchored main reinforcements in the absence of axial forces and is able to account for effects of compressive stresses and loading rates on concrete shear strength. In transforming the continuous structural element into the equivalent flexural SDOF system, variable response-dependent transformation factors are employed in lieu of traditional constant factors. Based on dynamic equilibrium of an elastic beam loaded at midspan, an expression for the dynamic reaction was derived and expressed in inertia terms. The dynamic material and structural models discussed are implemented into the P-I analyses of RC beams subjected to transient concentrated loading.

8.8 APPLICATION EXAMPLES FOR SDOF AND P-I COMPUTATIONAL APPROACHES

8.8.1 SDOF and P-I Computations for Reinforced Concrete Beams

The material and structural models discussed earlier were incorporated by Soh and Krauthammer (2004) into the numerical P-I analyses of actual reinforced concrete beams for which test data were available (Feldman and Siess, 1958). The approach described previously in this chapter is illustrated for beam 1-c in that report, as discussed next.

The beam geometry and idealized load pulse are shown in Figure 8.19 and Figure 8.20, respectively. Also, the properties of the tested beams are provided in Table 8.6 and the numerical parameters for the analysis are shown in Table 8.7. Based on the dynamic analysis approach presented in Chapter 5 and previously in this chapter, the moment–curvature and load deflection relationships for Beam 1-c are shown in Figure 8.21a and Figure 8.21b, respectively. Two curves were plotted for each relationship, one representing flexural response only, and the other including a correction for diagonal shear based on the discussion in Chapter 5 (Krauthammer et al., 1987).

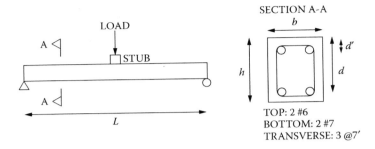

FIGURE 8.19 Reinforced concrete beam C1 (Soh and Krauthammer, 2004).

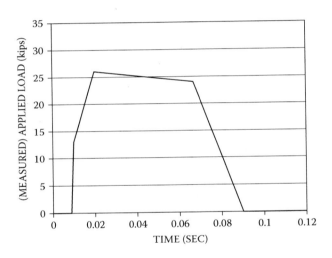

FIGURE 8.20 Load function for beam C1 (Soh and Krauthammer, 2004).

Figure 8.22 shows the corresponding (numerical) P-I diagram for beam 1-c subjected to the idealized load pulse. Note that the peak force and impulse were used to define the axes, as compared to the use of the traditional dimensionless approach. Such representation is more useful for design purposes because the results represent a specific set of materials and structural and loading conditions. No appreciable difference is noted for the impulsive asymptotes from the correction for diagonal shear. This was expected because the effect of diagonal shear on these beams was small (due to their slenderness and adequate shear reinforcement) and the reduction in flexural strain energy was also small. The quasi-static values differed slightly, corresponding to a reduction in static load capacity.

As mentioned earlier, traditional P-I curves for one-way spanning RC elements are obtained based on the assumption that flexure mode dominates the failure. The P-I curves generally have simple hyperbolic shapes as noted above. On the other hand, the threshold curve for direct shear response has not been documented in the

TABLE 8.6
Properties of Beam 1-c

	Properties	Symbol	Values				
			C-1	G-1	H-1	I-1	J-1
1	Beam span	L	106″				
2	Height	h	12″				
3	Width	b	6″				
4	Depth to tension steel	d	10.0625″				
5	Depth to compression steel	d'	1.5″				
6	Tension steel area	A_{st}	1.20 in³ (2 #7)				
7	Compression steel area	A_{sc}	0.88 in³ (2 #6)				
8	Shear reinforcement	A_{sv}	#3 @ 7″ U-stirrup		#3 @ 7″ closed welded		
9	Tension steel						
a	Yield stress (ksi)	f_{yt}	46.08	47.75	47.17	47.00	47.42
b	Yield strain (×10⁻³)	ε_{yt}	1.60	1.50*	1.40	1.40	1.50*
c	Strain hardening strain (×10⁻³)	ε_{sht}	14.4	14.0*	12.5	15.0	14.0*
10	Compression steel						
a	Yield stress (ksi)	f_{yc}	46.70	48.30	47.61	47.95	48.86
b	Yield strain (×10⁻³)	ε_{yc}	1.55*	1.55*	1.50	1.55*	1.60
c	Strain hardening strain (×10⁻³)	ε_{shc}	13.5*	13.5*	15.0	13.5*	12.0
11	Steel ultimate stress (ksi)	f_u	72.0*				
12	Steel ultimate strain	ε_{su}	0.15*				
13	Concrete comp. stress (psi)	f'_c	5,835	6,388	5,963	6,488	6,000
14	Stirrup yield stress	f_{ys}	48.67*	49.50	46.90	49.60	48.67*

* Estimated based on available test data.

Source: Soh and Krauthammer, 2004.

TABLE 8.7
Numerical Analysis Data

Beam Mass (lbs-sec²/ft)	M_t	0.0017
Flexural Damping Ratio (%)	x	2.0
Shear Damping Ratio (%)	ξ_s	0.0
Rate Enhancement Factor		1.25

Source: Soh and Krauthammer, 2004.

literature, and one of the objectives of this study was to investigate its nature in greater detail through the use of this numerical technique. The P-I curve for direct shear response has been added to the previous P-I curve, as shown in Figure 8.23. For the given load pulse with a peak of about 26.3 kips and impulse of about 1.6

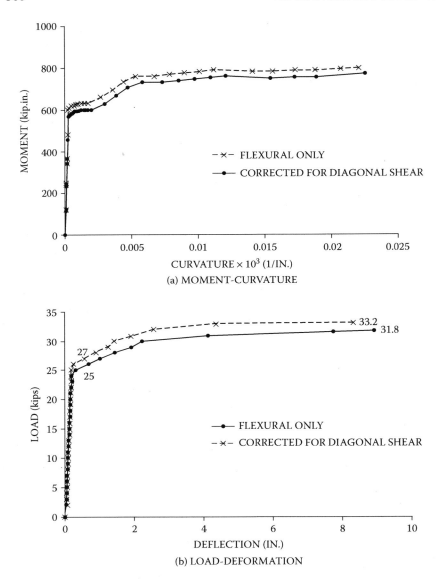

FIGURE 8.21 Moment versus curvature and load versus deflection for beam C1 (Soh and Krauthammer, 2004).

kips/sec, the beam will respond below the flexural curve and not fail (a residual deflection of 3 in. was reported by Feldman and Siess, 1958).

In the quasi-static loading regime, a significantly larger peak load (six times) is required to initiate simultaneous flexural and direct shear failures for a given load impulse. More interestingly, the numerical solution shows that direct shear will be the dominating failure mode for a peak load larger than about 320 kips

Pressure–Impulse Diagrams and Their Applications

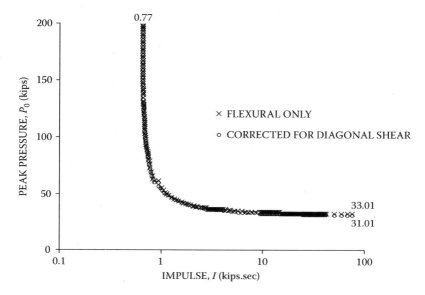

FIGURE 8.22 P-I Curve for flexural response including diagonal shear correction (Soh and Krauthammer, 2004).

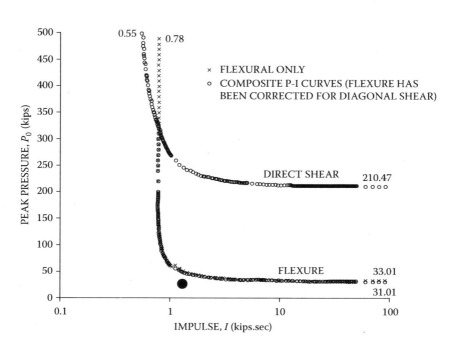

FIGURE 8.23 P-I diagrams for flexure, flexure with diagonal shear, and direct shear (Soh and Krauthammer, 2004).

and an impulse lower than 0.8 kips/sec. Further, direct shear cannot dominate the behavior for peak loads lower than about 200 kips. Loads with peak values larger than 200 kips and impulse values larger than about 0.8 kips/sec can cause a combined flexure and direct shear failure. In other words, if only a flexural mode of response is assumed, the P-I curve fails to detect failure of the RC beam due to direct shear that would be caused by high intensity short duration loads.

This limitation was overcome artificially by shifting the flexural curve to the left so that the impulsive asymptote reduces to the value for an observed direct shear failure, as noted by Oswald and Skerhut (1993). The current approach, however, does not require such artificial corrections because it can consider direct shear in the analysis.

Soh and Krauthammer (2004) also addressed the issue of dynamic reactions at the supports and showed that one could use both a previous approach (Biggs, 1964; Krauthammer et al., 1987) and a modified approach developed in the more recent study. Since the P-I curve for direct shear response depends on the time histories for the dynamic reactions, they showed that the accuracy of the numerical approach to compute dynamic reactions could affect the generation of the P-I curve. Although the approach by Krauthammer et al. (1987) provided slightly more accurate results, additional study of these approaches is required before a better conclusion can be reached.

The illustration of the application of the advanced SDOF computer code DSAS (Krauthammer et al., September 2005) for the analysis of the same beams (Table 8.6) is based on a study by Kyung and Krauthammer (June 2004). The peak values for the central deflections and support reactions are presented in Table 8.8. These results show clearly that the SDOF approach provided very good estimates for the peak responses and these can be used in support of design activities. Clearly the combination of SDOF time domain calculations and numerical P-I diagram assessments can provide accurate data for such activities.

TABLE 8.8
Peak Deflections and Support Reactions

Beam	Peak Test Deflection Δ_T (in.)	Peak Test Reaction Q_T (kips)	Peak Computed Deflection Δ_C (in.)	Peak Computed Reaction Q_C (kips)	Δ_C/Δ_T	Q_C/Q_T
1-c	3.0	18.0	2.9	16.2	0.97	0.90
1-g	4.1	19.3	4.0	16.6	0.98	0.86
1-h	8.9	18.0	8.6	16.8	0.97	0.93
1-I	10.0	18.0	10.0	17.1	1.00	0.95
1-j	9.5	19.0	8.5	17.0	0.90	0.86
Avg.					0.96	0.91
Std. Dev.					0.04	0.04

Source: Kyung and Krauthammer, 2004.

Pressure–Impulse Diagrams and Their Applications

TABLE 8.9
Test Case Parameters

Parameters	DS1-1	DS2-5	DS1-3	DS2-3
Length of slab (m)	1.2	1.2	1.2	1.2
Width of slab (m)	1.2	1.1	1.2	1.1
Span-to-effective-depth ratio	10	7	10	7
Approximate concrete strength (MPa)	30	50	30	50
Percentage steel at each face (%)	1	1.2	1	0.75
Charge density (kg/m^3)	21.9	25.6	14.6	18.3

8.8.2 SDOF AND P-I COMPUTATIONS FOR REINFORCED CONCRETE SLABS

Case studies were conducted by Ng and Krauthammer (2004) to test the proposed procedure's capability to evaluate reinforced concrete slabs subjected to complex transients loads. Data from experimental work carried out by Slawson (1984) on reinforced concrete roof slabs subjected to explosive loading were used. Four cases with parameters shown in Table 8.9 are illustrated next.

Test case DS1-1 — The slab in test DS1-1 was subjected to high load during the experiment and it was totally severed from its supports along vertical planes. Slawson (1984) noted that all the concrete except for the center 12 in. (0.3 m) was crushed such that it fell away from the reinforcement.

Figure 8.24 shows the generated P-I diagram of experimental results with an insert of the post-test view. The experimental P-I combination's position on the right side of the diagram agreed with the observations of the experiment. The numerical results showed that the slab failed in direct shear at 1.26 msec after being loaded beyond the threshold point (E on Figure 8.15). The displacement history for the direct shear degree of freedom in plotted in Figure 8.25(a) and the direct shear resistance of the slab is shown in Figure 8.25(b). The numerical analysis also indicated that the slab went into tensile membrane mode but did not fail in flexure.

Test case DS2-5 — This test was conducted on a slab with a smaller span-to-effective-depth ratio, higher concrete strength, and an extremely high load. Permanent deflection at the mid-span of the slab was approximately 12 in. (0.3 m). The concrete was broken up over the entire span and most of the concrete cover spalled from the bottom of the roof slab, exposing the reinforcement. The generated P-I diagram with the experimental results was plotted and is shown in Figure 8.26; the insert provides a post-test view. The position of the experimental P-I combination on the far right side agreed with the observation that the slab failed in direct shear with extensive flexural response. The displacement histories and resistance functions for both direct shear and flexural degrees of freedom are plotted in Figure 8.27(a) and Figure 8.27(b), respectively.

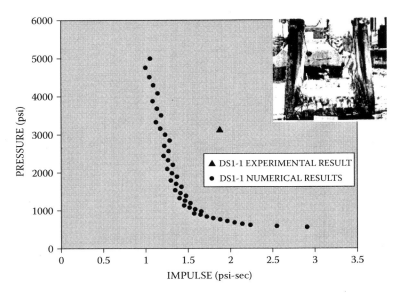

FIGURE 8.24 P-I diagram and damage state for case DS1-1.

Test case DS1-3 — This test was conducted on a slab similar to DS1-1, but at a relatively lower load. The permanent mid-span deflection was about 10 in. (0.25 m). The center 18 in. (0.46 m) of the slab remained relatively flat. The generated P-I diagram with experimental results is plotted in Figure 8.28; the insert shows the post-test view. The pressure and impulse combination measured during the test resided in the same position as the numerical P-I curve.

This case illustrates the condition on a slab subjected to the threshold loading of the numerical procedure. The numerical procedure evaluated the slab as "safe." The slab responded predominantly in shear but did not fail. The slab oscillated in shear about a permanent plastic deformation with a larger period of a weaker system. The slab went into compressive membrane mode but not tensile membrane mode. The displacement histories for shear and flexure are shown in Figure 8.29a and Figure 8.29b, respectively.

Test case DS2-3 — This test was conducted on a stronger slab than test DS2-5, but at a lower load. The structural response appeared to be predominantly in direct shear with a permanent mid-span deflection of only about 4 in. (0.1 m). Figure 8.30 is the generated pressure-impulse diagram; an insert shows the post-test view. The pressure and impulse combination measured just crossed the threshold curve to the right. The slab was assessed to have failed in direct shear. Although the slab went into the compressive membrane mode, the flexural resistance showed that its capacity was not fully utilized. The displacement histories in shear and flexure are shown in Figure 8.31a and Figure 8.31b, respectively.

Pressure–Impulse Diagrams and Their Applications

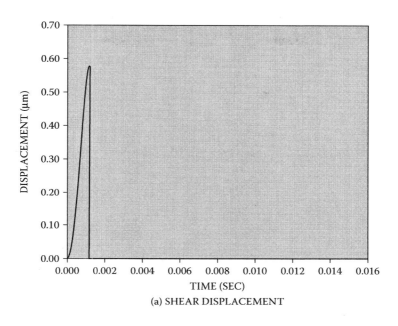

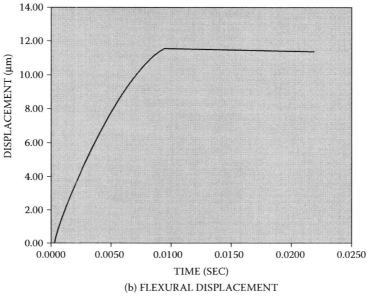

FIGURE 8.25 Direct shear response for case DS1-1.

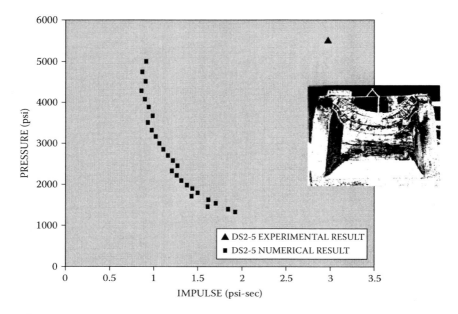

FIGURE 8.26 P-I diagram and damage state for case DS2-5.

8.8.3 SUMMARY

This chapter addressed the important topic of P-I diagrams and their potential for both design and damage assessment. The approaches presented here go beyond the traditional empirical derivations of such diagrams, and they enable one to derive engineering and behavioral based P-I diagrams for any structural system. The following conclusions were drawn by Soh and Krauthammer (2004), Ng and Krauthammer (2004), and Kyung and Krauthammer (2004):

- These studies illustrated the derivation of closed-form P-I solutions for a simple mechanical system subjected to simple transient loads based on structural dynamics principles.
- For idealized SDOF systems, the results from the numerical approach agreed very well with available closed-form solutions. It appears that the numerical approach was able to describe correctly and accurately the threshold curves for idealized SDOF system with reasonable computation effort.
- These studies demonstrated the ability of the present numerical approaches to analyze SDOF systems with complicated loads and highly nonlinear resistance functions.

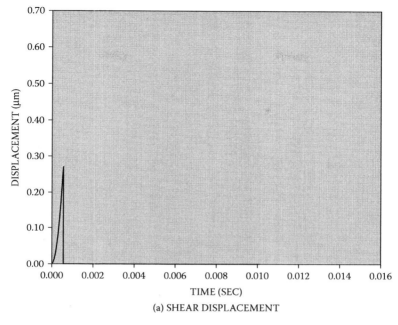
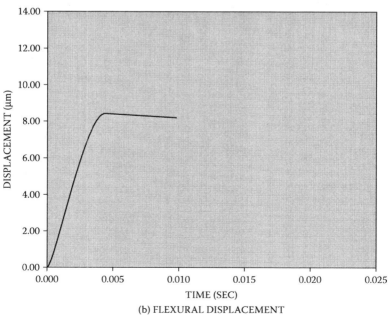

FIGURE 8.27 Direct shear and flexural response for case DS2-5.

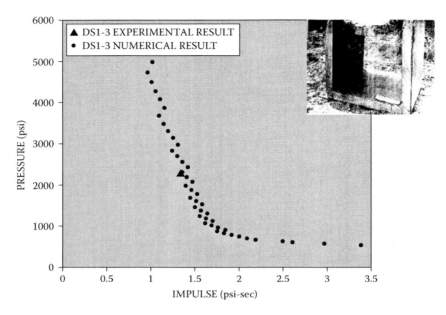

FIGURE 8.28 P-I diagram and damage state for case DS1-3.

- Unlike traditional P-I diagrams for RC structures for which one normally assumes a single response mode, usually flexure without shear effects, the present approaches are able to generate rational P-I diagrams with flexural, diagonal shear, and direct shear modes of responses. It was shown that for certain high intensity short duration loads, the assumption of a single flexural mode of response may be inadequate.
- The P-I curves for direct shear generally follow the same hyperbolic profiles as the threshold curves for flexural responses. However, the influence of a finite rise time on the P-I curves is significantly different for the two types of responses. The results of the responses to a triangular load with finite rise time for certain RC beams indicates a minimum loading rate below which no direct shear failure will occur.
- The proposed methodologies generate rational and physics-based P-I diagrams that are useful tools for structural design and assessment. Complicated time-dependent loadings and more than one non-linear structural response mechanism can be considered. The procedures are efficient and reasonably accurate. They offer alternative methods to the expensive practice of empirical P-I curve fitting.

Pressure–Impulse Diagrams and Their Applications

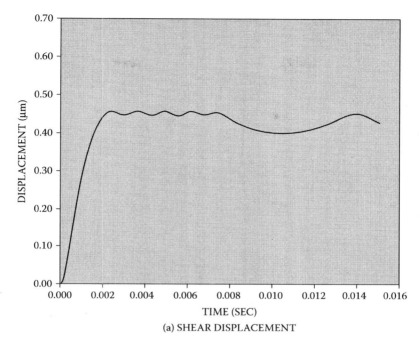

(a) SHEAR DISPLACEMENT

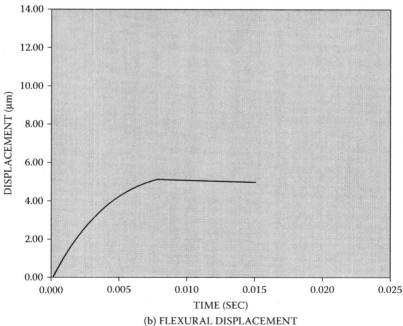

(b) FLEXURAL DISPLACEMENT

FIGURE 8.29 Direct shear and flexural response for case DS1-3.

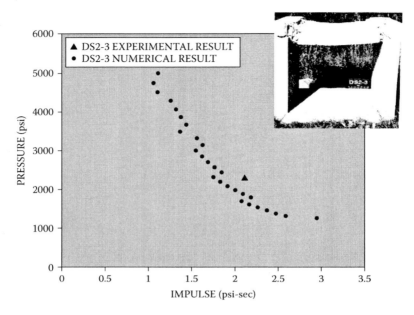

FIGURE 8.30 P-I diagram and damage state for case DS2-3.

- Although the methodologies are approximate, the position of a P-I combination in the diagram is a strong indication of the likely performance of a reinforced concrete beam or slab to a specific loading.
- The accurate approach for fully nonlinear SDOF computations has been validated.

Pressure–Impulse Diagrams and Their Applications

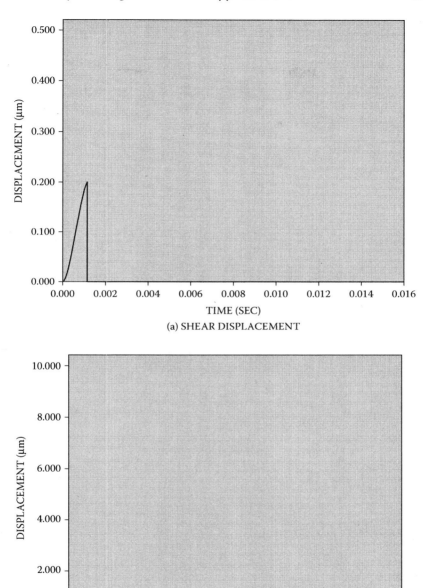

FIGURE 8.31 Direct shear and flexural response for case DS2-3.

9 Progressive Collapse

9.1 INTRODUCTION

9.1.1 Progressive Collapse Phenomena

Explosive loading incidents have become serious problems that must be addressed frequently. Besides the immediate and localized blast effects, one must consider the serious consequences associated with progressive collapse that could affect people and property in an entire building. Progressive collapse occurs when a structure has its loading pattern or boundary conditions changed such that structural elements are loaded beyond their capacity and fail. The residual structure is forced to seek alternative load paths to redistribute the load applied. As a result, other elements may fail, causing further load redistribution.

The process will continue until the structure can find equilibrium either by shedding load as a by-product of the failures of other elements or by finding stable alternative load paths. In the past, structures designed to withstand normal load conditions were over-designed and were usually capable of tolerating abnormal loads. Modern building design and construction practices enabled us to build lighter and more optimized structural systems with considerably fewer over-design characteristics.

Progressive collapse became an issue following the Ronan Point incident (HMSO, 1968; Griffiths et al., 1968), when a gas explosion in a kitchen on the 18th floor of a 22-story precast building caused extensive damage to the entire corner of that building, as shown in Figure 9.1. The particular type of joint detail used in the Ronan Point apartment building relied heavily on joint friction between precast panels. This resulted in a structure that has been termed a "house of cards," indicating that buildings with similar joint characteristics were particularly susceptible to progressive collapse (Breen, 1980). The failure investigation of that incident resulted in important changes in the United Kingdom building code (HMSO, 1976). It requires a minimum level of strength to resist accidental abnormal loading by (1) comprehensive "tying" of structural elements, (2) if tying is not possible, allow the "bridging" of loads over the damaged area (the smaller of 15% of the story area or 70 m^2); (3) if bridging is not possible, ensure that key elements can resist 34 kN/m^2. These guidelines have been incorporated in subsequent British standards (HMSO, 1991; BSI, 1996; BSI, 2000). Although many in the U.K. attribute the very good performance of numerous buildings subjected to blast loads to these guidelines, it is not always possible to quantify how close those buildings came to progressive collapse.

Progressive collapse is a failure sequence that relates local damage to large scale collapse in a structure. Local failure can be defined as a loss of the load-

FIGURE 9.1 Damage at Ronan Point apartment building.

carrying capacity of one or more structural components that are parts of the whole structural system. Preferably, a structure should enable an alternative load-carrying path after a structural component fails. After the load is redistributed, each structural component will support different loads. If any load exceeds the load-carrying capacity of any member, it will cause another local failure. Such sequential failures can propagate through a structure. If a structure loses too many members, it may suffer partial or total collapse. This type of collapse behavior may occur in framed structures such as buildings (Griffiths et al., 1968; Burnett et al., 1973; Ger et al., 1993; Sucuolu et al., 1994; Ellis and Currie, 1998; Bazant and Zhou, 2002), trusses (Murtha-Smith, 1988; Blandford, 1997), and bridges (Ghali and Tadros, 1997; Abeysinghe, 2002).

9.2 BACKGROUND

9.2.1 ABNORMAL LOADINGS

It is estimated that at least 15 to 20% of the total number of building failures are due to progressive collapse (Leyendecker and Burnett, 1976). A notable example of such a failure is the Ronan Point collapse (Griffiths et al., 1968). In the years following that incident, hundreds of engineering articles and reports on these subjects have been published (Breen, 1980).

Since an explosion caused the progressive collapse at Ronan Point, a number of studies were devoted to the relationships of abnormal loadings and progressive collapse (Astbury, 1969; Astbury et al., 1970; Burnett et al., 1973; Mainstone,

Progressive Collapse

1973 and 1976; Burnett, 1974 and 1979; Leyendecker et al., 1975; Leyendecker and Burnett, 1976; Mainstone et al., 1978). An abnormal load is any loading condition a designer does not include in the normal and established practice course design (Gross and McGuire, 1983). Abnormal loadings include explosions, sonic booms, wind-induced localized over-pressures, vehicle collisions, missile impacts, service system malfunctions, and debris resulting from incidents (Burnett, 1974).

Leyendecker and Burnett (1976) discussed plausible sources of abnormal loading events and estimated the risk of the loadings in residential building design with regard to progressive collapse. They concluded that gas explosions occurred with an annual frequency of 1.6 events per million dwelling units, bomb explosions occurred with an annual frequency of 0.22 events per million dwelling units, and vehicular collisions with buildings that caused severe damage occurred with an annual frequency of 6.8 events per million dwelling units.

For certain abnormal loading events, the probability of an event occurring in a building increases with building size. In particular, high-rise buildings tend to be at a higher risk for gas and bomb explosions. In contrast, vehicular collisions affect primarily ground story areas. It was found that the hazard due to a gas explosion exceeds that for a vehicular collision in a building more than about five stories in height (Leyendecker and Burnett, 1976). However, a structural designer has very little control over the probability of an abnormal event. Therefore, structural engineers should consider other approaches to prevent progressive collapse.

9.2.2 Observations

Recent developments in the efficient use of building materials, innovative framing systems, and refinements in analysis techniques may result in structures with lower safety margins. Both the Department of Defense (2002 and 2005) and the General Services Administration (2003) issued clear guidelines to address this critical problem. Nevertheless, these procedures contain assumptions that may not reflect accurately the actual conditions in a structure damaged by an explosive attack (see Figure 9.2) that highlight the very complicated state of damage that must be assessed before correct conditions can be determined. The structural behavior associated with such incidents involves highly nonlinear processes in both the geometric and material domains. One must understand that various important factors can affect the behavior and failure process in a building subjected to an explosive loading event, but these cannot be easily assessed.

The idea of removing a single column as a damage scenario adopted in various design requirements while leaving the rest of the building undamaged is unrealistic. An explosive loading event near a building (e.g., terrorist attacks in London and Oklahoma City) will cause extensive localized damage affecting more than a single column. The remaining damaged structure is expected to behave very differently from the ideal situation. Therefore, it is critical to assess accurately the post-attack behaviors of structural elements that were not removed from the

(a) ST. MARY AXE, APRIL 1992 (b) BISHOPSGATE, APRIL 1993

FIGURE 9.2 Post-incident view of building damage arising from London bombing incidents. (a) St. Mary Axe, April 1992. (b) Bishopsgate, April 1993.

building by the blast loads in their corresponding damaged states. This requires first a fully nonlinear blast–structure interaction analysis, then determining the state of the structural system at the end of this transient phase, and then proceeding with a fully nonlinear dynamic analysis for the damaged structure subjected to only gravity loads. Such comprehensive analyses are very complicated and time consuming, require extensive resources, and are not suitable for design office environments.

Damaged structures may have insufficient reserve capacities to accommodate abnormal load conditions (Taylor, 1975; Gross, 1983). We have few numerical examples of computational schemes to analyze progressive collapse. Typical finite element codes can only be used after complicated source level modifications to simulate dynamic collapse problems that contain strong nonlinearities and discontinuities. Several approaches have been proposed for including progressive collapse resistance in building design. The alternative load path method is a known analytical approach that follows the definition of progressive collapse (Yokel et al., 1989). It refers to the removal of elements that failed the stress or strain limit criteria.

In spite of their analytical characteristics, alternative load path methods are based on static considerations and may not be adequate for simulating progressive collapse behavior. Choi and Krauthammer (2003) described an innovative approach to address such problems by using algorithms for external criteria screening (ECS) techniques applicable to these types of problems. As a part of such ECS, element elimination methods were classified into direct and indirect approaches and compared with each other. A variable boundary condition (VBC) technique was also proposed to avoid computational instability that could occur while applying the developed procedure.

A more comprehensive study of progressive collapse by Krauthammer et al. (2004) addressed the behavior of both two-dimensional and three-dimensional multi-story structures under several damage conditions and highlighted conditions (i.e., combinations of geometric, material, and damage characteristics) that could lead to progressive collapse in moment-resisting steel frame structures.

These two studies presented preliminary theories to analyze progressive collapse of a multi-story steel frame structure. The procedures included stress–strain failure criteria of linear elastic and elastic–perfectly plastic materials. Also included were elastic–plastic with kinematic hardening material models as well as linear and nonlinear column buckling and several types of beam-to-column connections (e.g., hinge, semi-rigid, fixed, and variable moment capacities). That approach is currently being developed further to study the physical phenomena associated with progressive collapse and develop fast running computational algorithms for the expedient assessment of various multi-story building systems.

The 1994 Northridge earthquake highlighted troublesome weaknesses in design and construction technologies of welded connections in moment-resisting structural steel frames. As a result, the U.S. steel construction community embarked on an extensive R&D effort to remedy the observed deficiencies. Around the same time, domestic and international terrorist attacks have become critical issues that must be addressed by structural engineers. Here, too, it has been shown that structural detailing plays a very significant role in the response of a building to a blast (Krauthammer, 1999; Krauthammer et al., 2004).

In blast-resistant design, however, most of the attention during the last half century has been devoted to structural concrete. Since many buildings that could be targeted by terrorists are moment-resisting steel frames, their behaviors under blast are of great interest, with special attention to structural details. Typical structural steel welded connection details currently recommended for earthquake conditions underwent preliminary assessments for their performance under blast effects. The assessments also addressed current blast design procedures to determine their applicabilities for both the design and analysis of such details. The finding highlighted important concerns about the blast resistance of structural steel details and about the assumed safety in using current blast design procedures for structural steel details.

Obviously, one must address not only the localized effects of blast loads, and the idealized behavior of typical structural elements such as columns, girders, etc., but also the behavior of structural connections and adjacent elements that define the support conditions of a structural element under consideration. The nature of blast loads, the behavior of structural connections under such conditions, and progressive collapse are addressed in the following sections to provide the background for current and proposed research activities.

9.3 PROGRESSIVE COLLAPSES OF DIFFERENT TYPES OF STRUCTURES

Progressive collapse may occur in all types of framed structures including steel or concrete framed buildings, space trusses, and bridges. Progressive collapse is not necessarily related to a specific type of structural material. However, some types of materials may be more prone to progressive collapse in that they may have less ductility or weaker connection details than others (Breen and Siess, 1979). Although this chapter will focus on moment-resisting steel frames, several other structural systems will be discussed.

9.3.1 PRECAST CONCRETE STRUCTURES

The risk of progressive collapse of precast concrete large panel and bearing wall structures is greater than the risk for traditional cast-in-place structures (Breen, 1980). Because precast concrete structures are vulnerable to progressive collapse, the problem was studied to address structural integrity and design issues (Speyer, 1976; Burnett, 1979; Hendry, 1979; Pekau, 1982).

The collapse of the Ronan Point building (Griffiths et al., 1968) is an example of this type of failure. The report revealed several deficiencies in the British codes and standards of that time, particularly as they applied to multi-story construction and focused on the lack of redundancy or alternate load paths in the structure. As a consequence of that investigation, British building regulations were changed to require that the design of a multi-story structure either provide an "alternate path" in case of loss of a critical member or have sufficient local resistance to withstand the effects of a gas-type explosion (Breen, 1980).

A more recent British code (Steel Construction Institute, 1990) states that a building should be checked to see whether at each story in turn any single column or beam carrying a column could be removed without causing collapse of more than a limited portion of the building local to the member concerned. The ACI building code (2005) provides a separate chapter covering precast concrete structures. It requires that adequate horizontal, vertical, and peripheral ties be provided to link all structural elements to develop tensile continuity and ductility of the elements. This combination of system continuity and ductility should enable a structure to either absorb the abnormal loads with minimal damage or bridge localized damage as a result of the abnormal load. The provision of general structural integrity will bring the safety of precast large panel structures closer to that of the traditional cast-in-place reinforced concrete buildings (Fintel and Schultz, 1979).

9.3.2 MONOLITHIC CONCRETE STRUCTURES

Cast-in-place concrete construction provides better performance against progressive collapse than precast concrete construction (Fintel and Schultz, 1979) although reinforced concrete slab structures have undergone progressive collapses (Feld, 1964; Leyendecker and Fattal, 1973; Litle, 1975; Lew et al., 1982; Hueste and Wight, 1999). The mechanism most likely to trigger such a collapse under conventional loads is a punching shear failure at an interior column (Hawkins and Mitchell, 1979). The key to preventing such progressive collapse may be to design and detail slabs such that they are able to develop secondary load carrying mechanisms after initial failures occur. If the continuous reinforcement is properly

Progressive Collapse

anchored, the slab can develop tensile membrane action after initial failure. If the final collapse load is higher than the initial failure load, then a means of preventing progressive collapse has been provided (Mitchell and Cook, 1984).

The ACI building code (2005) states that even if top reinforcement is continuous over a support, it will not guarantee integrity without stirrup confinement or continuous bottom reinforcement. This is because top reinforcement tends to pull out of concrete when a support is damaged. Hawkins and Mitchell (1979) concluded that only continuous bottom reinforcement through a column or properly anchored in a wall or beam should be considered effective as tensile membrane reinforcement. If abnormal loads are considered, one must address the removal of one or more exterior or interior columns, probably combined with damage to horizontal structural members.

9.3.3 Truss Structures

Space trusses are highly redundant structures. This means that space truss structures are expected to survive even after losses of several members. However, the failure of the Hartford (Connecticut) Coliseum space roof truss in 1978 (Ross, 1984) showed that this assumption was not always correct. Progressive collapse can occur following the loss of one of several potentially critical members when a structure is subjected to full service loading (Murtha-Smith, 1988).

Force redistributions may cause members to exhibit nonlinear behavior and yield in the case of a tension member or buckle in the case of a compression member (Schmidt and Hanoar, 1979). However, because of strain hardening, a yielded tension member can typically absorb additional force, whereas a compression member will carry lower loads after reaching its buckling load. Thus, a compression member cannot resist additional force but must shed force, and cause additional force redistributions to other members.

The other members may also buckle and cause further force redistributions, and thus failure can progress through a structure to cause collapse. In addition, because the snap-through phenomenon is rapid, dynamic effects can increase the force redistribution intensity further (Murtha-Smith, 1988). Morris (1993) pointed out that neglect of the dynamic response due to member snap-through leads to a significant overestimate of a structure. Malla and Nalluri (1995) presented variations of natural frequencies of the truss structures due to member failure.

9.3.4 Steel Frame Buildings

Steel moment resisting frame structures have similar structural characteristics as truss structures but trusses do not resist moments in the members. Each steel member can have a different failure mode because it will be subject to different load, cross-sectional shape, and material property combinations. Traditionally, the anticipated failure modes were categorized in the following two general domains (Salmon and Johnson, 1996).

Material failures may be either ductile (e.g., plastic deformation or failure induced by high temperature) or brittle (e.g., fracture fatigue failures) and are expected mostly in tension. Steel members show ductile behavior before failure in an ideal condition such as a uniaxial tension test. However, steel members can fail in a brittle manner in a practical case due to an initial flaw or notch and they can lose stiffness in a high temperature induced by a fire. Repeated loading and unloading may eventually result in a fatigue failure even if the yield stress is never exceeded. Also, strain hardening in steel provides additional strength that makes material failure in compression an unlikely event in massive cross-sections. Therefore, steel tends to buckle before it reaches the material failure strain in compression.

Buckling types include flexural column (global) buckling, torsional buckling, lateral torsional buckling, and local buckling. A steel member will buckle if its critical load is exceeded. If a steel member such as an angle, tee, zee, or channel has a relatively low torsional stiffness, it may buckle torsionally while the longitudinal axis remains straight. A steel beam can also buckle under bending without a proper lateral restraint because the flange can be considered as a column when it is subject to compression by bending. Components such as flanges, webs, angles, and cover plates that are combined to form a column section may buckle locally before the entire section achieves its maximum capacity.

Nevertheless, the traditional characterization of failure ignored the role of connections in structural behavior. As a result, the behavior connections under extreme conventional and unconventional loads are not adequately addressed in typical structural design books. As noted previously, this lack of attention to structural steel connections was highlighted by surprising connection failures during the Northridge earthquake of 1994 and the development of new recommendations for steel connections (FEMA, 2000). The behavior of moment-resisting steel connection under blast was also addressed (Krauthammer et al., 2004) and uncovered additional deficiencies that require further design and construction measures to prevent such outcomes. Marchand and Alfawakhiri (2004) presented some information on blast effects and progressive collapse in steel buildings and referred readers to existing design guidelines, as presented next.

9.4 DEPARTMENT OF DEFENSE AND GENERAL SERVICES ADMINISTRATION GUIDELINES

9.4.1 DEPARTMENT OF DEFENSE (DoD) GUIDELINES

9.4.1.1 Design Requirements for New and Existing Construction

The DoD unified facilities criteria (UFC) on progressive collapse (2005) adopted recommendations in the British code cited earlier. The code requires a minimum level of strength to resist accidental abnormal loading be provided by (1) comprehensive "tying" of structural elements; (2) if tying is not possible, allowing

Progressive Collapse

the "bridging" of loads over the damaged area (the smaller of 15% of the story area or 70 m^2); and (3) if bridging is not possible, ensuring that key elements can resist 34 kN/m^2. Also, the UFC classify buildings according to their respective levels of protection (LOP). The progressive collapse design requirements employ two design and analysis approaches: tie forces (TFs) and alternate paths (APs). Details of design recommendations for each LOP for both new and existing construction are presented next.

Buildings with very low levels of protection (VLOP) — Provide adequate horizontal tie force capacity.

Buildings with low levels of protection (LLOP) — Provide both horizontal and vertical tie force capacities.

Buildings with medium and high levels of protection (MLOP and HLOP) — Perform AP analysis.

For all LOP, all multi-story vertical load-carrying elements must be capable of supporting the vertical load after the loss of lateral support at any floor level (i.e., a laterally unsupported length equal to two stories must be used in the design or analysis). In order to prevent slab failure due to upward blast pressure, the slab and floor system must be able to withstand a net upward load of:

$$1.0\,D + 0.5\,L \qquad (9.1)$$

where D is the dead load based on self-weight only and L is the live load, both in kilonewtons per square meter or pounds per square foot. This load is to be applied to each bay, one at a time (not concurrently to all bays), and one should use the appropriate strength reduction factors and over-strength factors.

The DoD UFC (2005) also includes detailed flow charts for the design processes that must be applied to buildings with the various levels of protection, from VLLOP through HLOP.

9.4.1.2 Design Approaches and Strategies

As noted above, the DoD design recommendations (2005) are based on the British approach and they require either the ability to develop sufficient TFs or the ability to bridge over the damaged area through an AP as discussed next.

9.4.1.2.1 Tie Forces

Tie force can be developed if the structural components assembled to form a building are tied together mechanically to enhance continuity, ductility, and, in the case of element failure, allow the development of alternate load paths. Such tie forces must be provided by the various structural elements and their connections based on the prevailing design procedures to carry the specified standard loads expected to be applied to the structure, as illustrated in Figure 9.3.

Several horizontal ties must be provided, depending on the construction type: internal, peripheral, and ties to edge columns, corner columns, and walls. Similarly, vertical ties are required in columns and load-bearing walls. Note, however,

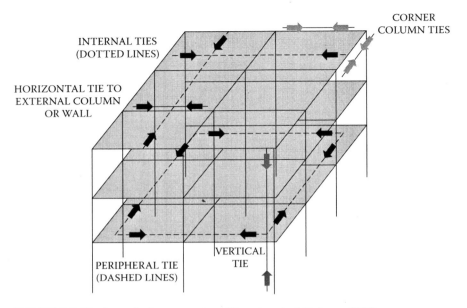

FIGURE 9.3 Tie forces in frame structure (Department of Defense, 2005).

that tie forces are not synonymous with reinforcement ties as defined in ACI 318-05 (2005) for reinforced concrete design. These specifications emphasize that peripheral ties must ensure continuous load paths around the plan geometry and internal ties must ensure continuous load paths from one edge to the other. Vertical ties must be continuous from the lowest level to the highest level of a structure. Although horizontal ties to edge columns and walls do not have to be continuous, they must be adequately anchored back into the structure. All structural members and the connections between such members that form any such load path must be able to resist the appropriate tie forces. Accordingly, the design tie strength (φR_n, in which φ is a strength reduction factor and R_n is the nominal tie strength as defined by the appropriate design code) must be not less than the required tie strength.

9.4.1.2.2 Alternate Path Method

The DoD guidelines (2005) specify use of the AP method in two situations. First, when a vertical structural member is either missing or cannot provide the required tie strength, one must ensure that the forces can be bridged over the damaged area. A designer may use the AP method to determine whether a structure can bridge the forces over the damaged area (e.g., by notionally removing the specific element and analyzing the remaining structure). If the structure does not have the required capacity to bridge forces, the designer must modify the design until such conditions can be satisfied. Second, the AP method must be applied for the removal of specific vertical load-bearing elements for structures that require MLOP or HLOP.

The AP method was developed to follow the load and resistance factor design (LRFD) philosophy (ASCE, 2002) by employing load factor combinations for extreme loading and resistance factors to define design strengths. Accordingly, the design tie strength (φR_n, in which φ is a strength reduction factor and R_n is the nominal tie strength, as defined by the appropriate design code) must be not less than the required tie strength. Although two-dimensional models may be used for a general response and three-dimensional effects can be adequately idealized, it is recommended that three-dimensional models be used. The three analysis procedures permitted are linear static, nonlinear static, and nonlinear dynamic. They are briefly summarized below.

Linear static — The damaged structure is subjected to the full load applied at one time and small deformations and linear elastic materials are assumed. Discrete hinges may be inserted to simulate limit states.

Nonlinear static — The damaged structure is subjected to a load history from zero to the full factored load, and both the material and geometry are treated as nonlinear. The following factored load combinations should be applied to bays immediately adjacent to the removed element and at all floors above the removed element for both linear and nonlinear static analyses:

$$2[(0.9 \text{ or } 1.2)D + (0.5L \text{ or } 0.2S)] + 0.2\,W \tag{9.2}$$

where D is the dead load, L is the live load, S is the snow load, and W is the wind load per Section 6 of ASCE 7 (2002). All are expressed in kilonewtons per square meter or pounds per square foot. In load-bearing wall systems, the adjacent bay is defined by the plan area that spans the removed wall and the nearest load bearing walls. The rest of the structure will be loaded as for nonlinear dynamic analysis, as defined below.

DoD (2005) recommends that at the time of failure (element removal), the load from the failed element defined by Equation (9.2) in linear or nonlinear static analyses be applied instantaneously to the structural area directly below the failed element. The loads from the area supported by the failed element should be applied to an area not larger than the area from which they originated.

Nonlinear dynamic — The selected damage is induced instantaneously in the fully loaded structure, a dynamic analysis is performed to analyze the response, and both the material and geometry are treated as nonlinear. The following factored load combinations should be applied to the entire structure:

$$(0.9 \text{ or } 1.2)D + (0.5L \text{ or } 0.2S) + 0.2\,W \tag{9.3}$$

where D is the dead load, L is the live load, S is the snow load, and W is the wind load per Section 6 of ASCE 7 (2002). All are expressed in kilonewtons per square meter or pounds per square foot.

DoD (2005) recommends that at the time of failure (element removal) the load from the failed element to be applied in nonlinear dynamic analyses be doubled for impact effects, and applied instantaneously to the structural area directly below the failed element. The loads from the area supported by the failed element should be applied to an area not larger than the area from which they originated.

The structural damage definitions (vertical load-bearing structural element removal) depend on the construction material (structural concrete, steel, wood, etc.), as specified in the appropriate chapters of DoD criteria (2005). Such induced damage includes an external column or load-bearing wall removal at recommended critical locations and internal column or load-bearing wall removal at recommended critical locations. Inducing damage at other locations may be required to address the effects of unique geometric features. The corresponding AP analyses are performed for each external column location and each floor individually.

9.4.1.3 Damage Limits

DoD's proposed damage limits (2005) for AP analyses are based on the British approach, as cited above. Accordingly, the following are recommended:

Removal of an external column or load-bearing wall — The collapse area directly above the removed element must be less than the smaller of 70 m² (750 ft²) or 15% of that floor; the floor directly under the removed element must not fail. Any collapse is not permitted to extend beyond the structure tributary to the removed element.

Removal of an internal column or load-bearing wall — The collapse area directly above the removed element must be less than the smaller of 140 m² (1500 ft²) or 30% of that floor; the floor directly under the removed element must not fail. Any collapse is not permitted to extend beyond the structure tributary to the removed element.

General requirements — All other damage, strength, and deformation parameters must comply with the appropriate design code for the specific structural system under consideration.

9.4.1.4 Other Topics

Following the general guidelines summarized above, the DoD publication (2005) contains additional chapters that specifically address design requirements for reinforced concrete, structural steel, masonry, and wood. These issues will not be discussed herein, and readers are advised to review the appropriate material in the original reference for further details.

9.4.2 GSA Guidelines

The GSA guidelines (2003) include a detailed procedure for evaluating the potential for progressive collapse resulting from an abnormal loading situation for existing reinforced concrete and steel framed buildings along with guidelines for the design of such new facilities. These procedures are summarized below.

9.4.2.1 Potential for Progressive Collapse Assessment of Existing Facilities

If an existing facility is at an extremely low risk for progressive collapse or if human occupancy is extremely low (as determined in that process), the facility may be exempt from further consideration of progressive collapse. The GSA process is performed in four steps that include detailed flow charts and tables, as summarized below.

Step 1 — Determine the potential for total exemption to the remaining methodology.

Step 2 — Determine the minimum defended standoff distance consistent with the construction type and required GSA level of protection of the facility under consideration. The defended standoff is considered only as one factor in determining whether a facility is exempt. If the type of construction is not listed in the appropriate table, go directly to Step 3. Otherwise, follow the steps in the flowchart for this step to determine the potential for total or partial exemption to the remaining methodology. If a facility is not exempt and analysis is required, the analysis process is threat-independent.

Step 3 — This is a more detailed consideration of the facility if the Step 2 requirements are not achievable or the construction type is not included in the appropriate table. The user shall use a specified flowchart to determine the potential for total exemption. The user will then continue to appropriate flowcharts for concrete structures or steel frame structures, as indicated.

Step 4 — The results determined in the exemption process (Steps 1 through 3) shall be documented by and submitted to the GSA project manager for review. This process is documented in all STANDGARD-generated progressive collapse assessment reports.

9.4.2.2 Analysis and Design Guidelines for Mitigating Progressive Collapse in New Facilities

The GSA guidelines (2003) for new facilities aim to reduce the potential for progressive collapse as a result of an abnormal loading event, regardless of the required level of protection. The outline consists of an analysis and redesign approach intended to enhance the probability that a structure will not progressively collapse (or be damaged to an extent disproportionate to the original cause of the damage) if localized damage is induced by an abnormal loading event. The GSA process is outlined in several flow charts and is accompanied by detailed equations, tables, and discussions.

Generally, the guidelines address both local and global conditions that must be handled during design and analysis. They are intended to ensure continuity along at least two full girder spans (i.e., a double-span condition consisting of two full bays) by requiring beam-to-beam structural continuity across a removed column and the ability of girders and beams to deform in flexure well beyond their elastic limits without inducing structural collapse. Designers are required to show that a proposed design ensures redundancy and beam-to-beam continuity across a removed column and that gravity loads can be redistributed for a multiple-span condition. Also, one must show that a proposed beam-to-column connection system can mitigate the effects of an instantaneous column loss through ductile behavior.

The selected beam-to-column-to-beam connection configuration must provide positive, multiple, and clearly defined beam-to-beam load paths. Only steel frame beam-to-column connection types that have been *qualified* by full-scale testing should be used in the design of new buildings to mitigate progressive collapse. The ability of a girder or beam to structurally accommodate a double-span condition created by a missing column scenario is considered *fundamental* in mitigating progressive collapse. The GSA (2003) included detailed discussions for calculating steel connection strength demands, the locations of plastic hinges, and the corresponding rotations. Redundant lateral and vertical force-resisting steel frame systems are preferred to promote structural robustness and ensure alternate load paths in the case of a localized damage state.

GSA (2003) permits one to apply linear elastic static analysis to assess the potential for progressive collapse in all new and upgraded construction. Other analysis approaches may also be used with appropriate analysis considerations. The recommendation is to use three-dimensional analytic models to account for potential three-dimensional effects and avoid overly conservative solutions. Nevertheless, two-dimensional models may be used provided that the general response and three-dimensional effects can be adequately idealized. The GSA (2003) provides specific guidelines on performing such analyses.

The analyses should cover all unique structural differences that could affect the outcome of predicting the potential for progressive collapse, for example, differences in beam-to-beam connection type (simple versus moment connection), significant changes in beam span and/or size, and significant changes in column orientation or strength (weak versus major axes), etc. GSA (2003), like DoD (2005), specifies the location of localized damage (i.e., column removal) in both reinforced concrete and steel structures for facilities that have relatively simple, uniform, and repetitive layouts (for both global and local connection attributes), with no atypical structural configurations. Facilities that have underground parking and/or uncontrolled public ground floor areas require different interior analyses.

The following vertical load is required to be applied downward to a structure under investigation for static analysis:

$$\text{Load} = 2(\text{DL} + 0.25\text{LL}) \tag{9.4}$$

where DL is the dead load and LL is the live load.

9.4.2.3 Analysis and Acceptance Criteria

The GSA (2003) requires that the maximum allowable extent of collapse resulting from the instantaneous removal of an exterior primary vertical support member one floor above grade shall be confined to (1) the structural bays directly associated with the instantaneously removed vertical member in the floor level directly above the instantaneously removed vertical member **or** (2) 1,800 ft^2 at the floor level directly above the instantaneously removed vertical member, whichever is the smaller area, as shown in Figure 9.4.

Similarly, the allowable extent of collapse resulting from the instantaneous removal of an interior primary vertical support member in an uncontrolled ground floor area and/or an underground parking area for one floor level shall be confined to the smaller area of either (1) the structural bays directly associated with the instantaneously removed vertical member **or** (2) 3,600 ft^2 at the floor level directly above the instantaneously removed vertical member, as shown in Figure 9.4.

The internal consideration is not required if there is no uncontrolled ground floor area and/or an underground parking area present in the facility under evaluation.

The linear elastic analysis results must be examined to identify the magnitudes and distributions of potential demands on both the primary and secondary structural elements for quantifying potential collapse areas. Following induced local damage, all beams, girders, columns, joints, or connections are checked to determine whether any exceeded their respective maximum allowable demands. The

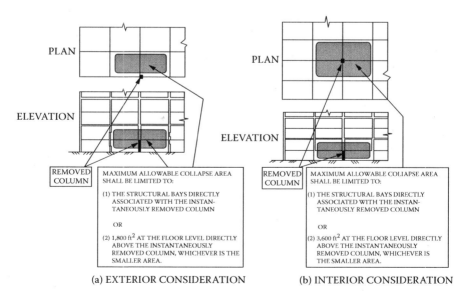

FIGURE 9.4 Maximum allowable collapse areas for structure using columns as primary vertical support system (GSA, 2003).

magnitude and distribution of demands will be indicated by demand-capacity ratios (DCRs), as defined in Equation (9.5).

$$DCR = Q_{UD}/Q_{CE} \qquad (9.5)$$

where Q_{UD} is the acting force or moment (i.e., demand) and Q_{CE} is the expected ultimate unfactored capacity of the component. Recommended limits for DCR values are provided by the GSA (2003). However, for other structural configurations, a value of 0.75 DCR should be used (factor of 0.75 for uncertainties), and a DCR below 1.0 shall not be required for any condition. Member ends exceeding their respective DCR values will then be released and their end moments redistributed. A five-step procedure for conducting the linear elastic static analysis (GSA, 2003) is provided and explained in detail as follows.

Step 1 — Remove a vertical support from the location being considered and conduct a linear static analysis of the structure. Load the model with 2(DL + 0.25LL).

Step 2 — Determine which members and connections have DCR values that exceed the provided acceptance criteria. If the DCR for any member end or connection is exceeded based upon shear force, the member is to be considered failed. In addition, if the flexural DCR values for both ends of a member or its connections as well as the span are exceeded (creating a three-hinged failure mechanism), the member is to be considered failed. Failed members should be removed from the model and all dead and live loads associated with failed members should be redistributed to other members in adjacent bays.

Step 3 — For a member or connection whose Q_{UD}/Q_{CE} ratio exceeds the applicable flexural DCR values, place a hinge at the member end or connection to release the moment. This hinge should be located at the center of flexural yielding for the member or connection. Use rigid offsets and/or stub members from the connecting member as needed to model the hinge in the proper location. For yielding at the end of a member, the center of flexural yielding should not be taken to be more than half the depth of the member from the face of the intersecting member, which is usually a column.

Step 4 — At each inserted hinge, apply equal but opposite moments to the stub or offset and member end to each side of the hinge. The magnitude of the moments should equal the expected flexural strength of the moment or connection, and the direction of the moments should be consistent with direction of the moments in the analysis performed in Step 1.

Step 5 — Rerun the analysis and repeat Steps 1 through 4. Continue this process until no DCR values are exceeded. If moments have been redistributed throughout the entire building and DCR values are still exceeded in areas outside of the allowable collapse region, the structure will be considered to have a high potential for progressive collapse.

9.4.2.4 Material Properties and Structural Modeling

GSA (2003) has specified material properties per various ASTM standards and the material strengths in the design may be increased by employing appropriate strength increase factors (e.g., as per TM 5-1300, Department of the Army, 1990). However, such material enhancement should be used only if the actual states of the materials in the facility are well defined.

Similarly, the analytic models used in assessing the potential for progressive collapse should be as accurate as possible to capture anticipated or existing conditions. This includes all material properties, design details, types of boundary conditions (fixed, simple, etc.). Any limitations or anomalies of the software adopted for the analysis should be known and well defined. The progressive collapse analysis is based on the appropriate removal of critical vertical elements (columns, bearing walls, etc.) and their removal should be instantaneous. Although the element removal rate does not affect static analyses, it may have an important effect on structural responses in a dynamic analysis. Therefore, the element should be removed over a time period no more than 1/10 of the period associated with the structural response mode for the vertical element removal. The element removal should be performed such that it does not affect the connection and joint or horizontal elements attached to the removed element at the floor levels.

9.4.2.5 Redesigns of Structural Elements

Structural configurations determined to have high potentials for progressive collapse should be redesigned to a level consistent with a low potential for progressive collapse. The following two-step procedure (GSA, 2003) is outlined for such redesign.

Step 1 — As a minimum, the structural elements and/or connections identified as deficient should be redesigned consistently with the redistributed loading determined in this process in conjunction with the standard design requirements of the project-specific building codes using well established design techniques. The redesign criteria for typical and atypical structural configurations follow:

Typical structural configurations — Structural elements and beam-to-column connections must meet the DCR acceptance criteria in the design of deficient components and connections. If an approved alternate analysis criterion is used, the deficient components should be designed, as a minimum, to achieve the allowable values associated with that criterion for the redistributed loading.

Atypical structural configurations — Structural elements and beam-to-column connections must meet the DCR acceptance criteria in the design of deficient components and connections. Note that a reduction factor of 3/4 must be multiplied by the DCR value for atypical structures. If an approved alternate analysis criterion is used, the deficient components should be designed, as a minimum, to achieve the allowable values associated with that criterion for the redistributed loading.

Step 2 — Upon the completion of Step 1, the redesigned structure shall be re-analyzed consistently with the previously described analysis procedure.

9.5 ADVANCED FRAME STRUCTURE ANALYSIS

9.5.1 BACKGROUND

In the previous sections, we reviewed progressive collapse phenomena and several analysis and design approaches proposed in the U.K. and U.S. Nevertheless, progressive collapse did not attract much attention between the mid 1970s and the mid 1990s. Limited consideration was given to this topic until the catastrophic explosive attack on the Alfred P. Murrah federal building in Oklahoma City and the resulting failure that involved some aspects of progressive collapse (FEMA, 1996; Hinman and Hammond, 1997). This type of failure attracted much more attention after the horrific terrorist attacks on the World Trade Center and the Pentagon in 2001 (FEMA, 2002; ASCE, 2003).

One of the comprehensive studies has been used as a foundation for the discussion of advanced research activities to elucidate progressive collapse in steel frame structures (Choi and Krauthammer, 2003; Krauthammer et al., 2004), as presented next.

When a change in the geometry of a structure or structural component under compression results in the loss of its ability to resist loadings, the condition is called instability. Advanced analysis is any method of analysis that sufficiently represents the strength and stability behavior such that separate specification member capacity checks are not required (Chen and Toma, 1994). The main distinction between advanced analysis and other simplified analysis methods is that advanced analysis combines, for the first time, the theories of plasticity and stability in the limit states design of structural steel frameworks. Other analysis and design methods treat stability and plasticity separately — usually through the use of beam–column interaction equations and member-effective length factors (Liew et al., 1991). The strength of a column may be expressed by the following modified Euler equation for critical buckling loads:

$$P_{cr} = \frac{\pi^2 E_t}{(KL/r)^2} A_g = F_{cr} A_g \qquad (9.6)$$

in which E_t is the tangent modulus of elasticity at a stress P_{cr}/A_g, A_g is the gross cross-sectional area of the member, KL is the effective (or equivalent pinned-end) slenderness ratio, K is the effective length factor, L is the actual member length, r is the radius of gyration $(I/A_g)^{0.5}$, and I is the cross-sectional moment of inertia. The tangent modulus of elasticity was used instead of the modulus of elasticity.

Failure modes of a column depend on the load, boundary condition, shape of cross-section, and especially the slenderness ratio. If a column is long enough (has a high slenderness ratio), it will buckle. If it is short enough (has a low

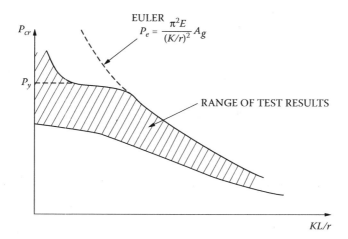

FIGURE 9.5 Typical range of column-strength-versus-slenderness ratio (Salmon and Johnson, 1996).

slenderness ratio), it will crush (experience a material failure). Figure 9.5 shows a typical range of column strength-versus-slenderness ratio. Note that a column fails at the Euler buckling load for a high slenderness ratio, while it fails at the material failure load for a low slenderness ratio. Because the slenderness ratio is the only factor that changes a failure mode, it is the most important factor in calculating column strength. As this figure shows, the accurate strength of a column can be obtained by considering the strength and stability of a member simultaneously. Load and resistance factor design (AISC, 1994) also provides design procedures to estimate critical column stress. Figure 9.6 depicts the critical column stress-versus-slenderness ratios for various yield stresses. It shows that a column fails at the yield stress if it has a low slenderness ratio and fails at a much lower stress if it has a high slenderness ratio. These two regions are connected by a smooth curve, implying that the transition area is affected by both the strength and stability of a member.

Another issue is inelastic buckling (Salmon and Johnson, 1996). Euler's theory pertains only to situations where compressive stress below the elastic limit acts uniformly over the cross-section when buckling failure occurs. However, in many cases, some of the fibers in the cross-section yield when the member buckles; this is inelastic buckling. If all fibers in the cross-section yield before the member reaches its critical load, a material failure has occurred. Inelastic buckling is affected by both strength and stability simultaneously. Therefore, inelastic buckling is in the transition area in Figure 9.5 and Figure 9.6. Since many structural members have slenderness ratios in the vicinity of the transition area, advanced analysis is essential for obtaining accurate results.

To perform a structural analysis under these complicated conditions, equilibrium equations must be written based on the deformed structural geometry. This is known as second-order analysis (Chen and Lui, 1987). The stiffness of a

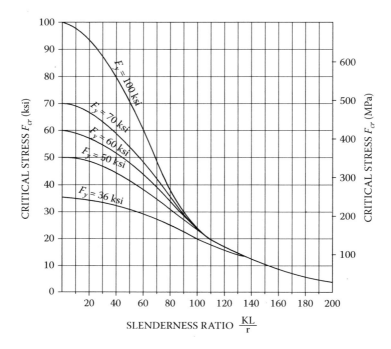

FIGURE 9.6 Critical column stress F_{cr} vs. KL/r according to load and resistance factor design for various yield stresses (Salmon and Johnson, 1996).

structure changes as the geometry changes. If the load and deformation keep changing or increasing, the stiffness of a structure may reach a point where the stiffness vanishes. This is the buckling condition.

In ordinary structural analysis, the original geometry does not change even if the load goes to an extreme value. Therefore, ordinary structural analysis cannot capture the buckling phenomenon. A second-order analysis is essential for stability analysis, as explained above. One also needs a plastic analysis approach for strength consideration. Therefore, a second-order inelastic analysis is required.

9.5.2 Semi-Rigid Connections

One very common engineering practice is to use rigid or pinned connections between steel members for analysis purposes, but experiments have shown that a real steel connection is neither rigid nor pinned (Kameshki and Saka, 2003). Furthermore, these experiments have shown that when a moment is applied to a ductile connection, the relationship between the moment and the beam column rotation is nonlinear.

Other studies were performed to evaluate the effective length with semi-rigid connections (Ermopoulos, 1991; Kishi et al., 1997 and 1998; Aristozabal-Ochoa, 1997; Kameshki and Saka, 2003; Liew et al., 2000). Figure 9.7 shows moment

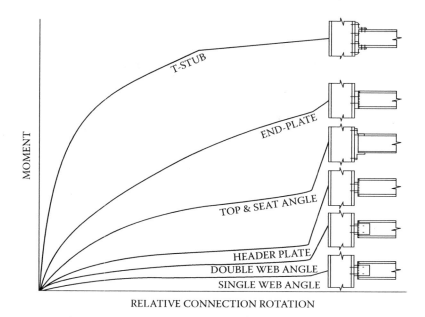

FIGURE 9.7 Connection moment rotation curves (Kameshki and Saka, 2003).

rotation curves for various steel connections. A connection should be considered as an individual member of a structure when its behavior is semi-rigid. Rotational springs can be adopted to simulate the behaviors of semi-rigid nonlinear connections, as shown in Figure 9.8.

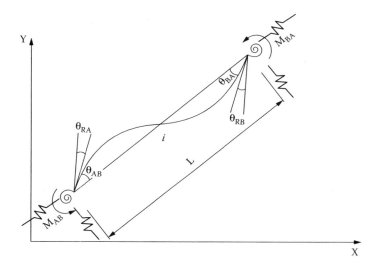

FIGURE 9.8 Semi-rigid plane member.

9.5.3 Computer Code Requirements

Two time integration techniques are available when nonlinear dynamic analysis is performed numerically (e.g., with finite element codes): the implicit and explicit methods (Bathe, 1996). For example, ABAQUS/Standard (implicit solver) is designed to analyze the overall static or dynamic response of a structure, in contrast to a wave propagation solution that can be performed with ABAQUS/Explicit that can address local responses in continua (Hibbitt, Karlsson, and Sorensen, 2005).

The implicit approach to obtain a solution in nonlinear analysis is achieved by iterative methods. The iterative equations in dynamic nonlinear analysis using implicit time integration have the same forms as the equations used for static nonlinear analysis, except that both the coefficient matrix and the nodal point force vector contain contributions from the inertia of the system (Bathe, 1996). If a system is nonlinear, an iterative process such as the Newton-Rhapson method should be used to obtain the displacement increment vector at each time increment. This procedure can be performed with ordinary structural dynamics problems. However, the situation is different in progressive collapse analysis.

The stiffness matrix is approaching a singular state if members of the system buckle so that convergence is hard to achieve. This means that the implicit method may not guarantee a solution when buckling problems are considered. Therefore, another method is required to perform progressive collapse analysis, and the explicit integration approach can be considered. The most common explicit time integration operator used in nonlinear dynamic analysis is probably the central difference operator (Bathe, 1996). The central difference integration operator is explicit in that the kinematic state can be advanced using known variable values from the previous time increment.

The key to the computational efficiency of the explicit procedure is the use of diagonal element mass matrices that enable the efficient inversion of the mass matrix used in the computation for the accelerations at the beginning of the time increment. The explicit procedure requires no iterations and no tangent stiffness matrix (Hibbitt, Karlsson, and Sorensen, 2005). A special treatment of the mean velocities, etc., is required for initial conditions, certain constraints, and presentation of results. The state velocities are stored as linear interpolations of the mean velocities for presentation of results. The central difference operator is not self-starting because the value of the mean velocity needs to be defined. The initial values of velocity and acceleration are set to zero or given specific values. The explicit method integrates through time by using many small time increments. It is conditionally stable and the stability limit for the operator (with no damping) is given in terms of the highest eigenvalue in the system. Explicit integration is fast, reliable, and well suited for problems with many degrees of freedom with extreme loads. Most of all, the explicit method does not require the inversion of the stiffness matrix that can cause matrix singularity problems associated with buckling.

9.5.4 EXAMPLES OF PROGRESSIVE COLLAPSE ANALYSIS

As in a recent study by Krauthammer et al. (2004), a 10-story steel frame building was selected as an example of progressive collapse analyses. It was designed as an ordinary office moment-resisting frame building. Its overall properties are listed below. Figure 9.9 shows the selected girders and columns for each frame and Figure 9.10 shows selected beams for the roof and floors. The building was designed based on ASCE 7 (2002) and AISC's *Manual of Steel Construction* (2003).

Building outline
10-story steel office building
Dimensions: 120 ft × 120 ft
Span length: 30 ft (4 × 4 bays in plan)
Height: 146 ft (first story: 20 ft; other stories: 14 ft)
Lateral load resisting moment frames (A–A and C–C planes in Figure 9.9)
Gravity frame (B–B plane in Figure 9.9) are non–moment-resisting frames

Location: low seismic and wind regions
Occupancy category (classification of building): II
Importance factor: 1.0 for occupancy category II
Seismic use group: I
Site class definition: D
Seismic design category: B
Exposure category: B
Basic wind speed: 90 mph

Materials
Beams, girders and columns: hot rolled shape steel (A992 with $F_y = 50\ ksi$)
Slab: normal weight concrete (145 *pcf* and $f'_c = 4\ ksi$)

9.5.4.1 Semi-Rigid Connections

Connections were designed as a part of the whole building design procedures according to requirements in AISC (2003). Two connection types were selected: moment and shear. The moment connection was set up as an exterior girder-to-column connection resisting lateral loads. Shear connections were used for the assemblage of interior building parts, and typical moment and shear connections were chosen. Loads were determined by the shear and flexural capacities of the girders, that is, the maximum forces applied to the connections were selected from the design shear and flexural strengths of the girders that transmit external loads. Figure 9.11 shows typical final designs of the moment and shear connections.

396   Modern Protective Structures

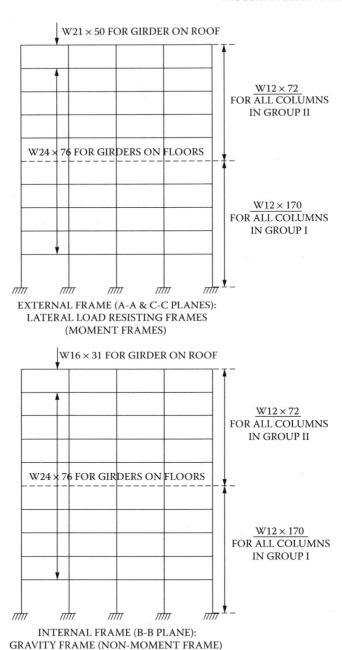

FIGURE 9.9 Selected girders and columns for each frame.

Progressive Collapse

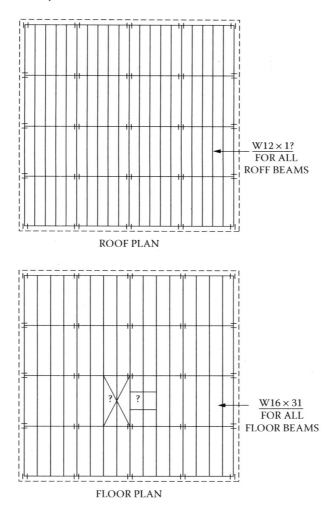

FIGURE 9.10 Selected beams for roof and floor.

9.5.4.2 Analyses

Connections are characterized by the amount of restraint: fully or partially restrained and simple connection. Moment rotation curves are generally assumed to be the best characterizations of connection behavior. The moment rotation curves of connections were derived from numerical simulations, as discussed in Krauthammer et al. (2004). The curves defined properties of connector elements in the macrosimulation of the whole structure. The main aim of this modeling was to analyze the behavior of the designed connection details, and define connector elements of the progressive collapse models in terms of moment-end rotation relationships. Connection details were positioned at the corners of this

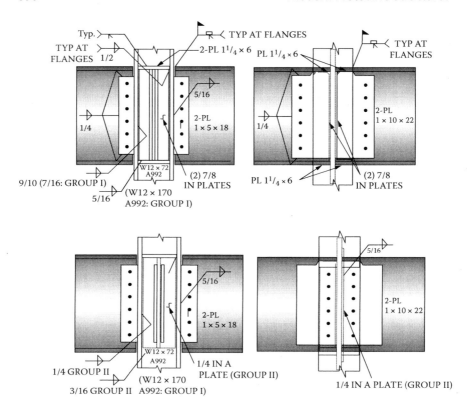

FIGURE 9.11 Designs of moment and shear connections.

theoretical room model as indicated in Figure 9.12. In the same way, the middle parts of girders and columns and the connection components were assembled as corner configurations. The finite element models for the moment and shear connections in Group II are illustrated in Figure 9.13.

The Young's modulus of structural steel is 2.9×10^9 psi and the yield stress is 5.0×10^4 psi. Each floor has a 3-in. thick concrete slab. Finite membrane strain and large rotation shell elements were used to model a concrete slab. The Young's modulus of concrete is 3.5×10^6 psi and yield stress is 4000 psi. The concrete slab and beam were fully connected so that the deflections were continuous between the slabs and the beams. Steel decks and reinforcements were not considered for simplicity of the analyses. The given loads were roof loads, floor loads, and exterior wall loads. Gravity acceleration was applied gradually for 8 sec so that the inertial effect could be avoided. Figure 9.14 shows an overview of the frame.

Two types of steel connections were used. The first type was an ideal connection. The ideal connections were hinges and rigid connections. A hinge

Progressive Collapse

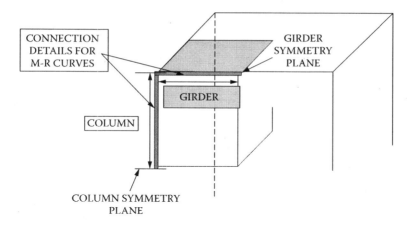

FIGURE 9.12 Geometric model used for moment-versus-rotation relationships.

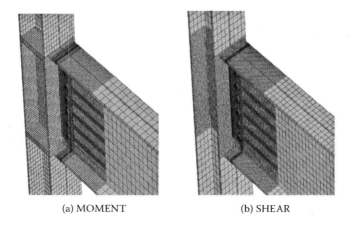

(a) MOMENT (b) SHEAR

FIGURE 9.13 Numerical model of designed moment and shear connections, Group II.

connection is incapable of transferring moments from one member to another; a rigid connection transfers moments between members without loss. The other type of connection was semi-rigid. These connections were designed as part of the building. The moment rotation relationships of the connections will be shown later. These relationships were adopted for the progressive collapse analyses of the 10-story steel frame building. There were six initial column failure cases that were applied to both the ideal connection building and the semi-rigid connection building. The initial column failure cases are shown in Figure 9.15.

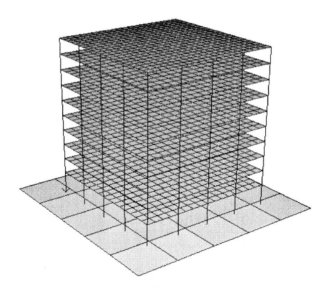

FIGURE 9.14 Overview of building model.

9.5.4.3 Results

The moment rotation relationships of the moment and shear connections in Group I and II from simulation results are shown in Figure 9.16. These relationships were used as semi-rigid connections in the progressive collapse simulations.

Twelve progressive collapse analyses cases of the 10-story steel frame building were performed. Only three initial column removal cases (Case 6) caused total collapse of the building. Figure 9.17 shows results for Case 6 with ideal connections. The first failure was initiated at the connection, as shown in Figure 9.17(b). As these connections broke, the floors above the removed columns started to collapse to the ground, and this caused column buckling of the sixth floor, as shown in Figure 9.17(c). This column buckling phenomenon initiated a horizontal failure propagation at the sixth floor so that the whole floor collapsed. After that, the columns in the first floor started to buckle because of the enhanced vertical impact load and it led to the total collapse of the building. Buildings with other failure cases did not show collapse, but several members had permanent damage due to yielding. As noted, this building was designed to resist lateral forces and it showed enhanced resistance to progressive collapse. Case 6 with semi-rigid connections showed similar results. However, the building collapsed differently, as shown in Figure 9.18.

9.5.4.4 Conclusions

The analyses described above showed interesting results. The three initial column removals (Case 6) with ideal and semi-rigid connection cases showed different types of total collapse. The only differences between the models were the types

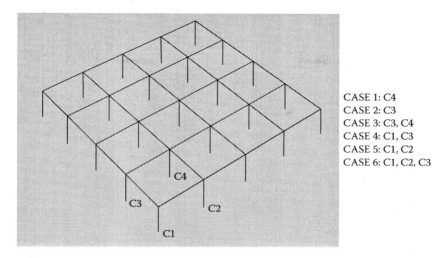

FIGURE 9.15 Initial column failure cases.

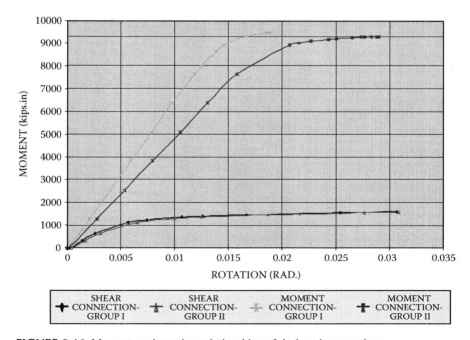

FIGURE 9.16 Moment–end rotation relationships of designed connections.

of connections adopted in the various cases. The collapse of the semi-rigid connection case was caused by a cascade of local failures such as connection failures and column buckling. However, the collapse of the ideal connection case

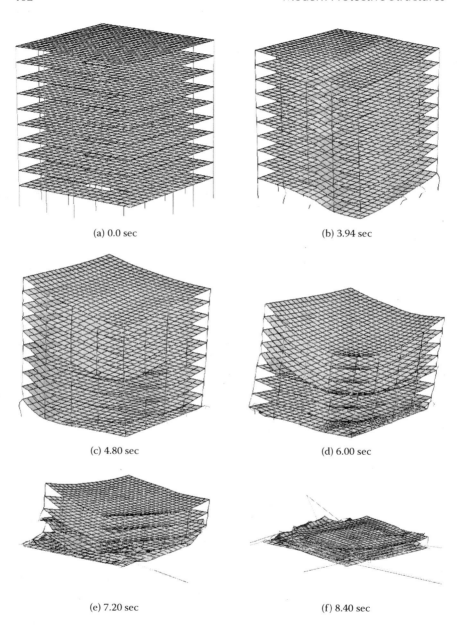

FIGURE 9.17 Case 6 with ideal connections.

was caused by column buckling in the first floor. Columns buckled on the sixth floor because the columns on that floor were smaller than the columns on the first to fifth floors. Column buckling on the sixth floor propagated through the

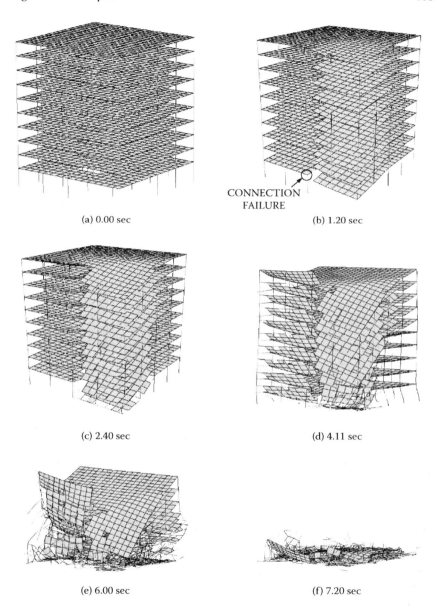

(a) 0.00 sec

(b) 1.20 sec — CONNECTION FAILURE

(c) 2.40 sec

(d) 4.11 sec

(e) 6.00 sec

(f) 7.20 sec

FIGURE 9.18 Case 6 with semi-rigid connections.

floor so that the floors from the sixth to the roof became an impact mass that caused the total collapse of the building.

Therefore, both the columns on the first floor and the columns on the sixth should be addressed carefully. One or two initial column removal cases did not cause collapse because the building was designed to resist lateral forces such as moderate seismic and wind loads. Only three initial column removal cases caused total collapse. A lateral load resistance design increased member sizes and the responses of the same size building designed without lateral loads is expected be much different. Therefore, the lateral load assumptions in the design process should be considered to improve the mitigation of progressive collapse.

9.6 SUMMARY

Current provisions for mitigating progressive collapse are based on single column removal assumptions, as discussed previously. Furthermore, they permit the designer to apply linear elastic, elastic–plastic, or fully nonlinear analysis methods. Although such approaches will lead to more robust buildings, they highlight two serious shortcomings:

- A single column removal approach does not represent typical damage from an explosive attack on a building. Previous incidents have shown that more than one column would be damaged and other structural members are also expected to be severely damaged. Therefore, realistic progressive collapse analyses should consider more accurately the expected damage states associated with explosive attack conditions on a facility.
- The use of linear elastic analysis tools will not provide results that represent the actual behavior of a damaged building. The application of elastic–plastic analysis tools may provide reasonable results, but only if no geometric nonlinear effects (i.e., buckling) are involved in the actual building response. Unfortunately, one cannot predict this situation without a fully nonlinear analysis. Therefore, progressive collapse mitigation should require one to use only fully nonlinear analysis.

10 A Comprehensive Protective Design Approach

10.1 INTRODUCTION

This chapter is aimed at reviewing the topics covered in the previous chapters and providing guidelines on a comprehensive protective design approach. This information is provided to ensure that individuals involved in this field will implement systematic and balanced approaches without overlooking issues that must be addressed.

10.2 BACKGROUND

Protective structural design requires a sound background in fortification science and technology, and must be performed within a comprehensive physical security plan. The reader should realize that loading environments associated with many relevant threats (impact, explosion, penetration, etc.) are extremely energetic and their durations are measured in milliseconds (i.e., about one thousand times shorter than typical earthquakes). Structural responses under short-duration dynamic effects can be significantly different from the much slower loading cases, requiring a designer to provide suitable structural details. Therefore, the designer must explicitly address the effects related to such severe loading environments, in addition to considering the general principles used for structural design to resist conventional loads.

As a starting point, the reader should review background material on security (AIA, 2004; ASCE, 1999, Department of the Army, 1994) and structural consideration and design (ASCE, 1985; Department of the Army, 1986, 1998, and 1990). Information from other sources can add significantly to this background (Allgood and Swihart, 1970; ASCE, 1997 and 1999; Crawford et al., 1971 and 1974; Drake et al., 1989; Gut, 1976; Schuster et al., 1987; Swedish Civil Defence Administration, 1978; Swiss Civil Defense Administration, 1971 and 1977; Department of Defense, 2003 and 2005; GSA, 2003). The works by Smith and Hetherington (1994), Mays and Smith (1995), and the National Research Council (1995, 2000, and 2001) also contain valuable information. They address the general issues of weapon and blast effects on buildings and how to protect against such events.

Biggs (1964) presented an introduction to the dynamic principles of structural analysis and design for resisting impulsive loads. More advanced treatments of general structural dynamics problems can be found in other texts (Clough and Penzien, 1993; Tedesco et al., 1999; Chopra, 2001). Baker et al. (1983) provided a comprehensive treatment of explosion hazard mitigation; Smith and Hetherington (1994) covered designs of structures to resist weapon effects.

More advanced considerations of a broader range of related problems can be found in various conference proceedings (Concrete Society, 1987; Clarke et al., 1992; Krauthammer, 1989 and 1993; Bulson, 1994; Yankelevsky and Sofrin, 1996; Lok, 1996 and 1997; Goering, 1997; Jones et al., 1998; Langseth and Krauthammer, 1998). Additional relevant information can be found in the proceedings of the bi-annual International Symposium on the Interaction of the Effects of Munitions with Structures sponsored since 1983 by the German Federal Ministry of Defense and the U.S. Department of Defense; the proceedings of the bi-annual Explosive Safety Seminar sponsored by the U.S. Department of Defense, Explosive Safety Board; the annual Shock and Vibration Symposium; and the bi-annual MABS conference. Specific recommendations on the design of industrial blast-resistant building are available in ASCE (1997). Recommendations for civil defense shelters can be found in publications by several European organizations (Swedish Civil Defence Administration, 1978; Swiss Civil Defense Administration, 1977).

Lessons learned from recent terrorist bombing incidents (Mays and Smith, 1995; FEMA, 1996; Goering, 1997; Hinman and Hammond, 1997) include information about the behaviors of attacked facilities, post-attack conditions, and forensic approaches used to study these incidents. Although the following resources were described in Chapters 1 and 5, brief summaries of their applications to protective design are provided next.

The **Tri-Service Manual TM 5-1300** (Department of the Army, 1990) is intended primarily for explosives safety applications. It is the most widely used manual for structural design to resist blast effects and is approved for public release with unlimited distribution. The manual addresses general background, loads, analysis, structural design, and planning for explosive safety. Allowable design response limits for the structural elements are given as support rotations based on an assumed three-hinge mechanism. This manual is currently under revision.

Army Technical Manual 5-855-1 (Department of the Army, 1986 and 1998) is intended for use by engineers involved in designing hardened facilities to resist the effects of conventional weapons and other explosive devices. It contains many of the issues covered in TM 5-1300 but focuses on their military aspects. This manual has been replaced by UFC 3-340-01 (Department of Defense, 2002).

ASCE Manual 42 (1985) was prepared to provide guidance in the design of facilities intended to resist nuclear weapon effects. It addresses general background, nuclear loads, analysis, structural design, and planning for various structural and nonstructural systems in protected facilities. Although the manual is an excellent source for general blast-resistant design concepts, it lacks specific design guidelines on structural details.

A Comprehensive Protective Design Approach

ASCE Guidelines for Blast-Resistant Buildings in Petrochemical Facilities (1997) contains detailed information and guidelines on the design of industrial blast-resistant facilities, with emphasis on petrochemical installations. It includes considerations of safety, siting, types of construction, material properties, analysis and design issues, and several detailed examples. The information is very useful for all aspects of blast-resistant design and can also be used to address physical security.

ASCE Structural Design for Physical Security (1999) was a state-of-the-practice report addressing a broad range of topics in this field. It contains procedure for threat assessment, approaches for load definitions, structural behavior assessment and design approaches, considerations of non structural systems, and retrofit options.

Department of Defense [UFC 4-010-01 (October 2003) and UFC 4-023-03 (January 2005)] and **GSA** (June 2003) criteria are intended to meet minimum antiterrorism standards for buildings along with the prevention of progressive collapse. Both DoD and GSA publications recognize that progressive collapse may be a primary cause of casualties in facilities attacked with explosive devices and include guidelines to mitigate them.

DOE Manual 33 is similar to TM 5-1300, as described above, but it contains updated material based on more recent data.

FEMA Guidelines (December 2003) do not constitute a technical or design manual, but contain clear and comprehensive guidelines on issues that need to be addressed and include simple explanations. The guidelines can serve as an effective starting point for those who wish to learn how to handle protective construction projects.

In addition to these references, one should consider specific modifications for addressing antiterrorist design (Mays and Smith, 1995; Drake et al., 1989; Clarke et al., 1992; National Research Council, 1995; Ettouney et al., 1996; Longinow and Mniszewski, 1996). Accordingly, the following steps must be taken. First, a facility's functions must be well defined. Then, the normal loading environment including anticipated corresponding risks and their occurrence possibilities must be specified. The hazardous (or abnormal) loading environment including anticipated corresponding risks and their occurrence possibilities must be defined. Site conditions must be evaluated in light of all such data, and influences on the design and construction approaches must be defined. Limits must be set on the various risks and the corresponding costs. Finally, methods must be set for ensuring quality control and facility performance requirements (for all loading environments under consideration).

10.3 PROTECTION APPROACHES AND MEASURES

The information in UFC 4-010-01 (Department of Defense, October 2003) indicates clearly the hazards to be addressed for the effective protection of personnel. Several key design strategies are applied throughout that standard. Although they do not account for all the measures that can be considered, the following strategies are the most effective and economical for protecting personnel from terrorist attacks (in order of anticipated effectiveness).

"**Maximize standoff distance** — The primary design strategy is to keep terrorists as far away from inhabited DoD buildings as possible. The easiest and least costly opportunity for achieving the appropriate levels of protection against terrorist threats is to incorporate sufficient standoff distance into project designs. While sufficient standoff distance is not always available to provide the minimum standoff distances required for conventional construction, maximizing the available standoff distance always results in the most cost-effective solution. Maximizing standoff distance also ensures that there is opportunity in the future to upgrade buildings to meet increased threats or to accommodate higher levels of protection." As noted in Chapters 2 through 4, the standoff distance R is a key parameter in defining blast environments. Larger values of (or z) = $R/W^{1/3}$ correspond to much lower blast environments that reduce the levels of hazards associated with explosive and/or ballistic attacks. If space is available around a protected facility, maximizing the standoff distance may be the most efficient protective measure to be implemented. These recommendations are illustrated in Figure 10.1 and Figure 10.2.

"**Prevent building collapse** — Provisions relating to preventing building collapse and building component failure are essential to effectively protecting building occupants, especially from fatalities. Designing those provisions into buildings during new construction or retrofitting during major renovations, repairs, restorations, or modifications of existing buildings is the most

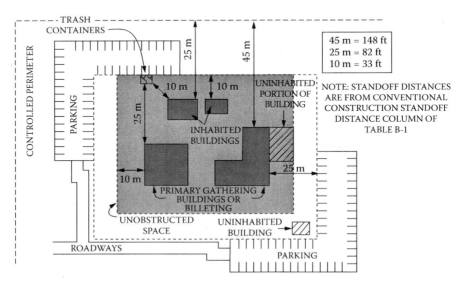

FIGURE 10.1 Standoff distance and buildings separation, controlled perimeter (Department of Defense, October 2003).

A Comprehensive Protective Design Approach

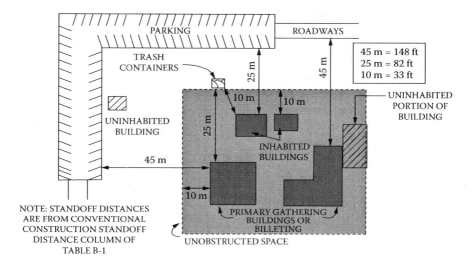

FIGURE 10.2 Standoff distances and building separation, no controlled perimeter (Department of Defense, October 2003).

cost-effective time to do that. In addition, structural systems that provide greater continuity and redundancy among structural components will help limit collapse in the event of severe structural damage from unpredictable terrorist acts." Building collapse is known to be associated with many casualties (e.g., Oklahoma City in 1995, Kenya in 1998, and others). Preventing such outcomes is critical. Clearly, both DoD (January 2005) and GSA (2003) guidelines for preventing progressive collapse are specifically aimed at achieving this objective, as discussed in Chapter 9.

"**Minimize hazardous flying debris** — In past explosive events where there was no building collapse, a high number of injuries resulted from flying glass fragments and debris from walls, ceilings, and fixtures (nonstructural features). Flying debris can be minimized through building design and avoidance of certain building materials and construction techniques. The glass used in most windows breaks at very low blast pressures, resulting in hazardous, dagger-like shards. Minimizing those hazards through reduction in window numbers and sizes and through enhanced window construction has a major effect on limiting mass casualties. Window and door designs must treat glazing, frames, connections, and the structural components to which they are attached as an integrated system. Hazardous fragments may also include secondary debris such as those from barriers and site furnishings." Although this book does not address the behavior and design of blast resistant glazing systems, one can find useful information, elsewhere (ASCE, 1999).

"**Provide effective building layout** — Effective design of building layout and orientation can significantly reduce opportunities for terrorists to target building occupants or injure large numbers of people." Besides general common sense-type measures, one could also employ more advanced considerations, for example, placing less important assets around a building perimeter, especially facing the side of the building that could be more exposed to an attack. Obviously, the more important assets should be placed farther from the building perimeter and/or a side that could face an attack. Other measures include attention to hallways and stairways to enable effective evacuation, rescue, and recovery.

"**Limit airborne contamination** — Effective design of heating, ventilation, and air conditioning (HVAC) systems can significantly reduce the potential for chemical, biological, and radiological agents being distributed throughout buildings."

"**Provide mass notification** — Providing a timely means to notify building occupants of threats and what should be done in response to those threats reduces the risk of mass casualties."

"**Facilitate future upgrades** — Many of the provisions of these standards facilitate opportunities to upgrade building protective measures in the future if the threat environment changes."

10.4 PLANNING AND DESIGN ASSUMPTIONS

Defining the various performance criteria that a facility must meet is important for all protective design activities. Such criteria correlate the types and magnitudes of structural responses with various anticipated loading environments and response specifications for the facility. These criteria are directly related to the protective approaches, selection of structural systems, and design methodology. Accordingly, the following general serviceability criteria are proposed as options, but the design team may select any combination of these for a specific facility:

1. The facility will be always fully operational. The user is willing to accept damage to the facility and/or its contents if the overall operational requirements can be met.
2. The facility will remain intact and respond elastically. The user requires that the facility will remain undamaged after the loading event.
3. The facility is permitted to have a specified range of permanent deformations, but not loss of integrity. The user is willing to accept a certain level of damage if the facility remains an integral system (e.g., no separation between building components, etc.).
4. The facility is permitted to have a specified range of permanent deformations, but not loss of overall robustness. The user is willing to accept a certain level of damage if the structure does not reach a state of imminent collapse. Here, stability is ensured if additional deformation does not correspond to lower structural resistance.
5. The facility is permitted to fail for loads that are greater than the peak design environment. The user is willing to accept total failure if the loads exceed a selected limit. For loads within the acceptable range, however, the user needs to specify one of the other serviceability criteria.

A Comprehensive Protective Design Approach

Each of these criteria may have secondary definitions that address various aspects of the response and performance of the structure and/or its contents. The issues of maintenance, protection of vital nonstructural systems, and post-event recovery operations should be included. Although these issues are beyond the scope of this chapter, they are addressed briefly. Additional information on these issues can be found in many of the publications cited earlier.

Differences would arise in design and cost considerations if a structure would be expected to remain fully operational during and after an anticipated loading event, as compared with another case in which a structure must remain stable during and after an anticipated event but permanent deformation and a specified loss in operational capacity are acceptable. The cost of recovery and restoration often are not included in design procedures, despite the fact that such considerations may affect specific design and construction choices. Obviously, considerations of structural stability and robustness must be central issues in the entire process (Clarke et al., 1992; Choi and Krauthammer, 2002 and 2003; GSA, 2003; Krauthammer et al., 2004; Department of Defense, 2005). Additionally, one must also integrate post-incident recovery to enhance the ability of casualty reduction into the design approach, as discussed by Miller-Hooks and Krauthammer (2002, 2003, and to appear).

One should emphasize that a comprehensive security plan must be established from the start. It should include consideration of all security measures in which structural hardening is one of the components. This should result in a layered defense approach that will enable a facility owner to intercept possible attacks before they can materialize. The enhanced structure should be considered as the last defensive layer of such a system whose role is to protect the facility contents and/or mission if an attack is carried out.

The following sections will address some of the topics noted above and related topics that should be considered during the planning, design, construction, and service lives of protected facilities.

10.5 SITING, ARCHITECTURAL, AND FUNCTIONAL CONSIDERATIONS

Similar to discussions in ASCE (1985, 1997, and 1999), siting, structural shape, and related parameters may also play important roles in designing for physical security. Specific guidelines on siting considerations and architectural and structural design issues are found in the recent DoD minimum antiterrorism standards for buildings (Department of Defense, July 2002). Although the chances of a terrorist attack directed against a facility may be very small, the consequences when one does occur can be devastating and include but are not limited to loss of life, property, and operations.

Certain types of buildings may attract terrorist activities more than others (government buildings, corporate headquarters, etc.). Because of the high costs associated with terrorist bombings, considering ways to incorporate protective design measures into the new buildings from the earliest design stages is worthwhile for the architect and engineer. The objective of these measures is not to create a bunker or bomb-proof building. That would be impractical for commercial structures. Instead, the goal is to find practical, cost-effective measures to mitigate the effects of an explosive attack. The designer's goal is to reduce the risk of catastrophic structural collapse, thus saving lives and facilitating evacuation and rescue efforts.

In discussing physical security measures implemented by architects and engineers, putting these countermeasures into the context of the overall security functions of a facility is worthwhile. Security measures can be divided into two groups: physical and operational. Physical or passive security measures (barriers, bollards, planters, structural hardening, site and structural three-dimensional geometries, etc.) do not require human intervention, whereas operational or active security measures (guards, sensors, CCTV, and other electronic devices) require human intervention. To achieve a balanced design, both types of measures must be implemented into the overall security of a facility.

Remember that any security system is only as strong as its weakest link. Architects and engineers can contribute to an effective physical security system that both augments the operational security functions and also simplifies them, for example, by providing a design that accommodates the inspection of pedestrians and vehicular traffic. If these factors are not considered in the design stages, congestion and unattractive makeshift security posts are the inevitable results. If properly implemented, security countermeasures will contribute toward achieving the following benefits.

Preventing attack or reducing its effectiveness — By making it more difficult to implement some more obvious attack scenarios (such as parking a truck along the curb beside a building), a would-be attacker may become discouraged from targeting a facility. In addition, if a vehicle bomb cannot be detonated near the facility, the corresponding blast load would be smaller. On the other hand, it may not be an advantage to make a facility too obviously protected. That may motivate a potential terrorist to escalate a threat to a higher level (another reason not to create a bunker-like facility).

Delaying attack — An architect and engineer can, through proper design, delay the execution of an attack by making it more difficult for the attacker to reach the intended target. Such a delay would give security forces and authorities time to mobilize their forces and ideally stop an attack before it is fully executed. This can be done by creating a buffer zone between the publicly accessible areas and the vital areas of the facility by means of an "obstacle course" consisting of a meandering path and/or a division of functions within the facility.

One effective way to implement these countermeasures is to create layers of security within a facility. The outermost layer is the perimeter. Interior to this line is the approach zone, then the building exterior, and finally its interior. The interior may be divided into successively more protected zones, starting with publicly accessible areas such as a lobby and retail space, then the more private office areas, and finally the vital functions such as the control room and emergency functions. The advantage of this approach is that breaching one line of protection does not completely compromise the facility. In addition, with this approach the design focus does not have to be the outer layer of protection, which would contribute to an unattractive, fortress-like appearance. Each protection zone is discussed below.

10.5.1 PERIMETER LINE

The perimeter line is the outermost boundary that can be protected by the security measures provided by the facility. In design, it is assumed that all large-scale explosive weapons such as car or truck bombs will be used outside this line of defense. This line should be defended by both physical and operational security methods. The recommendation is to locate a perimeter line as far away from a building exterior as is practical. This is an effective way to limit building damage because the pressure generated by an explosion is inversely proportional to the cube of the distance from the explosion. In other words, if we double the distance from the building to the explosive source, the over-pressure is reduced by roughly a factor of eight.

Many buildings thought to be vulnerable to terrorist attack are in urban areas where only the exterior walls stand between the outside world and the facilities. The options in these cases are limited. The perimeter line often can be pushed out to the edge of the sidewalk by means of bollards, planters, and other obstacles. To push this line even farther outward, restricting parking along the curb may be arranged with the local authorities or street closing may be an option. One might also consider the use of structurally composite blast-resistant and energy-absorbing panels to enhance and protect exterior walls (Krauthammer et al., 1997). The use of steel plates and/or special fabrics placed on the interiors of the external walls may be useful (Eytan and Kolodkin, 1993).

Off-site parking is recommended for facilities vulnerable to terrorist attack, but imposing such a restriction is often impractical. If on-site parking or underground parking is used, one should take precautions to limit access to these areas to only the building occupants and/or have all vehicles inspected. Parking should be located as far from a building as practical. If an underground area is used, one should consider a space next to the building rather than directly underneath it. Another measure is to limit the sizes of vehicles that can enter by imposing height or weight limitations.

Barrier walls designed to resist the effects of an explosion can sometimes reduce pressure levels acting on exterior walls. However, these walls may not enhance security because they prohibit observation of activities occurring on their other sides. An anti-ram knee wall with a fence may be an effective solution.

10.5.2 ACCESS AND APPROACH CONTROL

Access and approach control is the controlled access to a facility through a perimeter line. Architects and engineers can accommodate this security function by using design strategies that would make it difficult for a vehicle to crash onto the site. Example strategies include the use of barrier walls and other devices and ensuring that the location access points are oblique to oncoming streets so that vehicles cannot gain enough velocity to break through these stations. If space is available between the perimeter line and the building exterior, much can be done to delay an intruder. Examples include terraced landscaping, reflecting pools, staircases, circular driveways, planters, and any number of other obstacles that make it difficult to reach a building rapidly.

10.5.3 Building Exterior

The focus at a building exterior shifts from delaying intruders to mitigating the effects of an explosion. The exterior envelope of the building is the most vulnerable to an exterior explosive threat because it is the part of the building nearest the weapon. It also is a critical line of defense for protecting the occupants. The design philosophy to be used is "simpler is better." Generally a simple geometry with minimal ornamentation (which can become airborne during an explosion) is recommended. If ornamentation is used, it should consist of a lightweight material such as wood that is likely to become a less lethal projectile than bricks or metal.

The shape of a building can affect the overall damage to the structure. Airblast may be thought of as a wave that washes over and around a building, like a wave at the seashore washing over a box. As an example of the effect of shape on response, a U- or L-shaped building may trap a wave and exacerbate the effect of an airblast. Therefore, it is recommended that reentrant corners be avoided.

The issue of material to be used for exterior walls must be carefully evaluated. Depending on the level of protection selected, an exterior wall can be designed to survive or fail in response to an explosion. Cast-in-place reinforced concrete is an effective material for designing an exterior wall to survive relatively high pressure levels. Thick reinforced concrete walls with continuous reinforcement on both faces will afford a substantial level of blast protection. Wall thicknesses and the amounts and types of reinforcements must be determined for specific threats, as discussed in Chapter 4 of ASCE (1999). If blast-resistant exterior walls are used, the loads transmitted to the interior frame should be checked. If a frame structure is used, redundancy of the system will increase the likelihood of mitigating response. For instance, the use of transfer girders on exterior faces of a building should be carefully evaluated.

On the other end of the spectrum are curtain walls, which are highly vulnerable to breakage and not recommended. The responses of cladding panels to blast and their design considerations were studied by Pan and Watson (1998) and Pan et al. (2001). Their findings are useful in improving the performance of such systems. Although unreinforced masonry has substantial inertia, it also has little resistance to lateral forces. If masonry is used, it should be reinforced in both vertical and horizontal directions. Starr and Krauthammer (2005) reported on a study to assess impact load transfer between cladding panels and structural

A Comprehensive Protective Design Approach

frames. They found that a transmitted load could be modulated considerably by a cladding system and that the loads transmitted to a structural frame could be much smaller than the full impact loads. Clearly, the type of cladding and its attachment to the structural frame should be carefully addressed during the planning and design phases.

Windows are typically the most vulnerable portions of any building; this issue is addressed in more detail in Chapter 5 of ASCE (1999). While designing all windows to resist a large-scale explosive attack may be impractical, limiting the amount of glass breakage to reduce the injuries is desirable. Annealed glass breaks at low pressure levels (0.50 to 2.0 psi) and the shards created by broken windows are responsible for many injuries due to large-scale explosive attack.

To limit this danger, several approaches can be taken. One is to reduce the number and size of windows, thereby limiting the number of lethal shards created. In addition, smaller windows will generally break at higher pressures than larger windows, making them less prone to breakage. If blast-resistant walls are also used, fewer and/or smaller windows will cause less airblast to enter a facility. For example, one can limit the number of windows on lower floors where the pressures are expected to be higher because of an external explosive threat. Another idea is to use an internal atrium with windows facing inward rather than outward or clerestory windows near ceilings, above the heads of occupants. The information in ASCE (1999) can be used to enhance the protection provided by existing or new windows by selecting an appropriate glazing system and attaching it well into the window supports, using stronger window frames and attaching them well into the structural system, and using special catcher systems that can prevent window debris from flying into protected spaces.

In addition to studying the placements and sizes of windows, it is possible to use more resistant types of glass or a glass that fails in a less lethal mode. Options in this category include tempered glass, laminated glass, or glass/polycarbonate security glazing. Tempered glass breaks at higher pressure levels and also breaks into cube-shaped, possibly less lethal pieces. Laminated glass has been shown to hold shards together and deform before breaking. Glass/polycarbonate glazing is typically sold as bullet-resistant glazing and can resist high pressures. If polycarbonate is used on an exterior face, care must be taken to ensure that the surface is not prone to scratching, clouding, or discoloring. Gluing security films onto existing windows can enhance the level of protection by keeping the debris in larger pieces. Each of these products has advantages. Their manufacturers should be consulted before a decision is made.

Windows, once the sole responsibility of an architect, become structural issues when explosive effects are taken into consideration. When considering the installation of special window lights, the structural designs of mullions, frames, and supporting walls must be checked to ensure that they can hold the windows in place during an explosion. It is pointless to design windows that are more resistant than their supporting walls.

Two measures that may be considered for the retrofit of existing windows are polyester film coating on the inside faces combined with cross-bars to stop the retrofitted glass from entering protected spaces or blast curtains to stop shards from entering protected spaces. Again, window manufacturers must be consulted regarding the tendency of their materials to scratch, delaminate, cloud, or discolor.

10.5.4 BUILDING INTERIOR

The protection of a building interior can be divided into two categories (National Research Council, 1995; Mays and Smith, 1995): the functional and structural layouts. As for functional layouts, public areas such as the lobby, loading dock, and retail area must be separated from the more secured areas of the facility. For an effective design, the evaluation of the impact on the structural integrity of a building arising from the detonation of a weapon in the public areas is recommended. Spaces next to the exterior walls and on the lower floors must be examined as well. One way of protecting occupants is to place stairwells or corridors rather than office spaces beside the exterior walls to provide a buffer zone.

False ceilings, Venetian blinds, ductwork, air conditioners, and other equipment are vulnerable to becoming airborne in an explosion. A simple design that can limit these hazards should be adopted. Examples include placing heavy equipment such as air conditioners near floors rather than near ceiling, using curtains rather than blinds, and using exposed ductwork as an architectural device.

As for structural layouts, good engineering practice recommends that the bays of a building be kept to dimensions less than or equal to 10 m. Since outer bays are more vulnerable to collapse than inner bays, reducing the depth of outer bays is worthwhile. One should also use multiple interior bays to limit collapse hazards. Proper reinforcement for these members may be determined by using dynamic structural analysis techniques.

Progressive collapse measures should be implemented to ensure that damage is limited to the vicinity closest to an explosive source, as discussed in Chapter 9. Some methods for limiting progressive collapse in the design of new buildings include:

- Good plan layout: reduce spans throughout the width of a building.
- Returns on walls to increase structural stability.
- Two-way floor systems designed with dominant and secondary support systems. If the support in the dominant direction is lost, members can transmit the load in the secondary direction.
- Load-bearing internal partitions placed such that the slabs above them can adequately transmit the load to the partitions in their secondary direction.
- Enabling catenary action of floor slab, girders, and support beams. If the internal partitions are not available for support, the floor slab is designed to transmit the load as a catenary.
- Beam action of walls; walls are designed as beams to span an opening.

A Comprehensive Protective Design Approach

10.5.5 VITAL NONSTRUCTURAL SYSTEMS

Protection of nonstructural systems is often not considered when facilities are assessed or designed to resist explosive loads (National Research Council, 1995; Mays and Smith, 1995). This lack of consideration may produce very serious consequences. For example, elevator shafts can become chimneys in case of an explosion, transmitting smoke and heat from the explosion to all levels of the building. This situation can hinder evacuation and increase the risk of injury due to smoke inhalation. Similarly, emergency functions such as sprinkler systems and generators are critical for mitigating the effects of an explosion.

Elevator shafts and emergency functions should be located away from vulnerable areas such as underground parking facilities and loading docks. Furthermore, system redundancy and the separation of different types of utilities and systems should be regarded as critical issues. Obviously, all vital support systems can be damaged if placed in one location such as a utility room. Distributing, separating, and protecting such systems is recommended so that emergency services would be functional if an incident occurred. The attention to such systems must not be limited to the interior of a building. It is often possible to disrupt utilities (water, power, communications, ventilation) by damaging sites located outside a building (e.g., hook-up vaults and exhaust and intake shafts). All such systems should be included in a comprehensive physical security assessment and designed accordingly.

10.5.6 POST-INCIDENT CONDITIONS

Designers should consider the requirements to ensure effective post-attack rescue operations. All such operations include two important features: protection and evacuation. Persons in a building must be protected both during and after an incident. Typically, bombing incidents (or accidents) are unannounced and rescue operations (especially in large, heavily damaged facilities) must be conducted under very difficult circumstances. Persons who cannot exit without help must be sheltered until rescue personnel can remove them. Designers may wish to consider "safe havens" that can serve as temporary shelters. Such areas should be better protected than typical facility interiors and should include emergency supplies and services. For cost effectiveness, these areas could be designed to fulfill dual roles: to support normal activities and act as safe areas in an emergency. Additionally, evacuation routes and communication and other support equipment and services should be well designed to enable safe and efficient rescue operations.

10.6 LOAD CONSIDERATIONS

A structural design must be developed for selected design-based threats (DBTs) and should address an appropriate combination of relevant threats and corresponding loading environments. Detailed discussions of threats and loading environments are presented in Chapters 1 through 4 and in relevant references (Baker and Baker et al., 1983; Department of the Army, 1986 and 1990; Henrych, 1979; Johansson and Persson, 1970; Persson et al., 1994; Vrouwenvelder, 1986; Zukas and Walters, 1998). The load definition must be a preliminary step in the design process. Many such loads can be derived by employing the procedures contained in various design manuals. The application of computer codes for such purposes introduces many advantages.

One such code based on the procedures in TM 5-855-1 (Department of the Army, 1986) is ConWep (Hyde, 1992). Note that design manuals and even computer codes are approximations and that care should be taken when using these analysis tools. It also should be noted that the possibility of fire following an explosion must be considered in the design process, although this subject is not addressed herein. The following brief discussion of loading considerations is provided as a background summary to the overall design process.

An impulsive load has a short duration compared with a characteristic time T (i.e., natural period of the structural element) for a loaded structure. Such loads are often generated by high explosives or detonating gases. Blast effects, especially from explosive devices, are often accompanied by fragments, either from the explosion casement or miscellaneous debris engulfed by the blast wave. Blast parameters from spherical high-explosive charges in open air are well known and can be found in various handbooks and technical manuals, as explained in previous chapters. Such data are usually given for TNT charges; however, transformations to other explosives are also provided. Equivalence factors are provided for different kinds of explosives and for different effects with the same explosive. When an airblast is reflected by a nonresponding structure, a significant pressure enhancement is achieved (perhaps two to eight times the incident pressure) and such reflected pressures should be used as the load on the structure. For detonations in complex and/or nonresponding structural geometries, accurate determination of such parameters can be achieved with available computer programs. Currently, no accurate computer programs can predict load parameters for responding structural models.

Due to the lack of physical data, certain types of explosive loads cannot be determined accurately and the information available in various design manuals is based on estimates. This is particularly true when an explosive charge is placed very near or in contact with a target to produce combined effects of airblast and fragments and also applies to nonstandard explosive devices. Explosive charges of shapes other than spherical or cylindrical will not yield pressure distributions with rotational symmetry and the information provided in design manuals should be used cautiously for such cases. This difficulty is particularly true for close-in explosions; its importance diminishes significantly for more distant explosions. Usually this effect can be ignored for scaled distances larger than 1 ft/lb$^{1/3}$ (ASCE, 1985, 1997, and 1999). However, some experimental data for cylindrical charges suggest that this effect should be ignored for scaled distances larger than 2 ft/lb$^{1/3}$.

When addressing the effects of explosions, consideration must be given to whether a device is encased and to the position of the explosive device relative to the target structure. Variations of each parameter will affect the loading

FIGURE 10.3 Combined effects of blast and fragments (Department of the Army, 1990).

conditions, the responses of structural elements and systems, and the corresponding damage as shown in Figure 10.3.

The relative position of a device has three components: standoff distance, lateral position relative to the target (for certain conditions), and elevation or height of burst. Fragments will be generated when the explosive charge is encased and the fragmentation process will absorb part of the detonation energy (and somewhat lower the blast pressures). Fragments or debris also can be generated from damaged structural components, and they may impact other structural elements. The coupled effects of blast and fragments are not well understood.

Fragments and blast generated by a cased charge ordinarily propagate with different velocities. The blast travels faster than fragments near to the charge. However, because blast velocity decays faster than fragment velocity, the fragments will arrive first farther away from the charge. At a larger distance, because of drag effects, the fragment velocity may decrease below the blast velocity. Obviously, at close-in conditions, the coupling of blast and fragments may expose a structure to more than twice the impulse from the blast alone, and such coupling is a major concern to designers. However, with a close-in explosion, the damage is more likely to be localized (breaching) rather than global (flexural). Furthermore, if fragments load the structure first, the blast wave may load a damaged facility and a lower structural capacity should be considered.

In addition to damaging structural elements, fragments may penetrate them and affect people and/or property located in a facility. Penetration calculations can be done using the various design manuals mentioned earlier or with computational support tools. ConWep (Hyde, 1992) is one such code, but hydrocode calculations may be required for advanced analyses. These issues must be considered during project assessment and design.

Regardless of whether an explosive is encased, the effect of the standoff distance on blast loading is considerable. When a blast source is a considerable distance from a target, the blast wave will engulf the structure uniformly;

conversely, short standoff distances lead to significant pressure differentials across the structure. Furthermore, for a similar charge weight, the greater standoff distance causes a longer loading duration than the shorter standoff distance. The result of the increased standoff distance leads to a more global response of the structure.

A structure's lateral bracing system will also show greater response to this loading, and the nonresponding structural elements (walls, beams, and columns) will fail in more of a flexural manner. When an explosion is very close to a target, the loading characteristics change and become difficult to predict. The loading is very concentrated on the target surface nearest the point of detonation, and dissipates very quickly for locations away from that point. The effect of this loading leads to localized failures such as the breaching of some exterior wall sections or shear failures of beams or column sections.

Equally important for close-in explosions is the fact that the load durations are relatively short and highly impulsive. The lateral bracing system is unable to respond since the intense loading has dissipated by the time the structure reaches its peak response. One final point about close-in explosions is that if a bomb is large enough, a structure can experience an upward force, in essence putting the columns and their connections in tension. For this condition to occur, a bomb must be very large and a structure lightweight.

While it is understood that a bomb could be located at any position along a target face, its position relative to the lateral bracing system plays a significant role in determining the effectiveness of the bracing system. This is particularly true when the standoff distance is small; the localized load may not sufficiently engage the bracing system, causing considerable localized damage. With a large standoff distance, the blast wave is uniformly distributed across the target face and is thus more likely to engage the bracing system. When laying out a structure and performing analysis, proper placement of lateral bracing systems must be considered to ensure the structural integrity of a structure under these diverse loading conditions. As mentioned above, the effects of close-in explosions are difficult to predict due to lack of data.

The third scenario concerning standoff distance is the contact burst or zero standoff distance. Analysis of this condition requires a hydrocode calculation and detailed knowledge of material properties because data for this condition are not available in typical design manuals.

The last spatial component is the elevation or height of burst of an explosive device. The threats and loading environments presented in Chapters 3 and 4 are aimed primarily at above-ground facilities for which ground shock (explosive loads generated in and/or transmitted through the ground) is not a serious concern. Nevertheless, designers may need to consider such effects for specific cases. Ground shock data depend on various soil properties; information can be found in the manuals cited above. Krauthammer (1993) noted that the scatter in ground shock test data is higher, as compared with airblast data, and such scatters should be considered in the design process. This can be noted also when reviewing the treatment of ground shock in typical design manuals (ASCE, 1985; Department

A Comprehensive Protective Design Approach 421

of the Army, 1986 and 1998) in which large uncertainty values were assigned to the empirical approaches.

Free-field pressure from buried charges depends not only on the explosive charge but also on the properties of the soil (compressibility strength, stiffness, hysteretic behavior, moisture content). Existing data show considerable scatter, and this introduces difficulties in a structural analysis. The consideration of nonspherical charge shapes and casing adds to the complexity. Furthermore, soil–structure interaction considerations can lead to highly nonlinear problems and the use of advanced computer codes may be required (Department of Defense, June 2002).

The above discussion involves explosive attacks on a structure from the outside (external explosions). While many blast load definitions issues remain consistent for interior explosions, some points must be addressed. First, TM 5-1300 (Department of the Army, 1990) contains extensive guidelines on calculating pressure–time histories from contained explosions under different conditions. Another important issue is the reflection of a blast wave off of adjacent structural surfaces. Such surfaces are considered as nonresponding and the corresponding pressures would not be accurate. The reflective pressures can further damage a target element that has diminished structural capacity due to the initial blast wave. If the reflected standoff distance is larger than the direct target standoff distance, the net result can be little or no additional damage. The obvious solution for eliminating the potential for damage due to reflective loading is to provide sufficient venting for high-risk areas. Venting can be achieved through the creation of open space or with blast relief panels. Additional information on such features can be obtained also from various insurance companies who handle industrial customers with explosive safety hazards.

Various design manuals contain simple models for calculating explosive loads under containment conditions (i.e., explosions in closed, simple geometry volumes), as discussed in Chapter 3. Few data are available for enclosed explosions in complicated geometries and/or containment structures that are responding systems. It has been observed experimentally, however, that internal explosions cause longer duration pressure–time histories.

For such conditions, special computer codes must be used for deriving pressure–time histories at various locations within a facility. Typically, the pressure may not be uniformly distributed on the structure, and the spatial distribution must be considered in the design process. Failing to consider nonuniform pressure–time distributions may lead to nonconservative design. The design manuals cited earlier contain models for pressure distributions in tunnels and considerations for geometric variations along tunnels (i.e., cross-sectional changes, expansion chambers, bents). Such models were derived for long-duration blast waves. The corresponding information for short-duration blasts in complex geometries may not be as detailed. It should be noted, however, that conventional explosives can cause long-duration blast waves in tunnels since the pulse duration increases with propagation distance along the tunnel.

For gas explosions, an additional problem may be the difficulty to specify adequately the initial gas cloud geometry and conditions when a detonation of deflagration occurs. This problem becomes even more complicated for a gas explosion within a facility, when one must consider the geometrical effects of the detonating system, the pressure–structure interaction, and the effects of confinement.

For design purposes, the design manuals cited above should be used and care should be taken to select the case closest to the problem under consideration, as discussed in previous chapters. If the differences between the available information and the expected environmental conditions are significant, a designer should consider the use of computer simulations to obtain additional data. When contained explosions are under consideration, the application of controlled venting (i.e., by incorporating blowout panels that will enable the release of hot gas and explosion byproducts to the outside) should be explored.

The loads discussed above are dynamic in nature and can be loosely compared with the two other dynamic loading conditions, earthquake, and wind. Like seismic and wind loading, blast loads are high-energy, nonuniform events whose maximum force and timing remain difficult to predict. These events strongly differ in how they are applied to a structure. Both wind and seismic loads engage an entire structure, forcing the uniform global movement of the structure, whereas blast loadings typically remain localized and do not necessarily engage an entire structure.

10.7 STRUCTURAL BEHAVIOR AND PERFORMANCE

All the design manuals cited above use only approximate single-degree-of-freedom (SDOF) mass–spring–damper systems for the calculations of dynamic structural response parameters. Although such calculations can also be achieved with more advanced computational approaches, the current design procedures are still based only on SDOF calculations. Accordingly, the discussion in this book will emphasize the dynamic structural behaviors represented by SDOF models while addressing other structural analysis techniques in less detail.

This discussion is aimed at highlighting some important aspects of structural behavior under severe short-duration dynamic loads. Although considered an approximate approach, SDOF calculations have been shown to provide good accuracy in support of design activities (Krauthammer, 1993; Krauthammer et al., 1996 and 2004). Nevertheless, readers can find discussions about the applications of more advanced computational approaches in Pilkey (1995), in conference proceedings and books on structural dynamics as cited above, and in proceedings of specialized meetings (Krauthammer et al., 1996).

The main concept in approximate SDOF calculations is the need to establish a relationship between load and structural response (i.e., cause and effect). In the linear elastic static domain, the equation for an SDOF mass–spring–damper system as shown in Figure 10.4 is given by Hooke's law in which F is the magnitude of the applied load, K is the structural stiffness, and x is the corresponding deflection. The external force is resisted only by the spring, and the

A Comprehensive Protective Design Approach

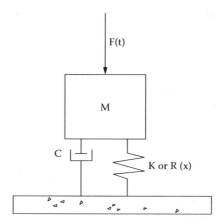

FIGURE 10.4 Single-degree-of-freedom model.

mass and damper do not contribute to this equation. In the dynamic domain, the corresponding equation of equilibrium is given by:

$$F(t) = M\ddot{x} + C\dot{x} + Kx \qquad (10.1)$$

F, K, and x were defined above (here, however, the force is time-dependent). M is the mass, C is the damping coefficient, and $\ddot{x}$ and $\dot{x}$ are the acceleration and velocity, respectively.

The differences between the static and dynamic cases arise from the effects of inertia $M\ddot{x}$ and damping $C\dot{x}$ that do not participate in the static response. Usually, the effect of damping is small, but the inertia effect may be significant and can dominate the response whenever loading durations are much shorter than structural response times.

Furthermore, unlike the static case where the magnitudes of force and stiffness determine directly the corresponding deflection, in the dynamic domain the response (deflection, velocity, and acceleration) is obtained by solving Equation (10.1). The system response will depend not only on the magnitude of the force, but also on the relationship between dynamic characteristics of the force and frequency characteristics of the structure. A detailed discussion of these issues is presented in Chapter 6 and other references (Biggs, 1964; Clough and Penzien, 1993; Tedesco, 1999; ASCE, 1985). The various design manuals cited above contain dynamic response charts and tables based on SDOF considerations, and these can be used for design, as discussed in previous chapters.

Based on this simplified approach, a designer must determine the following parameters as part of the design process: loading conditions, structural mass, structural damping, and structural stiffness (or resistance for nonlinear systems). These parameters are used for evaluation of the expected response and are varied until the design meets the present behavior criteria. It should be noted that damping effects can be ignored during initial analysis considerations, as noted in the various design manuals.

10.8 STRUCTURAL SYSTEM BEHAVIOR

Structural members, when subjected to extreme loads, are expected to respond inelastically. This inelastic response mobilizes the fully plastic or ultimate capacity of the section, at which point no additional load can be resisted. Indeterminate structures are better able to redistribute additional loads to the remaining undamaged stiffer elements, whereas determinate structures undergo unrestrained deformation. The nonlinear response depends on the redundancy of the structure, its ability to provide multiple load paths if a localized plastic behavior develops, and the capacity of the structure to deform until the load is removed. This inelastic deformation capacity is termed its ductility and may be calculated as the ratio of ultimate deformation to yield deformation (or the ratio of ultimate rotation to yield rotation).

The ductility of a member is typically associated with its capacity to dissipate significant amounts of inelastic strain energy and is an indicator of the extent to which a structure may deform before failure. Redundancy and ductility both contribute to the post-elastic capacity of a structure to resist extreme loads. The ability of a section to attain its ultimate capacity and continue to maintain this level of response while deforming plastically requires an attention to details. The section must be properly detailed to ensure that it can withstand the large deformations associated with plastic behavior while maintaining its level of resistance. Ultimately, if the loading persists and produces deformations that exceed the ductility limits or rotational capacities associated with the material and construction, the section will fail.

When subjected to extreme loads, structures may fail in a variety of ways. Depending on the characteristics of the loading, proximity, and intensity of the blast to a structural member, the response of a structure will determine the resulting mode of failure. A contact or close-in explosion produces a cratering effect on the near side of the element and spalling on the opposite face. These two damage mechanisms weaken the section, and when the zone of spall overlaps the cratered region the section is breached. The capacities of different materials to resist cratering, spalling, and ultimately breaching dictate the thickness required to maintain structural integrity. Typically, material behavior dictates the modes of deformation and the resulting patterns of failure. Other modes of local failure that may result from direct blast involve shear failures.

Typically, concrete materials are weak in tension. When subjected to a principal tension that exceeds the tensile capacity of the material, the reinforcing steel that crosses the failure plane holds the aggregate together, allowing shears to be transferred across the section. When the capacity of the reinforcing steel is reached, its ability to hold the sections together across the failure plane is diminished and the ability of the section to resist shear is lost. Similarly, flat plate slabs must transfer the loads to the columns through shear at the column–slab interface.

A Comprehensive Protective Design Approach

Although the area of this interface may be increased by means of drop panels, dowel action, and shear heads, the capacity of a slab to transfer the load to the columns is limited by the shear capacity of the concrete. Extreme loads on short deep members may result in a direct shear failure before the development of a flexural mode of response. This direct shear capacity is typically associated with a lower ductility and a brittle mode of failure.

Each of the preceding modes of failure is associated with a fundamental frequency of response and therefore influences the characteristics of the load that the member feels. The dynamic response of the structural element to the transient blast loads will determine how much deformation the element undergoes. The dynamic response analysis may be a multiple-degree-of-freedom finite element representation of the structure; however, in the design of conventional structures to respond to blast loads, this is often unnecessary.

As stated earlier, SDOF representations of the structural response, in which the analyst has postulated the anticipated mode of response, are effective and efficient methods of accounting for the transient nature of blast loads. Principles of elementary vibration analysis reveal the de-amplification of response that results when a short-duration load is applied to a flexible system. Highly impulsive loads are no longer acting on a structure by the time the structure reaches its peak response and the system may be idealized by a spring mass system set into motion by an initial velocity. Stiffer structural elements or modes of deformation associated with higher frequencies of response may respond to the same blast environment with a greater dynamic amplification factor. However, the characteristics of the higher frequency system will determine the peak deformations associated with the transient loading. Therefore, considering an elastic, rigidly plastic SDOF analogy for each of the various modes of failure will provide a reasonable means of determining the dynamic response of the element and its ability to deform within prescribed limits. Each analogy, however, depends on the proper detailing of the section to guarantee that the capacity can be achieved and the ultimate deformation can be withstood.

It was noted previously that the following four characteristic parameters must be known about a structure under consideration to employ an SDOF model of the structure: equivalent load F(t), mass M, stiffness K, and damping (ignored in most design manuals). Mass and stiffness parameters for the structural system under consideration are selected on the basis of the type of problem, load source, type of structure, and general conditions for load application to the structure, for example, localized load on a small part of the structure, a distributed load over a large part of the structure, etc., and the expected behavioral domain (linear elastic, elastic–perfectly plastic, nonlinear, etc.). The general approach for selecting such parameters has been discussed in detail by Biggs (1964), and most design manuals contain similar procedures. Neither Biggs nor the design manuals provide information on the treatment of fully nonlinear systems by SDOF simulations. Using such modified SDOF approaches has been shown to enhance the recommended design manual procedures, and they can be employed for analysis and design (Krauthammer et al., 1990).

One should distinguish between structural elements that are sensitive to pressure and those that are sensitive to impulse, as addressed in Chapters 6 and 8. This leads to the introduction of the pressure–impulse (P-I) diagram concept. The basic idea of a P-I relationship is not new. It is a direct outcome of applying a pressure pulse to a linear SDOF oscillator. One can compare the structural responses of different elements with the ratio t_d/T in which t_d is the duration of the applied load and T is the natural period of the element. The reader is reminded that $T = 2\pi/\omega$, in which $\omega = (K/M)^{0.5}$. K and M are the stiffness and mass for the equivalent SDOF, respectively.

It can been shown that if $t_d \gg T$, a structure will reach its peak displacement well before the load has diminished. Here, one can use the principle of energy conservation and show that the limiting peak displacement will be equal to twice the static displacement ($x_{max}/x_{st} = 2$). This behavioral domain has been called quasi-static or pseudo-static. If, however, $t_d \ll T$, the load will diminish well before the structure reaches its peak displacement. Again, one can use the principle of energy conservation and show that the limiting displacement ratio will be $x_{max}/x_{st} = 0.5 \omega t_d$. This behavioral domain is defined as impulsive. When $t_d \approx T$, the behavior is defined as dynamic and one needs to perform a dynamic analysis for deriving the structural response values. Furthermore, one can combine the information presented above into a normalized P-I diagram, as illustrated in Figure 10.5, and it can be used to define the type of expected response in a particular structure.

This approach enables designers to select appropriate analysis and design approaches for a particular case. This basic concept can be expanded by selecting different linear oscillators, each representing a different type of structural element, for deriving specific P-I diagrams for each case. Then, each element in the structure under consideration could be evaluated independently, and such individual behavior characteristics also could be used for high-explosive damage assessment.

Original P-I diagrams were based on deriving two theoretical asymptotes for the impulsive and quasi-static domains, and drawing arbitrary hyperbolas tangent to those asymptotes. Oswald and Skerhut (1993) describe the computer code FACEDAP, which is based on this approach and provides an approximate method to determine building vulnerability to explosive events. Damage calculations were made in two steps. First, damage in each structural component (beams, columns, walls, etc.) is calculated using P-I graphs that define various levels of damage.

These P-I curves are developed to predict component blast damage based on element type, structural properties, and blast loading environment. The levels of damage can be correlated with specific levels of protection, as discussed in other references (ASCE, 1999; Marchand et al., 1991). The damage calculated for each structural element can be combined in a weighted manner to derive a percentage of building damage. Building repairability and reusability are computed in a similar process. A similar approach was presented and discussed by Smith and Hetherington (1995). It is based on various approximate assumptions and should not be used for final design or for cases that require high degrees of accuracy.

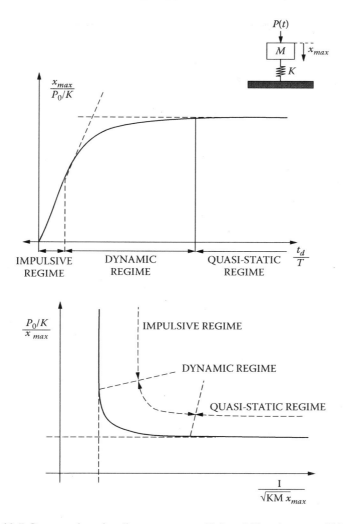

FIGURE 10.5 Pressure–impulse diagram concept (Soh and Krauthammer, 2004).

Nevertheless, the advanced treatment for deriving P-I diagrams, as presented in Chapter 8, provides innovative and accurate approaches for applying this concept in analysis and design.

Although each element within a structure responds locally to the blast load, the vertical and transverse loads imparted to the various structural elements must ultimately be carried to the foundation. This global capacity of a structure to locally resist direct loads and transfer them from element to element to the foundation is a less intense but not less demanding design criterion.

An earthquake shakes a structure over its entire foundation, developing inertial loads in proportion to the mass distributed throughout the building. Blast loads are typically applied to a facade with a local intensity that diminishes with range. The overall base shears to be resisted may be lower for the blast loading. However, the demands of the locally intense pressures may excite torsional modes and may exceed the local capacity of diaphragms to transfer the lateral loads to the primary lateral load-resisting systems. These locally intense effects may produce failures at base shears that are lower than would be associated with earthquake response. As a result, it is dangerous to equate damage levels associated with earthquake loads to those associated with blast loads. The desirable features of earthquake-resistant design (ductility, connection details, redundancy, and load redistribution) are equally desirable in blast design. The provision of seismic detailing to maintain the capacity of a section despite development of plastic hinges is equally desirable for resisting blast effects. The engineer has an obligation to guarantee that the full capacity of a section will be realized and that no premature failure will occur.

Blast pressures, particularly from short standoff bomb blasts that enter a building through openings in its facade, typically apply reflected loads to the undersides of floor slabs, with lower incident loads applied to the top side. This pressure imbalance will apply an uplift force to the columns and, depending on the magnitude of dead loads compared with the uplift forces, the columns may briefly be put into tension. In addition to cracking the concrete in direct tension, the tensile forces reduce the shear capacity of the columns. These columns must be designed and detailed to withstand the tensile forces and can resume carrying the downward gravity loads after the explosion.

The slabs must be detailed to resist upward forces by providing continuous reinforcement, both top and bottom. Although upward blast pressures must overcome the dead and live gravity loads that the slabs normally support, an analysis is required on a case-by-case basis to see whether the upward loaded slab capacity is sufficient. Similarly, prestressed and pretensioned structural systems are designed to amplify the capacity of a section to resist gravity loads; however, upward blast pressures may load these structural elements in modes for which they have little ultimate capacity. Prestressed concrete structures are therefore inherently dangerous in a blast environment in which the loads may be applied in any conceivable orientation. Precast construction typically lacks the kind of continuity that enhances the capacity of a structure to redistribute loads to fewer damaged elements. Precast construction also relies on mechanical connectors at discrete locations that may be damaged in a blast environment, leaving the panel detached from the remaining structure.

Designing a section to resist high-intensity localized blast effects may not always be reasonable, particularly if the location of the charge is not known. Therefore, a concept of limited and confined damage must be accepted. According to this concept, structural members immediately opposite the blast are allowed to fail if the failure is confined and does not progress throughout the building. The ability of a structure to redistribute loads as a structural element fails depends on the redundancy of the structural system. Provided the structural members in

A Comprehensive Protective Design Approach

way of the redistribution, (elements that provide the alternate load path) can withstand the additional load, the failure will be arrested.

The containment of failure to a localized region is the objective of a progressive collapse analysis. Once the concept of a tolerable level of failure is accepted, this analysis is essential. Although a transient dynamic inelastic analysis of the structural system is required to determine the extent of inelastic deformation and to trace the redistribution of loads throughout a structure, this is seldom done. The more common approach is to remove a single member, typically a column, and re-analyze the structure in the modified configuration. There are many scenarios in which the damaged structure will redistribute the forces. The removal of a column may require a horizontal member to carry the full weight of the columns above it and to span the failure. It is much more likely, however, that each floor will deform a significant amount and carry its own weight to the adjacent column line. The true redistribution of forces will depend on the relative stiffness and capacity of the elements surrounding the failure.

The means by which loads are redistributed to an adjoining structure may be through flexural behavior and the development of fully plastic hinges, and/or through catenary action. In this extreme scenario, the load carrying capacity of a member transforms from its flexural capacity to its axial capacity, as if the member transformed into a cable spanning between the remaining columns. Catenary behavior is therefore associated with very large deformations as the geometry of the deformed structure is required to satisfy equilibrium. It is essential that these elements be adequately detailed so that they will be able to carry the redistributed loads in the manner intended by the analyst.

This requirement has profound implications for reinforced concrete structures in which reinforcement is lap spliced. Making assumptions regarding the ability for the forces to redistribute is not sufficient unless provision is made for these forces to be carried by the surviving structural elements. In a progressive collapse analysis, a structure must be able to redistribute the dead weight and part of the live load; however, the failure of a column will be sudden and the redistribution of the loads will be dynamic. Therefore, dynamic amplification factors accounting for this sudden transfer of the load must be considered in the progressive collapse analysis.

Transfer girders and the columns supporting transfer girders are particularly vulnerable to blast loading. This form of construction poses a significant impediment to the safe redistribution of load in the event a girder or the columns supporting a girder are damaged. Typically, a transfer girder spans a large opening such as a loading dock or provides the means for shifting the location of column lines at a particular floor. Damage to a girder may leave several lines of columns that end at the girder from above totally unsupported. Similarly, the loss of a support column from below will create a much larger span that must carry the critical load-bearing structure. Transfer girders, therefore, create critical sections in which a loss results in a progressive collapse.

If a transfer girder is required and may be vulnerable to an explosive loading, it should be continuous over several supports. Furthermore, a substantial structure framing into the transfer girder to create a two-way redundancy and thus an alternate load path in case of failure is recommended. Finally, alternatives such as Vierendeel trusses or inclined columns that may be less vulnerable to a widespread, progressive collapse should be considered.

Designers should be concerned with the responses of structural materials (such as steel or structural concrete) to high loading rates, such as those induced by impact and blast. Under these conditions, the strength could increase significantly; current design manuals (Department of the Army, 1986 and 1990) provide recommendations on practical values of strain rate enhancements for such materials, as discussed in Chapter 5. It has been suggested that one may apply the average strain rate to calculate enhancement factors for the stress parameters of the material models, then employ the modified stress-strain relationships for deriving the resistance functions used in the dynamic analysis.

10.9 STRUCTURAL SYSTEM AND COMPONENT SELECTION

Different loading environments may require the applications of different types of structural systems, and the selection would be typically based on a facility's estimated effectiveness under the expected conditions. Such consideration would include the following general possibilities: above-ground, partially buried, or fully buried structures. For each of these cases, selecting one of several structure types (frame, box, shell, or a combination) would be possible. Construction materials must be selected for the particular application, considering both the structure and the backfill around it. Obviously, many factors will be involved in such a selection and the considerations of appearance, efficiency, and cost should be included. One of the most important issues that should be addressed is the problem of facility robustness or structural redundancy. A designer should ensure that the loss of a single structural component would not lead to extensive loss of function. Architectural features and the type of building construction also play a role in determining explosive scenarios and element loadings. Some of these effects are discussed next.

The tributary loading area for primary structural members can be reduced, for example, by designing walls that can break away from the columns and thus lessen loads on columns. Another option is to shield critical structural elements with sacrificial energy-absorbing panels. This approach enables the pressure to "flow" around the critical element. By the time the shielding panel is destroyed, the pressures on the front and rear faces of the critical element are approximately equal, thereby reducing the risk of its failure.

The magnitude of the quasi-static part of a blast load can be greatly reduced by venting the detonation products outside or into adjacent portions of a structure to limit the explosive loading to that associated only with the shock front. Venting can be achieved in a variety of ways such as using nonstructural walls that quickly break, using frangible secondary structural elements for the walls, floors, and

A Comprehensive Protective Design Approach

roofs that can "easily" break away from primary structural elements, or using a relatively narrow building cross-section with many "soft" walls so that a blast can quickly vent to the outside.

The sizes and shapes of structural members can influence their loadings and chances of survival. For example, one can use steel columns instead of reinforced concrete to reduce loaded areas, fill the spaces between the flanges of a steel beam with concrete (to provide added mass and reduce risks of flange damage), or use circular rather than rectangular cross-sections (to provide less load transfer to members).

Compared with other construction materials, well designed, cast-in-place reinforced concrete structures generally provide the highest levels of protection against explosive loads. To be effective, the following measures are recommended (ASCE, 1999):

- Two-way slab systems supported by beams on four sides
- At a minimum, seismic detailing at supports; blast-resistant details should be preferred
- Adequate shear reinforcement
- Continuous top and bottom reinforcement in slabs
- Progressive collapse provisions

For cast-in-place reinforced concrete design, the following minimum properties are recommended (actual dimensions should be determined by proper analysis):

- 28-day compressive strength of concrete: 4,000 psi
- Yield strength of reinforcing steel: 60,000 psi

Minimum dimensions recommended for cast-in-place reinforced concrete components are:

- Floor slab thickness: 8 in. minimum
- Rectangular column width: 12 in. minimum
- Joist width: 4 in. minimum
- Exterior wall thickness: 10 in. minimum
- Floor-to-floor height: 16 ft

It cannot be overemphasized that detailing such as connections between structural members (beams and columns, beams and slabs, columns and slabs) is critical for ensuring a monolithic response of the structural system. Proper anchorage of reinforcing bars and adequate shear reinforcement are necessary to enable the load transfer between the individual members. Although these issues continue to be studied (Krauthammer, 1998), existing information such as that contained in TM 5-1300 (Department of the Army, 1990) should be used to address such details.

Somewhat less effective, but also providing a significant level of protection, are other support systems using cast-in-place reinforced concrete, including one way joint systems, waffle slabs, and flat slabs with drop panels. Again, proper anchorage and shear reinforcement are required to make these systems effective. Reinforced masonry and precast concrete can also be designed to achieve modest levels of blast protection (up to 10 psi loading). TM 5-1300 (Department of the Army, 1990) contains provisions for these materials.

While this list is not exhaustive, it does show that some nontraditional thinking in the design process can improve the survivability of structural elements affected by a blast loading. Although typical design manuals do not provide much help on this process, it can be best illustrated by the following brief discussion.

A multistory office building to be located in a center-city business district that could be a target to terrorist activities will typically be a frame-type structure, possibly combined with shear walls. The structural appearance could require significant amounts of glass and several entrances. Parking facilities would usually be located under the structure. The design must include considerations of securing the parking facility against internally placed explosive devices, and ensuring sufficient robustness of the structural systems in that area. The use of energy-absorbing structural components for protecting the main structural frame components and/or other parts of the structure could be considered.

Although access control may not be very desirable for facilities that must be open to the public, some measures of monitoring human and vehicle traffic could be implemented successfully. Furthermore, externally placed explosive devices can cause significant damage to a fragile building envelope and measures must be taken to reduce such effects. This protection can be achieved by using special glazing materials and structural components connecting the glazing to the structure. The structural geometry (shape) and the incorporation of shielding devices with the external face of the building (considering both structural and aesthetic factors) can affect the outcomes of such incidents. Additional attention must be given to the protection of utilities (electric, telephone, water, etc.), computer systems, and areas with valuable assets (records, currency, etc.). Underground parking systems should enable blast venting to release the pressure quickly into the atmosphere. Furthermore, shielding critical structural components from the combined effects of blast and projectiles is important. The use of sacrificial barriers (expected to absorb a large amount of energy before failing) could be very cost-effective since they can be readily replaced.

The design of a control room in a petrochemical plant requires different considerations (ASCE, 1997). A control room is usually much smaller and contains considerably fewer assets. Nevertheless, control rooms contain personnel and equipment that are critical for the operation of the entire plant. The anticipated hazardous loading environment is usually related to an accidental explosion in the plant. Many physical security difficulties highlighted for the office building can be eliminated and the design would probably be based on an above-ground box-type structure. Here too, the use of sacrificial panels could be an attractive approach for enhancing existing structures or serving as integral components in

A Comprehensive Protective Design Approach

new construction. A third case could be the design of storage facilities for valuable assets that could be targets of either accidents or hostile activities (depending on the stored materials). The structure could be a system of underground boxes, shells, and tunnels. Access control would clearly be a key element for protection against external threats, while explosive safety measures would protect the facility against localized internal events. The design of above-ground petrochemical storage facilities was discussed in substantial detail by Thielen et al. (see Concrete Society, 1987) and Thielen and van Breugel (see Clarke et al., 1992).

Selection of materials for each of these cases should include safety and physical security aspects. For example, the type of glass to be used in the office building and the method of glazing installation are important (this issue, for example, is discussed in more detail in Chapter 5 of ASCE, 1999). Expecting that such structures must exhibit significant amounts of ductility is reasonable (designing such facilities for the elastic range is uneconomical), and the selected materials and structural details must fit these requirements. Furthermore, the performance criteria for each of these structures could be different, as discussed earlier. The selection of proper detailing is a very important issue, and the weak points in many designs are the connection details.

Analysis procedures would also be different. Usually, control rooms (i.e., box-type structures) can be designed and evaluated by advanced SDOF models (as described in Chapters 6 and 8) that can simulate the responses of associated structural components. Office buildings, production facilities, and complex underground storage systems composed of many different structural elements whose behavior may not be included in SDOF models could require the application of advanced numerical programs such as finite element codes. However, sophisticated computational tools cannot always replace sound engineering judgments, and designers should verify such computations by simple checks of force equilibrium and deformation compatibility.

Although this chapter is primarily aimed at structural engineering aspects, other important factors must be considered in designs for physical security. One such area outside the scope of this chapter is human tolerance to the threats and loading environments discussed in TM 5-1300 (Department of the Army, 1990). Besides the physiological considerations (human responses to pressure, shock, motion, heat, etc.) one must look also into psychological issues and their combined effects on personnel.

Another such area not defined as a structural engineering issue is the design of utilities entering a structure. This topic is very important. Even a structurally well designed facility can be rendered useless if certain utilities (power, communications, etc.) are disrupted. Designers should pay attention to how the facility is connected and serviced by such features.

10.10 MULTI-HAZARD PROTECTIVE DESIGN

Physical security requirements may not be the only loading environments to be considered. Additional loading (wind, seismic, fire, etc.) requirements often must be incorporated into a design. These different loading environments should not be treated separately. A comprehensive effort must be made to develop a "loading envelope" that includes the various conditions for a specific case. Obviously, such a combined consideration of the entire spectrum of conditions will enable the development of an efficient, cost-effective design. If a facility is designed for seismic effects, the incremental cost of adding specific physical security capabilities is low, compared with the cost of adding the same features to a structure without seismic resistance capabilities. This synergetic effect represents the concept of multi-hazard design and designers are encouraged to employ this approach whenever possible.

10.11 OTHER SAFETY CONSIDERATIONS

This chapter emphasizes design for physical security. However, many other safety-related issues could be considered, for example, the safety of industrial facilities subjected to accidental explosions. Although the loading environments are different, many structural design considerations are very similar. It is recommended that designers who wish to specialize in the general field of blast-resistant design be familiar with all such design literature (the manuals and specialized books cited in the various chapters).

Finally, an issue that can complicate physical security design requirements and other issues requiring protection against severe short-duration dynamic effects are access requirements for the disabled. Conflicts exist between access requirements aimed at making it easy to enter a facility and physical security demands aimed at making it difficult. This conflict must be resolved in the very early stages of the design process to avoid costly delays during the construction process or after the facility is ready for operation.

10.12 DEVELOPMENT AND IMPLEMENTATION OF EFFECTIVE PROTECTIVE TECHNOLOGY

The information presented in this book about our understanding of anticipated hostile environments and facility responses raised several practical issues regarding current methods for hardened systems, analysis, and design. Clearly, significant knowledge has been gained since World War II and events since then have taught us that threats from conventional and unconventional weapons should be taken very seriously. Nevertheless, additional research in many areas is badly needed and such needs must be adjusted to reflect changes in weapon systems, local and international conflicts, strategic and tactical concepts, and the capabilities of experimental and analytical tools. Unfortunately, both geopolitical and weapons technology changes outpace progress in the development of protective measures. It would be wise to address such issues actively and continuously in order to avoid tragic outcomes.

A Comprehensive Protective Design Approach 435

The following sections focus on specific technical issues and on the various analysis and design activities required for blast, shock, and impact mitigation. Chapter 1 contained several recommended actions related to the comprehensive development of effective protective science and technology. Since these actions are critical for supporting national defense and homeland security needs, they are summarized below and specific research recommendations based on observations in previous chapters are presented.

10.12.1 RECOMMENDED ACTIONS

As noted in Chapter 1, the following actions are recommended:

- To mobilize the international scientific community (including government, private, and academic sectors).
- To establish comprehensive and complementary long-and short-term R&D activities in protective technology to ensure the safety of international government, military, and civilian personnel, systems, and facilities under evolving terrorist threats.
- To develop innovative and effective mitigation technologies for the protection of government, military, and civilian personnel, systems, and facilities from terrorist attack.
- To launch effective technology transfer and training vehicles that will ensure that the required knowledge and technologies for protecting government, military, and civilian personnel, systems, and facilities from terrorist attack will be fully and adequately implemented.
- To establish parallel and complementary programs that address the nontechnical aspects of this general problem (culture, religion, philosophy, history, ethnicity, politics, economics, social sciences, life science and medicine, etc.) to form effective interfaces between technical and nontechnical developments and implement a comprehensive approach to combat international terrorism.

Adopting these steps will initiate a sequence of activities that address critical scientific and technical needs that are fully compatible with various other ongoing activities at land-, air-, and sea-based systems and facilities. Such compatibility is essential for ensuring that the required protective technologies address a broad range of critical national needs. The process is expected to involve a sequence of complementary activities, from basic research through implementation as described below.

10.12.1.1 Basic and Preliminary Applied Research

- Define and characterize design threats and corresponding loads; perform hazard and/or risk analyses, and determine corresponding consequences on assets, missions, and people.

- Predict and/or measure facility and/or system response to design threat loads, including considerations of characteristic parameters (material properties, geometry, structure/system type, assets, mission requirements, etc.).
- Develop theoretical and/or numerical retrofit and/or design concepts to bring a facility and/or system response and consequences within acceptable levels.

10.12.1.2 Applied Research, Advanced Technology Development, Demonstrations, and Validation Tests

- Develop retrofit and/or design concepts to bring the facility/system response and consequences within acceptable levels.
- Verify retrofit and/or design concept through laboratory and/or field tests and develop final design specifications.

10.12.1.3 Demonstration, Validation, and Implementation

- Conduct certification and/or validation tests.
- Implement retrofit and/or design technology transfer, guidance, and/or training.

These activities should be conducted internationally through national centers for protective technology research and development (NCPTR&D). These centers will direct, coordinate, and be supported by collaborative government, academic, and industry consortia that will perform various parts of the activities mentioned above. National academic support consortia (NASC) should be established to engage in this critical effort through both research and education activities. These NASCs will identify and mobilize faculty members from universities with appropriate scientific and technical capabilities, and lead some of the required R&D.

The NCPTR&Ds and consortia members will be staffed by a unique team of internationally known experts in all scientific and technological areas relevant to protective technology and will have access to advanced research facilities at all sites. Team members should have documented extensive experience in protective technology, developing and managing major research initiatives, developing and implementing innovative protective technologies, and training of both military and civilian personnel in the applications. Each collaborative effort should be conducted under the general guidance and oversight of an advisory committee consisting of internationally recognized technical experts and senior public, government, and military leaders in relevant fields. Specific technical guidance will be provided through combined government–academic–industry management teams with subgroups formed to focus on key S&T areas.

10.12.2 EDUCATION, TRAINING, AND TECHNOLOGY TRANSFER NEEDS

As technology is developed, it transitions to testing and evaluation to determine whether a technology is applicable for a given application. After several iterations, such technology is transferred to operational testing for evaluation under realistic conditions. Upon completion of this evaluation phase, acquisition and operational training occur. Training for known threats relies on a predetermined course of action. Some adversarial actions may be anticipated and countermeasures can be practiced during training. Criticism followed certain incidents because of failure to anticipate threat evolution and failing to train for it. This is also a shortcoming of conventional training for first responders; they are trained to respond to known conditions and may not be able to respond adequately under different conditions. This must be corrected by educating personnel to understand the possible threats and the ability of available technology to deal with them. The appropriate people should be able to modify their actions to address such threats intelligently, and hopefully develop preemptive measures.

This structured approach does not exist yet in the general field of protection from weapons of mass destruction (WMD). Furthermore, military solutions are often incompatible with civilian modes of operation and may be too rigid or too expensive to implement in nonmilitary organizations. Leaving this process to commercial vendors may be another option, but quality control and costs for commercial technology are frequently controversial. Further, the time available for training appropriate persons (engineers, security specialists, emergency and rescue operations staff, etc.) is limited compared to time available to train military personnel. The Department of Homeland Security (DHS) is expected to address these issues through collaboration with industry and academic institutions. One would expect similar collaborations with other government agencies. Universities will have to be involved to an extent beyond basic research. They will also have to play an integral role in functioning as think tanks and transferring the developed knowledge and technology to the end users.

Many government agencies and their supporting industrial organizations have a critical need to attract and/or develop employees with experience in protective science and technology as their current workforces age and reach retirement age. Since the mid-1980s, a gradual decline has taken place in academic protective technology-related R&D activities, along with the involvement of academicians in these R&D efforts. As a result, very few eminent academicians in this field are still available in the U.S., most are not supported by Department of Defense R&D funding, and no formal engineering training exists in the area of protective science and technology at U.S. universities. The situation is similar in most other developed countries.

Establishing government–academic–industry consortia in various countries with a mandate to develop new and cost-effective protective technologies, and train current and future engineers and scientists should be seriously considered. Such programs foster the input of fresh ideas and provide students with first hand experience from their work on real programs that can benefit national and international needs. This will allow nations to maintain protective technology knowledge centers that will ensure the development and broad dissemination of relevant technologies to all appropriate parties (law enforcement, military, various government agencies, as well as private organizations). The proposed consortia will address WMD threats through a multi-step science and technology (S&T) plan that provides an end-to-end solution, from R&D through system development, design certification, guidance, education and training, and technology transfer in all relevant areas.

10.12.3 Recommended Long Term Research and Development Activities

The following list of recommended long-term research activities has been developed as a result of many discussions with colleagues and sponsors. These R&D activities can be conducted over the next 3 to 5 years. They are needed to develop much more effective solutions to problems that can be currently addressed only with empirical and conservative approaches. These expected contributions will not only benefit specific projects, but also make a profound contribution to various critical national facilities, national and international defense, and homeland security.

The investment in the proposed approach will enable both meaningful technological enhancements and large cost savings in providing the required protection to society. Furthermore, these cost savings are estimated to be far larger than the cost of the recommended R&D. As noted earlier, essential research must be conducted also in several important areas related to terrorism and low intensity conflicts (history, sociology, politics, culture, economics, information transfer and media, religion, life sciences and medicine, psychology, etc.). These areas, however, will not be addressed here. The following material focuses on R&D related to scientific and technical issues directly related to enhancing the survivability of critical infrastructure systems and facilities. Recommended experimental and testing capabilities are identified for each of the following items by underlying the R&D topics that require such support.

1. Protection methodology, threat, risk assessment, its mitigation and resource allocation.
2. Load and environment definition (each of the following requires testing support):
   - Address time scales from microseconds to days and more.
   - Address explosive, fire, nuclear, chemical, and biological (NCB), and combined effects.
   - Study effects from site conditions.
   - Address both external and internal attack conditions for typical buildings, considering a broad range of blast wave propagation, complex geometries, and blast–structure interaction effects.

A Comprehensive Protective Design Approach

- Define the combined load arising from blast, primary and secondary fragments.
- Define and characterize load transfer through various single or multiphase media, including facility envelopes.
- Study and define short-duration loading conditions on critical structural elements (columns, etc.) as a function of the interaction between a blast and a typical building system.

3. Material behaviors under single and combined loading environments (each of the following requires testing support):
   - Address constant and/or variable loading rate effects.
   - Study behaviors of typical construction materials under blast loads and their constitutive formulations.
   - Address scaling and size effects.
   - Study the combined effects of loading rate and structural size.
   - Address high and low temperature effects.
   - Define and characterize material aging effects, and how they can influence any of the issues noted above.

4. Computational capabilities:
   - Precision impact tests for obtaining well defined data on the behaviors of simple structural elements in support of computer code validation.
   - Use of precision test data for the validation of computer codes that could be used for the numerical simulation of loading on buildings and building responses under blast effects. Define the accuracy of such computer codes and select those that are suitable for application.
   - Deterministic and probabilistic numerical capabilities.
   - Time and frequency domain computations.
   - Artificial intelligence and neural network capabilities.
   - Hybrid and/or coupled symbolic-numeric computational capabilities.
   - Computational structural dynamics (CSD) simulations and computer code validation.
   - Computational fluid dynamic (CFD) simulations and computer code validation.
   - Coupled CFD–CSD–thermomechanical simulations and required computer code validation.
   - Applications of Eulerian, Lagrangian, and/or combined numerical approaches.

5. Study behavior and effects of building enclosure (each of the following requires testing support):
   - Study behaviors of typical building envelopes under blast, shock, and impact loads.
   - Develop innovative building enclosure technologies to enhance protection from blast, shock, and impact.

- Study the performance of joints, particularly two-stage control joints, in building enclosures.
- Develop design criteria for building enclosures to enhance their performance under explosive loads.
- Study behaviors of closures, attachments, and penetrations in typical buildings under blast loads.
6. Building and structural science and behavior (each of the following requires testing support):
    - Study the behavior of structural detailing in buildings (structural concrete, structural steel, and composite construction) for enhancing their performance under blast loads.
    - Study and develop innovative structural retrofit options for enhancing the performance of typical buildings under blast loads.
    - Study how to shield and protect critical structural elements from blast effects for enhancing building robustness.
    - Develop effective measures for typical buildings to enhance their explosive safety under internal detonations.
    - Study behaviors and detailing of typical nonstructural elements (partitions, ceilings, ducts, lighting, etc.) under blast loads.
    - Develop design and construction recommendations for nonstructural elements.
    - Develop effective protection and integration procedures for mechanical and electrical systems in buildings subjected to blast effects.
    - Study combined effects of blast, primary and secondary fragments, and missiles on structural and nonstructural systems and techniques to protect such systems against such effects.
    - Study effects of openings in floors and walls in typical buildings under blast loads on their behavior and performance; develop appropriate design guidelines.
    - Determine structural behavior, design, construction and detailing, performance, and safety.
    - Devise perimeter protection systems (gates, walls, fences, landscape features, etc.).
    - Understand, characterize, and prevent progressive collapse.
    - Study nonstructural element behavior under blast, shock, and impact effects.
    - Study building and structural behavior under internal detonations.
    - Address structural systems unique to transportation infrastructure facilities (bridges, tunnels, ports, etc.).
    - Improve design codes.
7. Facility and system behavior under WMD environments: combine knowledge gained from R&D on the topics noted above to address complete facility behavior and performance.
8. Address multi-facility conditions; consider all the issues noted above to address large-scale WMD attack conditions that can affect many

facilities in various geographic areas (parts of cities, industrial complexes, etc.).
9. Facility assessment:
   - Development of both rigorous and practical guidelines for assessing structural damage and robustness before and after explosive incidents.
   - Pre- and post-incident condition and/or damage assessments in support of evacuation, rescue, and recovery (ERR) activities. These require consideration of both temporary shoring and demolition activities, rescue and construction equipment utilization, emergency supply chain applications, etc.; include applications of intelligent computer support tools and instrumented monitoring to provide expert advice during emergency operations.
10. Environmental effects (each of the following items requires testing support):
    - Study material and structural performance under various threat conditions, and controlling external and internal facility environments.
    - Study possible effects of low or high temperature and moisture on explosive loads, material and structural response, and behavior under blast, shock, and impact.
    - Study effects of time on the performance of building components and systems under blast effects under normal service conditions.
11. Technology transfer, education, and training:
    - Develop effective education, technology transfer, and training for the areas noted above that will ensure that the knowledge gained from the proposed research can be used and implemented by engineers, designers, and security personnel involved in force protection and industry.
    - Implement and integrate these activities with basic and advanced engineering education programs to ensure the long term supply of qualified personnel in this important field.
12. Use the knowledge gained from these R&D efforts to augment design recommendations for seismic and wind effect mitigation. This will result in the development of a uniform multi-hazard protection design approach for facilities subjected to abnormal loading conditions.

These recommendations should be reviewed periodically during the life of the proposed program and modified based on the findings and developments.

10.13 SUMMARY

This book is focused primarily on scientific and engineering issues in the general area of critical infrastructure facility survivability. It provides background on analysis and design capabilities in protective science and technology and recommendations for long-term research to address serious needs in this general area.

We must develop much more effective solutions to problems that can be currently addressed mainly with conservative and/or empirical approaches. The recommended R&D should be carried out by a well coordinated partnership of government, universities, and industry and is expected to be an effective approach for addressing critical topics that require innovative solutions.

The primary focus in the survivability of physical facilities is the protection of people and critical equipment. The emphasis should not be on simple fortifications (although consideration of modern hardened facilities should not be abandoned), but on the wide range of typical facilities used to support the needs of a modern society. The structural robustness and integrity of such facilities and the safety provided by them with regard to abnormal loadings such as explosion and impact (especially incidents involving deliberate, well planned efforts to maximize damage and consequence) are of great concern. Of particular interest are multi-unit, multi-story (five or more floors) civilian buildings (housing, offices, laboratories, related infrastructure systems, etc.). Of course these considerations are important for all building types but multi-unit buildings usually of a particular form of layout and construction are susceptible to progressive collapse and are inviting targets for terrorist attacks.

The U.S. has a strong need for innovative design and construction guidelines for enhancing the blast resistance of building systems (enclosure components or systems and corresponding structural frames) for above-grade walls. Of particular interest is the development of new blast-resistant wall systems with effective performance under normal environmental conditions (rain, sun, wind, etc.) or innovative retrofit measures for existing walls. Considerable research interest relates to wall configurations that have the potential for shielding and/or dissipating the effects of explosions. Additionally, we must develop a better understanding of the dynamic interactions of various combinations of structural and nonstructural building components, their effects on building performance, and the effects on personnel protection. The minimization of debris is an important consideration in any attempt to control the consequence of an explosion. The use of innovative structurally composite systems, and/or materials such as corrugated core metallic panels, polycarbonate, polypropylene, and various woven and non-woven fabrics in both new construction and retrofits of existing buildings has considerable potential for protection from external and internal explosive loading.

Although much information can be obtained from laboratory and/or field testing, relying only on such an approach would be too costly and not very efficient. The use of advanced computer simulation techniques along with well selected tests can significantly lower costs and enhance the overall efficiency and productivity of the required R&D.

References

Abeysinghe, R.S., Gavaise, E., Rosignoli, M., and Tzaveas, T., Pushover Analysis of Inelastic Seismic Behavior of Greveniotikos Bridge, *Journal of Bridge Engineering*, 7, 115, 2002.

Achenbach, J.D., *Wave Propagation in Elastic Solids*, North-Holland Publishing, Amsterdam, 1973.

Allgood, J.R. and Swihart, G.R., Design of Flexural Members for Static and Blast Loading, American Concrete Institute Monograph 5, 1970.

ACI 318-05, Building Code Requirements for Structural Concrete and Commentary, American Concrete Institute, Farmington Hills, MI, 2005.

ACI-ASCE Committee 352, Design of Beam–Column Joints in Monolithic Reinforced Concrete Structures, *ACI Structural Journal*, 82, 266, 1985.

ACI-ASCE Committee 352, Design of Slab–Column Connections in Monolithic Structures, *ACI Structural Journal*, 85, 675, 1988.

ACI-ASCE Committee 352, Recommendations for Design of Beam–Column Joints in Monolithic Reinforced Concrete Structures, Report ACI 352R-91, American Concrete Institute, Farmington Hills, MI, June 1991.

AIA, Security Planning and Design, American Institute of Architects, Washington, 2004.

AISC, Load and Resistance Factor Design: Structural Members, Specifications and Codes, American Institute of Steel Construction, Chicago, 1994.

AISC, *Steel Construction Manual*, 13th ed., American Institute of Steel Construction, Chicago, 2005.

AISC, *Resisting Blast Effects and Progressive Collapse in Steel Structures: Design Guide*, American Institute of Steel Construction, Chicago, 2006.

Aristizabal-Ochoa, J.D., Story Stability of Braced, Partially Braced, and Unbraced Frames: Classical Approach, *Journal of Structural Engineering*, 123, 799, 1997.

ASCE, *Design of Structures to Resist Nuclear Weapons Effects*, Manual 42, American Society of Civil Engineers, Reston, VA, 1985.

ASCE, Design of Blast-Resistant Buildings in Petrochemical Facilities, American Society of Civil Engineers, Reston, VA, 1997.

ASCE, Design for Physical Security: State of the Practice Report, American Society of Civil Engineers, Reston, VA, 1999.

ASCE, Minimum Design Loads for Buildings and Other Structures, Standard ASCE 7-02, American Society of Civil Engineers, Reston, VA, 2002.

ASCE, The Pentagon Building Performance Report, American Society of Civil Engineers, Reston, VA, January 2003.

Astbury, N.F., Brickwork and Gas Explosions, Technical Note 146, British Ceramic Research Association, London, November 1969.

Astbury, N.F., West, H.W.H., Hodgkinson, H.R., Cubbage, P.A., and Clare, R., Gas Explosions in Load-Bearing Brick Structures, Special Publication 68, British Ceramic Research Association, London, 1970.

Baker, W.J., Wave Propagation Studies in Laterally Confined Columns of Sand, AFWL-TR-66-146, Air Force Weapons Laboratory, June 1967.

Baker, W.E., *Explosions in Air*, Wilfred Baker Engineering, San Antonio, 1983.

Baker, W.E., Cox, P.A., Westine, P.S., Kulesz, J.J., and Strehlow, R.A., *Explosion Hazards and Evaluation*, Elsevier, Amsterdam, 1983.

Balint, P.S. and Taylor, H.P.J., Reinforcement Detailing of Frame Corner Joints with Particular Reference to Opening Corners, Technical Report 42.462, Cement and Concrete Association of Australia, February 1972.

Ball, J.W., Weathersby, J.H., and Kiger, S.A., Equipment Response to In-Structure Shock, U.S. Army Engineer Waterways Experiment Station, Final Report SL-91-4, March 1991.

Ball, J.W., Woodson, S.C., and Kiger, S.A., Conventional Weapon-Induced Shock Environment in a Prototype Structure, Proceedings of International Symposium, Interaktion Konventioneller Munition mit Schutzbauten, NATO, Mannheim, Germany, April 1991, p. N-35.

Bangash, M.Y.H., *Impact and Explosion: Analysis and Design*, CRC Press, Boca Raton, FL, 1993.

Bathe, K., *Finite Element Procedures*, Prentice Hall, New York, 1996.

Batsanov, S.S., *Effects of Explosions on Materials*, Springer-Verlag, Berlin, 1994.

Baûant, Z.P. and Planas, J. *Fracture and Size Effect in Concrete and Other Quasibrittle Materials*, CRC Press, Boca Raton, FL, 1998.

Baûant, Z.P. and Zhou, Y., Why Did the World Trade Center Collapse? Simple Analysis, *Journal of Engineering Mechanics*, 128, 2, 2002.

Beck, J.E. et al., Single-Degree-of-Freedom Evaluation, Final Report AFWL-TR-80-99, Air Force Weapons Laboratory, Kirtland AFB, NM, March 1981.

Ben Dor, G., *Shock Wave Reflection Phenomena*, Springer Verlag, Berlin, 1992.

Berg, G.V., *Elements of Structural Dynamics*, Prentice Hall, New York, 1989.

Biggs, J.M., *Introduction to Structural Dynamics*, McGraw Hill, New York, 1964.

Blandford, G.E., Review of Progressive Failure Analyses for Truss Structures, *Journal of Structural Engineering*, 123, 122, 1997.

Blasko, J.R. and Krauthammer, T., *Pressusre Impulse Diagrams of Structural Elements Subjected to Dynamic Loads*, Final Report to U.S. Army ERDC, PTC-TR-002-2007, Protective Technology Center, Penn State University, January 2007.

Bounds, W., Ed., Concrete and Blast Effects, SP-175, American Concrete Institute, Farmington Hills, MI, 1998.

Breen, J.E. and Siess, C.P., Progressive Collapse: Symposium Summary, *ACI Journal*, pp 997–1004, Vol. 76, No. 9, Sept. 1979.

Breen, J.E., Developing Structural Integrity in Bearing Wall Buildings, *PCI Journal*, Jan.-Feb. 1980.

Britt, J.R. and Ranta, D.E., Blast-X Code, Ver. 4.2, User's Manual, Final Report ERDC/GSL TR-01-2, U.S. Army Corps of Engineer Research and Development Center, Vicksburg, MS, March 2001.

BSI, Loading for Buildings, Part I: Code of Practice for Dead and Imposed Loads, BS6399, British Standards Institution, London, 1996.

BSI, Structural Use of Steelwork in Building, Part I: Code of Practice for Design of Rolled and Welded Sections, BS5950, British Standards Institution, London, 1996.

BSI, Structural Use of Steelwork in Building, Part I: Code of Practice for Design of Rolled and Welded Sections, BS5950, British Standards Institution, London, 2000.

Bulson, P.S., Ed., *Structures under Shock and Impact*, Elsevier, Amsterdam, 1989.

Bulson, P.S., Ed., *Structures under Shock and Impact*, Thomas Telford, London, 1992.

Bulson, P.S., Ed., *Structures under Shock and Impact*, Computational Mechanics Publications, Southampton, 1994.

References

Bulson, P.S., *Explosive Loading of Engineering Structures*, E. & F.N. Spon, London, 1997.

Burnett, E.F.P., Somes, N.F., and Leyendecker, E.V., Residential Buildings and Gas-Related Explosions, Center for Building Technology Institute for Applied Technology, National Bureau of Standards, Washington, June 1973.

Burnett, E.F.P., Abnormal Loadings and the Safety of Buildings, Structures Section, Center for Building Technology Institute for Applied Technology, National Bureau of Standards, Washington, 1974.

Burnett, E.F.P., Blast-Resistant Design of Precast Concrete Structures, Nonlinear Design of Concrete Structures, CSCE-ASCE-ACI-CEB International Symposium, University of Waterloo, Ontario, Canada, August 1979.

Chandra, D. and Krauthammer, T., Rate-Sensitive Micromechanical Models for Concrete, in Bounds, W., Ed., *Concrete and Blast Effects*, ACI SP-175, American Concrete Institute, Farmington Hills, MI, 1998, p. 281.

Chen, W.F. and Lui, E.M., *Structural Stability: Theory and Implementation*, Prentice Hall, New York, 1987.

Chen, W.F. and Toma, S., *Advanced Analysis of Steel Frames*, CRC Press, Boca Raton, FL, 1994.

Choi, H.J. and Krauthammer, T., Addressing Progressive Collapse in Multi-Story Steel Buildings, Proceedings of 73rd Shock and Vibration Symposium, Shock and Vibration Information Analysis Center, Newport, RI, November 2002.

Choi, H.J. and Krauthammer, T., Investigation of Progressive Collapse Phenomena in a Multi-Story Building, 11th International Symposium on Interactions of the Effects of Munitions with Structures, Mannheim, Germany, May 2003.

Chopra, A.K., *Dynamics of Structures*, 2nd ed., Prentice Hall, New York, 2001.

Clarke, J.L., Garas, F.K., and Armer, G.S.T., Eds., *Structural Design for Hazardous Loads*, E.& F.N. Spon, London, 1992.

Clough, R.W. and Penzien, J., *Dynamics of Structures*, 2nd ed., McGraw Hill, New York, 1993.

Collins, M.P. and Mitchell, D., *Prestressed Concrete Structures*, Prentice Hall, New York, 1991.

Coltharp, D.R., Vitayaudom, K.P., and Kiger, S.A., Semihardened Facility Design Criteria Improvement, Air Force Engineering and Services Center, Final Report ESL-TR-85-32, September 1985.

Concrete Society, Concrete for Hazard Protection, London, 1987.

Cook, R.D., Malkus, D.S., and Plesha, M.E., and Witt, R.J., *Concepts and Applications of Finite Element Analysis*, 4th ed., John Wiley & Sons, New York, 2002.

Cooper, P.W. and Kurowski, S.P., *Introduction to the Technology of Explosives*, Wiley-VCH, New York, 1996.

Crawford, E.R. et al., Protection From Nonnuclear Weapons, Air Force Weapons Laboratory, Technical Report AFWL-TR-70-127, February 1971.

Crawford, E.R., Higgins, C.J., and Bultmann, E.H., A Guide for the Design of Shock Isolation Systems for Ballistic Missile Defense Facilities, U.S. Army Construction Engineering Research Laboratory, Technical Report S-23, August 1973.

Crawford, E.R., Higgins, C.J., and Bultmann, E.H., The Air Force Manual for Design and Analysis of Hardened Structures, Air Force Weapons Laboratory, Technical Report AFWL-TR-74-102, October 1974.

Critchley, J., *Warning and Response*, Crane, Russak & Company, New York, 1978.

Crowl, D.A. and Louvar, J.F., *Chemical Process Safety: Fundamentals with Applications*, Prentice Hall, New York, 1990.

Das, B.M., *Fundamentals of Soil-Dynamics*, Elsevier, Amsterdam, 1983.

Davis, J.L., *Wave Propagation in Solids and Fluids*, Springer Verlag, Berlin, 1988.

Defense Civil Preparedness Agency, Protective Construction, TR-20, Vol. 4, Washington, May 1977.

Defense Intelligence Agency, Review of Soviet Ground Forces, RSGF 2-80, Washington, March 1980.

Department of the Army, Leavenworth Assessment of the Warsaw Pact Threat in Central Europe, Threats Division, Concepts and Force Design Directorate, April 1977.

Department of the Army, Soviet Army Operations, U.S. Army Intelligence and Threat Analysis Center, 1978.

Department of the Army, Fundamentals of Protective Design for Conventional Weapons, Technical Manual TM 5-855-1, November 1986.

Departments of the Army, Air Force, and Navy and Defense Special Weapons Agency, Design and Analysis of Hardened Structures to Conventional Weapon Effects, Technical Manual TM 5-855-1/AFPAM 32-1147(I)/NAVFAC P-1080/DAHSCWEMAN-97, August 1998 (official use only).

Department of the Army, Response Limits and Shear Design for Conventional Weapons Resistant Slabs, ETL 1110-9-7, Washington, 1990.

Department of the Army, Structures to Resist the Effects of Accidental Explosions, TM 5-1300, November 1990.

Departments of the Army, and Air Force, Security Engineering Project Development, TM 5-853-1/AFMAN 32-1071, May 1994 (official use only).

Department of Defense, Soviet Military Power, U.S. Government Printing Office, 1985.

Department of Defense, Ammunition and Explosive Safety Standards, Directive 6055.9-STD, July 1999.

Department of Defense, Unified Facilities Criteria: Design and Analysis of Hardened Structures to Conventional Weapon Effects, UFC 3-340-01, June 2002 (official use only).

Department of Defense, Unified Facilities Criteria: Minimum Antiterrorism Standards for Buildings, UFC 4-010-01, July 2002.

Department of Defense, Unified Facilities Criteria: Minimum Antiterrorism Standards for Buildings, UFC 4-010-01, October 2003.

Department of Defense, Unified Facilities Criteria: Design of Building to Resist Progressive Collapse, UFC 4-023-03, January 2005.

Department of Energy, Manual for the Prediction of Blast and Fragment Loadings on Structures, DOE/TIC-11268, July 1992.

Dobratz, B.M., Ed., Properties of Chemical Explosives and Explosive Simulants, Technical Report UCRL-51319, Rev. 1, Lawrence Livermore Laboratory, Berkeley, July 1974.

Dobratz, B.M., LLNL Handbook of Explosives, Technical Report UCRL-52997, Lawrence Livermore Laboratory, Berkeley, March 1981.

Dove, R.C., Evaluation of In-Structure Shock Prediction Techniques for Buried RC Structures, Defense Nuclear Agency, Technical Report DNA-TR-91-89, March 1992.

Downing, W.A., Force Protection Assessment of USCENTCOM AOR and Khobar Towers, Report of the Downing Assessment Task Force, August 1996.

Drake, J.L. et al., Protective Construction Design Manual, Final Report, Air Force Engineering and Services Center, ESL-TR-87-57, November 1989.

DTRA, Workshop on Coupled Eulerian and Lagrangian Methodologies, Defense Threat Reduction Agency, San Diego, CA, January 22–24, 2001.

References

Elfahal, M., Krauthammer, T., Ohno, T., Beppu, M., and Mindess, S., Size Effects for Normal Strength Concrete Cylinders Subjected to Axial Impact, *International Journal of Impact Engineering*, 31, 461, 2005.

Elfahal, M.M. and Krauthammer, T., Dynamic Size Effect in Normal and High Strength Concrete Cylinders, *ACI Materials Journal*, 102, 77, 2005.

Ellis, B.R. and Currie, D.M., Gas Explosions in Buildings in the UK: Regulation and Risk, *Structural Engineer*, 76, 19, 1998.

Engelhardt, M.D., Sabol, T.A., Aboutaha, R.S., and Frank, K.H., Overview of the AISC Northridge Moment Connection Test Program, Proceedings of the National Steel Construction Conference, San Antonio, May 17–19, 1995, pp. 4-1 and 4-22.

Ermopoulos, J.Ch., Buckling Length of Framed Compression Members with Semirigid Connections, *Journal of Constructional Steel Research*, 18, 139, 1991.

Ettouney, M., Smilowitz, R., and Rittenhouse, T., *Blast-Resistant Design for Commercial Buildings: Practice Periodical on Structural Design and Construction*, American Society of Civil Engineers, Reston, VA, 1996, p. 31.

Eytan, R. and Kolodkin, A., Practical Strengthening Measures for Existing Structures to Increase their Blast Resistance: Walls and Ceilings, Proceedings of 6th International Symposium on Interactions of Non-nuclear Munitions with Structures, Panama City Beach, FL, May 1993.

Feld, J., Lessons from Failures in Concrete Structures, American Concrete Institute, Monograph Series 1, 1964, p. 30.

Feldman, A. and Siess, C.P., An Investigation of Resistance and Behavior of Reinforced Concrete Members Subjected to Dynamic Loading, Part III, SRS Report 165, Department of Civil Engineering, University of Illinois, Urbana, IL, 1958.

FEMA, The Oklahoma City Bombing, Report 277, Federal Emergency Management Agency, Washington, August 1996.

FEMA, Recommended Seismic Design Criteria For New Steel Moment-Frame Buildings, Report 350, Federal Emergency Management Agency, Washington, July 2000.

FEMA, World Trade Center: Building Performance Study, Report 403, Federal Emergency Management Agency, Washington, May 2002.

FEMA, Reference Manual to Mitigate Potential Terrorist Attacks Against Building, Report 426, Federal Emergency Management Agency, Washington, December 2003.

Fickett, W., *Introduction to Detonation Theory*, University of California Press, Berkeley, 1985.

Fintel, M. and Schultz, D.M., Structural Integrity of Large Panel Buildings, *ACI Journal*, Vol. 76, No. 5, 583–620, May 1979.

Furlong, R.W. et al., Shear and Bond Strength of High-Strength Reinforced Concrete Beams under Impact Loads: First Phase. Technical Report AFWL-TR-67-113, Air Force Weapons Laboratory, Albuquerque, NM, 1968.

Gamble, W.J., Hendron, A.J. Jr., Rainer, H.J., and Schnobrich, W.C., *A Study of Launch Facility Closures*, SAMSO TR-67-15, Norton Air Force Base, CA, Space and Missile Systems Organization, November 1967.

Ger, J., Cheng, F.Y., and Lu, L., Collapse Behavior of Pino Suarez Building during 1985 Mexico City Earthquake, *Journal of Structural Engineering*, 119, 852, 1993.

Ghali, A. and Tadros, G., Bridge Progressive Collapse Vulnerability, *Journal of Structural Engineering*, 123, 227, 1997.

Ghosh, S.K. and Cohn, M.Z., Ductility of Reinforced Concrete Sections in Combined Bending and Axial Load, *Inelasticity and Non-Linearity in Structural Concrete*, University of Waterloo Press, Waterloo, Ontario, 1972, p.147.

Glasstone, S. and Dolan, P., The Effects of Nuclear Weapons, 3rd ed., Department of Defense and Department of Energy, Washington, 1977.

Goering, K.L., Blast Threats to Buildings and Blast Mitigation Technology, Proceedings of Architectural Surety, Sandia National Laboratories, Albuquerque, NM, May 1997, p. 47.

Grechko, A.A., The Armed Forces of the Soviet State, Moscow, 1975 (translated and published under the auspices of the U.S. Air Force), 1975.

Griffiths, H., Pugsley, A., and Saunders, O., Collapse of Flats at Ronan Point, Canning Town, Her Majesty's Stationery Office, London, 1968.

Gross, J.L. and McGuire, W., Progressive Collapse-Resistant Design, *Journal of Structural Engineering*, 109, 1, 1983.

GSA, Progressive Collapse Analysis and Design Guidelines for New Federal Office Buildings and Major Modernization Projects, General Services Administration, Washington, June 2003.

Gut, J., Waffenwirkungen und Schutzraumbau, Forschugsinstitute für militärische Bautechnik Zürich, FMB 73–11, 1976.

Han, Z. and Yin, X., *Shock Dynamics*, Kluwer Academic, Dordrecht, 1993.

Hardwick, E.R. and Bouillon, J., *Introduction to Chemistry*, Saunders College Publishing, Philadelphia, 1993.

Harris, K.L. and Tucker W.D., Power System Report for MUST-IV (Multi-Unit Structure Test, Full-Size Rectangular Shelter), Final Report, AFWL-TR-85-125, Air Force Weapons Laboratory, August 1986.

Harris, C.M. and Piersol, A.G., Eds., *Shock and Vibration Handbook*, 5th ed., McGraw Hill, New York, 2002.

Haselkorn, A., The Evolution of Soviet Security Strategy 1965–1975, National Strategy Information Center, New York, 1978.

Hawkins, N.M., The Strength of Stud Shear Connections, in *Civil Engineering Transactions*, IE, Australia, 1974, p. 39.

Hawkins, N.M. and Mitchell, D., Progressive Collapse of Flat Plate Structure, *ACI Journal*, Vol. 76, No. 7, pp. 775–809, June, 1979.

Hendry, A.W., Summary of Research and Design Philosophy for Bearing Wall Structure, *ACI Journal*, Vol. 76, No. 6, pp. 723–737, June, 1979.

Henrych, J., *The Dynamics of Explosions and Its Use*, Elsevier, Amsterdam, 1979.

Hibbitt, Karlsson & Sorensen, *ABAQUS/Standard User's Manual*, Ver. 6.6, Pawtucket, RI, 2005.

Hines, J.G. and Petersen, Ph.A., Strategies of Soviet Warfare, *Wall Street Journal*, January 7, 1983, p. 8.

Hinman, E.E. and Hammond, D.J., *Lessons from the Oklahoma City Bombings*, American Society of Civil Engineers, Reston, VA, 1997.

HMSO, Report of Inquiry in the Collapse of Flats at Ronan Point, Canning Town, Her Majesty's Stationery Office, London, 1968.

HMSO, Building and Buildings: The Building Regulation, Statutory Instrument 1676, Her Majesty's Stationery Office, London, 1976.

HMSO, Building and Buildings: The Building Regulation, Statutory Instrument 2768, Her Majesty's Stationery Office, London, 1991.

Hognestad, E., *A Study of Combined Bending and Axial Load in Reinforced Concrete Members*, Bulletin 399, Engineering Experiment Station, University of Illinois, Urbana, p. 128, 1951.

References

Hueste, M.B.D. and Wight, J.K., Nonlinear Punching Shear Failure Model for Interior Slab–Column Connections, *Journal of Structural Engineering*, 125, 997, 1999.

Humar, J.L., *Dynamics of Structures*, Balkema Publishers, Exton, PA, 2002.

Hyde, D., ConWep: Application of TM5-855-1, Structural Mechanics Division, Structures Laboratory, USAEWES, Vicksburg, MS, August 1992.

Jane's Strategic Weapon Systems, Jane's Information Group, Coulsdon, U.K., latest edition.

Jarrett, D.E., Derivation of British Explosives Safety Distances, *Annals of the New York Academy of Sciences*, 152, 18, 1968.

Johansson, C.H. and Persson, P.A., *Detonics of High Explosives*, Academic Press, New York, 1970.

Jones, N., *Structural Impact*, Cambridge University Press, Cambridge, 1989.

Jones, N., Talaslidis, D.G., Brebbia, C.A., and Manolis, G.D., Eds., *Structures Under Shock and Impact V*, Computational Mechanics Publications, Southampton, 1998.

Kameshki, E.S. and Saka, M.P., Genetic Algorithm Based Optimum Design of Nonlinear Planar Steel Frames with Various Semi-Rigid Connections, *Journal of Constructional Steel Research*, 59, 109, 2003.

Kani, G.N.J., Basic Facts Concerning Shear Failure, *ACI Journal*, 63, 675, 1966.

Kiger, S.A. et al., Vulnerability of Shallow-Buried Flat Roof Structures, Technical Report SL-80-7, U.S. Army Engineer Waterways Experiment Station, Vicksburg, MS, 1980–1984.

Kishi, N., Chen, W.F., Goto, Y., and Komuro, M., Effective Length Factor of Columns in Flexibly Jointed and Braced Frames, *Journal of Constructional Steel Research*, 47, 93, 1998.

Kolsky, H., *Stress Waves in Solids*, Dover Publications, New York, 1963.

Koos, R., Die kombinierte Luftstoss-splitter-Last, eine quantitiv bestimmbare Grösse für die Schutzbaubemessung, *Internationales Symposium, Interaktion Konventioneller Munition mit Schutzbauten, Band I*, Mannheim, Germany, March 1987, p. 348.

Krauthammer, T. and Hall, W.J., Modified Analysis of Reinforced Concrete Beams, *Proceedings of ASCE*, ST2 (108), 1982, p. 457.

Krauthammer, T., Shallow-Buried RC Box-Type Structures, *Journal of Structural Engineering*, 110, 637, 1984.

Krauthammer, T., Hill, J.J., and Fares, T., *Enhancement of Membrane Action for Analysis and Design of Box Culverts*, Transportation Research Record, No. 1087, pp. 54–61, 1986.

Krauthammer, T., Bazeos, N. and Holmquist, T.J., Modified SDOF Analysis of RC Box-type Structures, *Journal of Structural Engineering*, 112, 726, 1986.

Krauthammer, T., A Numerical Gauge for Structural Assessment, *Shock and Vibration Bulletin*, 56, Part 1, 179, 1986.

Krauthammer, T., Analysis of Reinforced Concrete Structures under the Effects of Localized Detonations, *Shock and Vibration Bulletin*, 57, 9, 1987.

Krauthammer, T., Shahriar, S., and Shanaa, H.M., Analysis of Reinforced Concrete Beams Subjected to Severe Concentrated Loads, *Structural Journal*, 84, 473, 1987.

Krauthammer, T. and Chen, Y., Free Field Earthquake Ground Motions: Effects of Various Numerical Simulation Approaches on Soil–Structure Interaction Results, *Engineering Structures*, 10, 85, 1988.

Krauthammer, T., Flathau, W. J., Smith, J.L., and Betz, J.F., Lessons from Explosive Tests on RC Buried Arches, *Journal of Structural Engineering*, 115, 810, 1989.

Krauthammer, T. and DeSutter, M.A., Analysis and Design of Connections Openings and Attachments for Protective Structures, Final Report WL-TR-89-44, Air Force Weapons Laboratory, Kirtland AFB, NM, October 1989.

Krauthammer, T., Ed., *Structures for Enhanced Safety and Physical Security*, American Society of Civil Engineers, Reston, VA, 1989.

Krauthammer, T., Shahriar, S., and Shanaa, H.M., Response of RC Elements to Severe Impulsive Loads, *Journal of Structural Engineering*, 116, 1061, 1990.

Krauthammer, T., Structural Concrete Slabs under Impulsive Loads, Fortifikatorisk Notat 211/93, Norwegian Defence Construction Service, Oslo, June 1993.

Krauthammer, T., Assadi-Lamouki, A., and Shanaa, H.M., Analysis of Impulsively Loaded RC Structural Elements, Part 1: Theory, *Computers and Structures*, 48, 851, 1993a.

Krauthammer, T., Assadi-Lamouki, A., and Shanaa, H.M., Analysis of Impulsively Loaded RC Structural Elements, Part 2: Implementation, *Computers and Structures*, 48, 861, 1993b.

Krauthammer, T., Shanaa, H.M., and Assadi-Lamouki, A., Response of Reinforced Concrete Structural Elements to Severe Impulsive Loads, *Computers and Structures*, 53, 119, 1994a.

Krauthammer, T., Stevens, D.J., Bounds, W., and Marchand, K.A., Reinforced Concrete Structures: Effects of Impulsive Loads, *Concrete International*, pp. 57–63, October, 1994b.

Krauthammer, T. and Ku, C.K., A Hybrid Computational Approach for the Analysis of Blast-Resistant Connections, *Computers and Structures*, 61, 831, 1996.

Krauthammer, T., Jenssen, A., and Langseth, M., Precision Testing in Support of Computer Code Validation and Verification, Fortifikatorisk Notat 234/96, Norwegian Defence Construction Service, Oslo, May 1996.

Krauthammer, T., O'Daniel, J.L., and Zineddin, M., Innovative Blast-resistant Corrugate Core Steel Systems, Proceedings of 2nd Asia–Pacific Conference on Shock and Impact Loads on Structures, Melbourne, November 1997.

Krauthammer, T., Blast Mitigation Technologies: Development and Numerical Considerations for Behavior Assessment and Design, *Proceedings of International Conference on Structures under Shock and Impact*, Computational Mechanics Publications, Southampton, 1998, p. 3.

Krauthammer, T., Structural Concrete and Steel Connections for Blast-Resistant Design, Proceedings of International Symposium on Transient Loading and Response of Structures, Norwegian University of Science and Technology, Trondheim, May 1998.

Krauthammer, T., Explosion Damage Assessment, Proceedings of First Forensic Engineering Seminar on Structural Failure Investigations, University of Toronto, January 1999.

Krauthammer, T., Jenssen, A., and Langseth, M., Follow-Up Workshop on Precision Testing in Support of Computer Code Validation and Verification, Fortifikatorisk Notat 276/99, Norwegian Defence Construction Service, Oslo, May 1999.

Krauthammer, T., Structural Concrete and Steel Connections for Blast-Resistant Design, *International Journal of Impact Engineering*, 22, 887, 1999.

Krauthammer, T. and Altenberg, A.E., Assessing Negative Phase Blast Effects on Glass Panels, *International Journal of Impact Engineering*, 24, 1, 2000.

Krauthammer, T., Numerical Support Capabilities for Blast Mitigation, Behavior Assessment and Design, Proceedings of Blast-Resistant Design Workshop, Structural Engineers Association of New York, June 2000a.

References

Krauthammer, T., Structural Concrete Behavior and Design, Proceedings of Blast-Resistant Design Workshop, Structural Engineers Association of New York, June 2000b.

Krauthammer, T., Lim, J., and Oh, G.J., Lessons from Using Precision Impact Test Data for Advanced Computer Code Validations, Proceedings of 10th International Symposium on Interaction of the Effects of Munitions with Structures, San Diego, CA, May 2001.

Krauthammer, T., Lim, J.H., and Oh, G.J., Moment Resisting Steel Connections under Blast Effects, Proceedings of 17th International Symposium on Military Aspects of Blast and Shock, Las Vegas, June 2002.

Krauthammer, T., Zineddin, M., Lim, J.H., and Oh, G.J., Structural Steel Connections for Blast Loads, Proceedings of 30th Explosive Safety Seminar, Department of Defense Explosive Safety Board, Atlanta, GA, August 2002.

Krauthammer, T., Zineddin, M., Lim, J.H., and Oh, G.J., Three-Dimensional Structural Steel Frame Connections under Blast Loads, Proceedings of 73rd Shock and Vibration Symposium, Shock and Vibration Information Analysis Center, Newport, RI, November 2002.

Krauthammer, T., Schoedel, R.M., and Shanaa, H.M., An Analysis Procedure for Shear in Structural Concrete Members Subjected to Blast, Final Report to U.S. Army, ERDC, PTC-TR-001-2002, Protective Technology Center, Pennsylvania State University, December 2002.

Krauthammer, T., Frye, M.T., Schoedel, R., Seltzer, M., Astarioglu, S., A Single-Degree-of-Freedom (SDOF) Computer Code Development for the Analysis of Structures Subjected to Short-Duration Dynamic Loads, Technical Report PTC-TR-002-2003, Protective Technology Center, Pennsylvania State University, August 2003.

Krauthammer, T., Elfahal, M., Ohno, T., Beppu, M, and Markeset, G., Size Effects for High Strength Concrete Cylinders Subjected to Axial Impact, *International Journal of Impact Engineering*, 28, 1001, 2003.

Krauthammer, T., Lim, J.H., Yim, H.C., and Elfahal, M., Three-Dimensional Steel Connections under Blast Loads, Final Report to U.S. Army Engineering Research and Development Center, PTC-TR-004-2004, Protective Technology Center, Pennsylvania State University, June 2004a.

Krauthammer, T., Lim, J.H., Choi, J.H., and Elfahal, M., Evaluation of Computational Approaches for Progressive Collapse and Integrated Munitions Effects Assessment, Final Report to Defense Threat Reduction Agency and U.S. Army Engineering Research and Development Center, PTC-TR-002-2004, Protective Technology Center, Pennsylvania State University, June 2004b.

Krauthammer, T., Seltzer, M., and Astariolglu, S., Dynamic Structural Analysis Suite, Ver. 1.0, User Manual, Final Report to U.S. Army Engineering Research and Development Center and Defense Threat Reduction Agency, PTC–TR-008-2004, Protective Technology Center, Pennsylvania State University, September 2005.

Ku, C.K. and Krauthammer, T., Numerical Assessment of Reinforced Concrete Knee Joints under Explosively Applied Loads, *ACI Structural Journal*, 96, 239, 1999.

Kyung, K.H. and Krauthammer, T., A Flexure–Shear Interaction Model for Dynamically Loaded Structural Concrete, Final Report to U.S. Army Engineering Research and Development Center, PTC–TR-003-2004, Protective Technology Center, Pennsylvania State University, June 2004.

Lakhov, G.M. and Polyakova, N.I., Waves in Solid Media and Loads on Structures, Report FTD-MT-24-1137-71, Foreign Technology Division (translated), 1967.

Lambe, T.W. and Whitman, R.V., *Soil Mechanics*, John Wiley & Sons, New York, 1969.

Langefors, U. and Kihlström, B., *The Modern Technique of Rock Blasting*, 3rd ed., Halsted Press, New York, 1978.

Langseth, M. and Krauthammer, T., Eds, Proceedings, Transient Loading and Response of Structures, Norwegian University of Science and Technology and the Norwegian Defence Construction Service, Trondheim, May 1998.

Lew, H.S., Carino, N.J., and Fattal, S.G., Cause of the Condominium Collapse in Cocoa Beach, Florida, *Concrete International: Design and Construction*, 4, 64, 1982.

Leyendecker, E.V. and Fattal, S.G., Investigation of the Skyline Plaza Collapse in Fairfax County, Virginia, Report BSS 94, Centre for Building Technology, Institute for Applied Technology, National Bureau of Standards, Washington, June 1973.

Leyendecker, E.V., Breen, J.E., Somes, N.F., and Swatta. M., *Abnormal Loading on Buildings and Progressive Collapse: An Annotated Bibliography*, National Bureau of Standards, May 1975.

Leyendecker, E.V. and Burnett, E.F.P., The Incidence of Abnormal Loading in Residential Buildings, National Bureau of Standards, December 1976.

Liberman, M.A. and Velikovich, A.L., *Physics of Shock Waves in Gases and Plasma*, Springer Verlag, Berlin, 1986.

Liew, J.Y.R., White, D.W., and Chen, W.F., Beam–Column Design in Steel Frameworks: Insight of Current Methods and Trends, *Journal of Constructional Steel Research*, 18, 269, 1991.

Liew, J.Y.R., Chen, H., Shanmugam, N.E., and Chen, W.F., Improved Nonlinear Plastic Hinge Analysis of Space Frame Structures, *Engineering Structures*, 22, 1324, 2000.

Litle, W.A., The Boston Collapse at 2000 Commonwealth Avenue, Symposium on Progressive Collapse, American Concrete Institute Annual Convention, Boston, April 1975.

Lok, T.S., Ed., Proceedings of First Asia–Pacific Conference on Shock and Impact Loads on Structures, Singapore, January 1996.

Lok, T.S., Ed., Proceedings of 2nd Asia–Pacific Conference on Shock and Impact Loads on Structures, Melbourne, Australia, November 1997.

Longinow, A. and Mniszewski, K.R., Protecting Buildings against Vehicle Bomb Attacks, Practice Periodical on Structural Design and Construction, American Society of Civil Engineers, Reston, VA, 1996, p. 51.

Love, A.E.H., *Some Problems of Geodynamics*, Dover Publications, New York, 1967, p. 160.

MacGregor, J.G. and White, J.K., *Reinforced Concrete Mechanics and Design*, 4th ed., Prentice Hall, New York, 2005.

Mainstone, R.J., The Hazard of Internal Blast in Buildings, Building Research Establishment Current Paper, Building Research Station, Garston, U.K., April 1973.

Mainstone, R.J., The Response of Buildings to Accidental Explosions, Building Research Establishment Current Paper, Building Research Station, Garston, U.K., March 1976.

Mainstone, R.J., Nicholson, H.G., and Alexander, S.J., Structural Damage in Buildings Cused by Gaseous Explosions and Other Accidental Loadings, 1971–1977, Department of the Environment, Her Majesty's Stationery Office, London, 1978.

Malla, R.B. and Nalluri, B.B., Dynamic Effects of Member Failure on Response of Truss-Type Space Structures, *Journal of Spacecraft and Rockets*, 32, 545, 1995.

Malvar, L.J., Crawford, J.E., Wesevich, J.W., and Simon, D., A Plasticity Concrete Material Model for DYNA3D, *International Journal of Impact Engineering*, 19, 847, 1997.

References

Marchand, K.A., Cox, P.A., and Peterson, J.P., Blast Analysis Manual, Part I: Level of Protection Assessment Guide, Key Asset Protection Program Construction Option, Southwest Research Institute, Final Report for U.S. Army Corps of Engineers, Omaha District, July 19, 1991.

Marchand, K.A. and Alfawakhiri, F., *Blast and Progressive Collapse*, American Institute of Steel Construction, Chicago, 2004.

Marquis J.P. and Berglund, J.W., General Guidelines for Predicting and Controlling the Response of Critical Equipment, Final Report WL-TR-90-20, Air Force Weapons Laboratory, Kirtland AFB, NM, July 1990.

Marquis, J.P., Morrison, D., and Hasselman, T.K., Development and Validation of Fragility Spectra for Mission-Critical Equipment, Final Report PL-TR-91-1017, Phillips Laboratory, Kirtland AFB, NN, August 1991.

Mays, C.G. and Smith, P.D., *Blast Effects on Buildings*, Thomas Telford, London, 1995.

McCarthy, G., Combined Blast and Fragment Loading: Definition and Influence on Structural Response, M.S. Thesis, Department of Civil and Environmental Engineering, Pennsylvania State University, August 2005.

McCormac, J.C. and Nelson, J.K., *Structural Steel Design*, Prentice Hall, New York, 2003.

Meamarian, N., Krauthammer, T., and O'Fallon, J., *Analysis and Design of One-Way Laterally Restrained Structural Concrete Members*, Structural Journal, American Concrete Institute, Vol. 91, No. 6, pp. 719–725, Nov.–Dec. 1994.

Meley, J.L. and Krauthammer, T., Finite Element Code Validation Using Precision Impact Test Data on Reinforced Concrete Beams, Final Report to U.S. Army ERDC, PTC–TR-001-2003, Protective Technology Center, Pennsylvania State University, April 2003.

Mett, H.G., Untersichug der Schocksicheren Lagerung, Aufstellung und Befestigung von Ausstattungsteilen in Schutzbauten bei Hoher, Kurtzzeitliger Belastung, *Internationales Symposium, Interaktion Konventioneller Munition mit Schutzbauten, Band I*, Mannheim, Germany, April 1991, p. 454.

Meyer, C. and Okamura, H., Eds., *Finite Element Analysis of Reinforced Concrete Structures*, American Society of Civil Engineers, Reston, VA, 1986.

Miller-Hooks, E. and Krauthammer, T., Intelligent Evacuation, Rescue and Recovery Concept, Proceedings of 30th Explosive Safety Seminar, Department of Defense Explosive Safety Board, Atlanta, GA, August 2002.

Miller-Hooks, E. and Krauthammer, T., Intelligent Evacuation, Rescue and Recovery Concept: A Functional System Description, 11th International Symposium on the Interaction of the Effects of Munitions with Structures, Mannheim, Germany, May 2003.

Miller-Hooks, E. and Krauthammer, T., An Intelligent Evacuation, Rescue and Recovery Concept, Fire Technology, (to appear).

Mitchell, D. and Cook, W.D., Preventing Progressive Collapse of Slab Structures, *Journal of Structural Engineering*, 110, 1513, 1984.

Morris, N.F., Effect of Member Snap on Space Trusses for Progressive Collapse, *Journal of Engineering Mechanics*, Vol. 119, No. 4, April 1993.

Murtha-Smith, E., Alternate Path Analysis of Space Trusses for Progressive Collapse, *Journal of Structural Engineering*, 114, 1978, 1988.

Mossberg, W.S., NATO Chief Warns of New Soviet Strategy to Deny the West Use of Its Nuclear Punch, *Wall Street Journal*, October 13, 1982.

Namburu, R.R. and Mastin, C.W. Eds., Recent Advances in Computational Structural Dynamics and High-Performance Computing, Waterways Experiment Station, Vicksburg, MS, November 1998.

National Research Council, *Protecting Buildings from Bomb Damage*, National Academy Press, Washington, 1995.

National Research Council, *Blast Mitigation for Structures*, National Academy Press, Washington, 2000.

National Research Council, *Protecting People and Buildings from Terrorism*, National Academy Press, Washington, 2001.

Nawy, E.G., *Reinforced Concrete: A Fundamental Approach*, Prentice Hall, Upper Saddle River, NJ, 2003.

NCEL, Shock User's Manual, Ver. 1, Naval Civil Engineering Laboratory, Port Hueneme, CA, January 1, 1988.

NDRC, Effects of Impacts and Explosion, Summary Technical Report of Division 2, Vol. 1, 1946.

Newmark, N.M., Hansen, Holley, and Biggs, J.M., Protective Construction Review Guide: Hardening, Vol. 1, Department of Defense, Washington, June 1961.

Newmark, N.M. and Haltiwanger, J.D., Air Force Design Manual, AFSWC-TDR-62-138, Air Force Special Weapons Center, December 1962.

Newmark, N.M., A Method of Computation for Structural Dynamics, *Transactions, ASCE*, 127, 1406, 1962.

Newmark, N.M. and Rosenblueth, E., *Fundamentals of Earthquake Engineering*, Prentice Hall, New York, 1971.

Ng, P.H. and Krauthammer, T., Pressure–Impulse Diagrams for Reinforced Concrete Slabs, Final Report to U.S. Army, ERDC, PTC–TR-007-2004, Protective Technology Center, Pennsylvania State University, November 2004.

Nilsson, I.H.E., Reinforced Concrete Corners and Joints Subjected to Bending Moment, Document D7, National Swedish Building Research, Stockholm, 1973.

O'Daniel, J.L., Krauthammer, T., Koudela, K.L., and Strait, L.H., An UNDEX Response Validation Methodology, *International Journal of Impact Engineering*, 27, 919, 2002.

Oswald, C.J. and Skerhut, D., FACEDAP User's Manual, Southwest Research Institute and U.S. Army Corps of Engineers, Omaha District, April 1993.

Otani, R.K. and Krauthammer, T., Assessment of Reinforcing Details for Blast Containment Structures, *ACI Structural Journal*, 94, 124, 1997.

Pahl, H., Penetration of Projectiles into Finite Thick Reinforced Concrete Targets, Proceedings of 4th International Symposium on Interaction of Non-Nuclear Munitions with Structures, Panama City Beach, FL, April 1989, p. 55.

Pan, Y. and Watson, A., Effect of Panel Stiffness on Resistance of Cladding Panels to Blast Loading, *Journal of Engineering Mechanics*, 124, 414, 1998.

Pan, Y.G., Watson, A.J., and Hobbs, B., Transfer of Impulsive Loading on Cladding Panels to the Fixing Assemblies, *International Journal of Impact Engineering*, Vol. 25, pp. 949–964, Nov. 2001.

Park, R. and Paulay, T., *Reinforced Concrete Structures*, John Wiley & Sons, New York, 1975.

Park, R. and Gamble, W.L., *Reinforced Concrete Slabs*, 2nd ed., John Wiley & Sons, New York, 2000.

Pekau, O.A., Structural Integrity of Precast Panel Shear Walls, *Canadian Journal of Civil Engineering*, Vol. 9, No. 1, March 1982.

References

Persson, P.A., Holmberg, R., and Lee, J., *Rock Blasting and Explosive Engineering*, CRC Press, Boca Raton, FL, 1994.

Pilkey, W. and Pilkey, B., Eds., Shock and Vibration Computer Programs, SVM-13, Shock and Vibration Information Center, Arlington, VA, 1995.

Pounder, C.C., *Pounder's Marine Diesel Engines*, 7th ed., Butterworth-Heinemann, London, 1998.

Proctor, R.V. and White, T.L., *Earth Tunneling with Steel Supports*, Commercial Shearing, Inc. 1977.

Rakhmatulin, Kh.A. and Dem'yanov, Yu.A., *Strength under High Transient Loads*, Israel Program for Scientific Translations Ltd., distributed by Daniel Davey & Co., New York, 1966.

Reddy, J.N., *An Introduction to the Finite Element Method*, 3rd ed., McGraw Hill, New York, 2006.

Rinehart, J.S., *Stress Transient in Solids*, Hyper-Dynamics, 1975.

Ross, C.A., Sierakowski, R.L., and Schauble, C.C., Concrete Breaching Analysis, Technical Report AFATL-TR-81-105, Air Force Armament Laboratory, Eglin AFB, FL, 1981.

Ross, C.A. and Rosengren, P.L., Expedient Nonlinear Dynamic Analysis of Reinforced Concrete Structures, *Proceedings of 2nd Symposium on the Interaction of Non-Nuclear Munitions with Structures*, University of Florida Graduate Engineering Center, Eglin AFB, FL, 1985, p. 45.

Ross, C.A., Kuennen, S.T., and Strickland, W.S., High Strain Rate Effects on Tensile Strength of Concrete, *Proceedings of 4th International Symposium on the Interaction of Non-Nuclear Munitions with Structures*, University of Florida Graduate Engineering Center, Eglin AFB, FL, 1989, p. 302.

Ross, S., *Construction Disasters: Design Failures, Causes and Prevention*, McGraw Hill, New York, 1984.

Ross, T.J., Direct Shear Failure in Reinforced Concrete Beams under Impulsive Loading, Technical Report AFWL-TR-83-84, Air Force Weapons Laboratory, Kirtland AFB, NM, 1983.

Russo, G., Zingone, G., and Puleri, G., Flexure-Shear Interaction Model for Longitudinally Reinforced Beams, *ACI Structural Journal*, 88, 60, 1991.

Russo, G. and Puleri, G., Stirrup Effectiveness in Reinforced Concrete Beams under Flexure and Shear, *ACI Structural Journal*, 94, 227, 1997.

Salmon, C.G. and Johnson, J.E., *Steel Structures: Design and Behavior*, 4th ed., Harper Collins College Publishers, 1996.

Schlaich, J., Schäfer, K., and Jennewein, M., Toward a Consistent Design of Structural Concrete, *PCI Journal*, pp. 227–238, May–June 1987.

Schmidt, L.C. and Hanoar, A., Force Limiting Devices in Space Trusses, *Journal of the Structural Division of ASCE*, 105, 939, 1979.

Schuster, S.H., Sauer, F., and Cooper, A.V., The Air Force Manual for Design and Analysis of Hardened Structures, Final Report, AFWL-TR-87-57, June 1987.

Seabold, R.H., Dynamic Shear Strength of Reinforced Concrete Beams, Part III, Technical Report R-695, Naval Civil Engineering Laboratory, Port Hueneme, CA, 1970.

Slawson, T.R., Dynamic Shear Failure of Shallow-Buried Flat-Roofed Reinforced Concrete Structures Subjected to Blast Loading, Final Report SL-84-7, U.S. Army Engineer Waterways Experiment Station, Vicksburg, MS, 1984.

Smith, J., *A Law Enforcement and Security Officer's Guide to Responding to Bomb Threats*, Charles C Thomas, Springfield, IL, 2003.

Smith, P.D. and Hetherington, J.G., *Blast and Ballistic Loading of Structures*, Butterworth-Heinemann, London, 1994.

Soh, T.B., and Krauthammer, T., Load-Impulse Diagrams of Reinforced Concrete Beams Subjected to Concentrated Transient Loading, Final Report to U.S. Army, ERDC, PTC–TR-006-2004, Protective Technology Center, Pennsylvania State University, April 2004.

Somerville, G. and Taylor, H.P.J., The Influence of Reinforcement Detailing on the Strength of Concrete Structures, *Structural Engineer* (London), 50, 7, 1972; discussion in 50, 309, 1972.

Soroushian, P. and Obaseki, K., Strain Rated-Dependent Interaction Diagrams for Reinforced Concrete Section, *American Concrete Institute Journal*, 83, 108, 1986.

Sozen, M.A., Hysteresis in Structural Elements, *Journal of Applied Mechanics*, 8, 63, 1974.

Speyer, I.J., Considerations for the Design of Precast Concrete Bearing Wall Buildings to Withstand Abnormal Loads, *PCI Journal*, March–April 1976.

Starr, C.M. and Krauthammer, T., Cladding-Structure Interaction under Impact Loads, *Journal of Structural Engineering*, 131, 1178, 2005.

Steel Construction Institute, Structural Use of Steelwork in Building, BS5950, Steel Construction Institute, Ascot, U.K., 1989.

Stevens, D.J. and Krauthammer, T., Nonlocal Continuum Damage/Plasticity Model for Impulse-Loaded RC Beams, *Journal of Structural Engineering*, 115, 2329, 1989.

Stevens, D.J. and Krauthammer, T., Analysis of Blast-Loaded, Buried Arch Response Part I: Numerical Approach, *Journal of Structural Engineering*, 117, 197, 1991.

Stevens, D.J., Krauthammer, T., and Chandra, D., Analysis of Blast-Loaded, Buried Arch Response Part II: Application, *Journal of Structural Engineering*, 117, 218, 1991.

Sucuolu, H., Çitipitiolu, E., and Altm, S., Resistance Mechanisms in RC Building Frames Subjected to Column Failure, *Journal of Structural Engineering*, 120, 765, 1994.

Swedish Civil Defence Administration, Technical Regulations for Shelters, TB 78E, Stockholm, 1978.

Swiss Civil Defence Administration, Technical Regulations for Private Shelters, TWP 1966, Bern, Switzerland, updated in November 1971 (in German).

Swiss Civil Defence Administration, Technical Regulations for Shelters (in German), Bern, 1977.

Taylor, D.A., Progressive Collapse, *Canadian Journal of Civil Engineering*, Vol. 2, No. 4, December 1975.

Tedesco, J.W., McDougal, W.G., and Ross, C.A., *Structural Dynamics*, Addison-Wesley, Reading, MA, 1999.

Terzaghi, K. and Peck, R.B., *Soil Mechanics*, John Wiley & Sons, New York, 1948.

Timoshenko, S.P., On the Correction for Shear of the Differential Equation for Transverse Vibrations of Prismatic Bars, *Philosophical Magazine* (London), 41, 744, 1921.

Timoshenko, S.P., On the Transverse Vibration of Bars of Uniform Cross-Section, *Philosophical Magazine* (London), 43, 125, 1922.

Timoshenko, S.P. and Goodier, J.N., *Theory of Elasticity*, 3rd ed., McGraw Hill, New York, 1970.

Triandafilidis, G.E., Calhoun, D.E., and Abbott, P.A., Simulation of Airblast-Induced Ground Motion at McCormic Ranch Test Site, AFWL-TR-68-27, October 1968.

Ulsamer, E., Military Power is the Root of Soviet Expansionism, U.S. Air Force, March 1982, p. 38.

U.S. Army, Design of Structures to Resist the Effects of Atomic Weapons, Corps of Engineers Manual EM 1110-345-415, 1957.

References

U.S. Army, Selected US and Soviet Weapons, Key Equipment, and Soviet Organization, USACGSC RB 30–2, U.S. Army Command and General Staff College, Fort Leavenworth, KS, July 1977.

U.S. Army, Suppressive Shields: Structural Design and Analysis Handbook, Corps of Engineers, Huntsville Division, HNDM-1110-1-2, November 1977.

U.S. Army, Security Engineering, Volumes 1–3, TM 5-853, December 1988.

USCNS/21, New World Coming, United States Commission on National Security, 21st Century, September 15, 1999.

USCNS/21, Seeking a National Strategy, United States Commission on National Security. 21st Century, April 15, 2000.

USCNS/21, Road Map for National Security, United States Commission on National Security, 21st Century, March 15, 2001.

U.S. Department of State, Patterns of Global Terrorism 2002, April 2003.

U.S. Department of State, Patterns of Global Terrorism 2003, April 2004.

U.S. Department of State, Country Reports on Terrorism 2004, April 2005.

U.S. Department of State, Country Reports on Terrorism 2005, April 2006.

U.S. Navy, Terrorist Vehicle Bomb Survivability Manual, Naval Civil Engineering Laboratory, Port Hueneme, CA, July 1988.

Vrouwenvelder, A., Explosion Loading on Structures, Instituut TNO Bouwmaterialen en Bouwconstructies, Report BI-86-43/4.8.2501, Amsterdam, 1986.

Wager, P. and Connett, J., FRANG User's Manual, Naval Facilities Laboratory, Port Hueneme, CA, May 1989.

Walters, W.P. and Zukas, J.A., *Fundamentals of Shaped Charges*, CMC Press, 1998.

Weidlinger, P. and Matthews, A.T., Shock and Reflection in Nonlinear Medium, *Journal of the Engineering Mechanics Division*, June 1965.

Weidlinger, P. and Hinman, E., Analysis of Underground Protective Structures, *Journal of Structural Engineering*, 114, 1658, 1988.

Whitham, G.B., *Linear and Nonlinear Waves*, John Wiley & Sons, New York, 1974.

Woodson, S.C., Effects of Shear Reinforcement on the Large-Deflection Behavior of Reinforced Concrete Slabs, Technical Report SL-94-18, U.S. Army Corps of Engineers, Waterways Experiment Station, September 1994.

Woodson, S.C., Shear Reinforcement in Deep Slabs, Technical Report SL-94-24, U.S. Army Corps of Engineers, Waterways Experiment Station, November 1994.

Yandzio, E. and Gough, M., *Protection of Buildings against Explosions*, Publication 244, Steel Construction Institute, Ascot, U.K., 1999.

Yokel, F.Y., Wright, R.N., and Stone, W.C., Progressive Collapse: U.S. Office Building in Moscow, *Journal of Performance of Constructed Facilities*, Vol. 3, No. 1, February 1989.

Index

A

ABAQUS/Explicit, 270, 273, 394
Academic consortia, *see* Collaboration, multidisciplinary
Acceleration
    dynamic response and analysis, 252
    ground shock, 151
    SDOF analysis, 240
Acceptance criteria, progressive collapse, 387–389
Access and approach control, 4, 408, 413–414, 432, 433
Accidental incidents, 3, 13, 14, 51, 432
Accumulator, in-line, 315, 317
Acoustic impedance, 113, 132
Acting pressure, 171
Adiabatic reactions, 59
Advanced analysis and computation methods
    conventional loads on buried structures, 169, 170
    design approach, comprehensive, 419
    dynamic response and analysis, 238–239, 247–250
    membrane behavior, 224–225
    nuclear loads on buried structures, 184
    penetration effects, 90
    planning and design, 8
    progressive collapse, frame structure analysis, 390–404
        computer code requirements, 394
        examples of, 395–404
        semi-rigid connections, 392–394, 395–397, 398
    SDOF, 247–250
        comprehensive design approach, 433
        DSAS, 362
    shock waves in three-phase media, 64
    size effects and combined size and rate effects, 31–32
    stress-time histories, 151
Advanced technology development, 38, 436
Aerial attacks, *see* Air power
Aggregate interlock, 238
Aggregate size, concrete, 88
Aging effects
    and compressive strength of concrete, 199–200
    research needs, 439, 441
Air, explosions in, *see* Overhead explosions
Airblast
    conventional weapons and high-explosive devices, 68–79
        blast fragment coupling, 175
        external explosions, 68–74
        internal explosions, 75–77, 78
        leakage blast pressure, 77–79
        loads on above-ground structures, 170
        loads on mounded structures, 175
        loads on surface-flush structures, 175
    design approach, comprehensive, 420
    ground shock
        coupling factor as function of scaled depth, 131
        surface-burst, 149, 150
    head waves, 148
    nuclear devices, 79–83, 84–90, 177
    peak airblast pressure estimation, 73, 74
    piping system protection, 315, 317
    planning and design considerations, 5, 16, 19
    superseismic conditions, 147, 150
    transmission through tunnels and ducts, 319, 322–324
Airborne contamination, minimizing, 6, 410
Airburst, 71, 72
Aircraft cannon projectiles, 42, 43, 44, 45
Airforce Weapons Laboratory (AFWL), 25
Air intake
    blast valves, 314
    door design air tightness, 311
Air power, 20–21, 23, 24
Airslap-induced waves, 148, 149, 151, 152
Alternate path (AP) analysis
    damage limits, 384
    DOD design recommendations, 382–383
Alternative load paths
    connections, 295
    progressive collapse, 377
    transfer of load, 429–430
Ambient pressure

blast design load derivation procedures,
  blast load on rear wall, 193
  pressure-time variation in external
    explosions, 68–69
American Society of Civil Engineers (ASCE),
  24
American Society of Civil Engineers (ASCE)
    guidelines/manuals/publications,
    6, 27, 28, 407, 411
  Manual 42, 27, 195, 197–198, 406
  progressive collapse, 395
  shear resistance, 206
Ammonium nitrate (AN), 58
Ammonium nitrate fuel oil devices (ANFO), 31,
    58, 66
Analysis/analytical solutions: *see also*
    Advanced analysis and
    computation methods
  connections, 294
  dynamic response and analysis
    computational systems, 282–287;
      see also Dynamic response and
      analysis
  general requirements and capabilities,
      30–31
    computational analysis, 30
    experimental analysis, 30–31
  P-I diagrams, 331–339, 340
  P-I diagrams, closed form, 331–335, 336
    rectangular load pulse, response to, 331,
      333–334
    triangular load pulse, response to,
      334–335, 336
  P-I diagrams, energy balance method,
      335–339
    approximating dynamic regions,
      336–338, 339
    continuous structural elements, 338, 340
  progressive collapse
    advanced frame structure analysis,
      390–404
    General Services Administration (GSA)
      guidelines, 383–389
Anchors: *see also* Connections and supports
  comprehensive design approach, 431
  dynamic response and analysis, 357
  knee joints, 301
  principal reinforcement, 207–208
  progressive collapse, concrete structures,
      378
  structural element behavior
    direct shear response, 228
    membrane behavior, 221–222

ANFO (ammonium nitrate fuel oil devices), 31,
    58, 66
Angle of impact
  design approach, comprehensive, 418–419
  and effectiveness of projectile, 43
Angle of incidence
  airburst blast environment, 72
  and penetration, concrete, 91, 94, 95
  and penetration path, 97
  reflected shock wave, 70
  slant bursts on buried structures, 106
Annihilation, Soviet doctrine, 22
Antirunway bombs, 47
Antitank weapons
  dispenser and cluster bombs, 47
  explosive device characteristics, 42, 44, 45,
      46
  penetration, shaped charge devices,
      104–105
  threat assessment, 14
AP devices, *see* Armor-piercing (AP) devices
Applied research, 436
Approach to buildings, *see* Access and approach
    control
Approximate methods
  dynamic response and analysis, 262,
      263–265
  multi-segmental forcing functions, 281,
      288–289
  P-I diagram, energy balance methods,
      336–338, 339
  SDOF, 422–423
Arches, 186
  connections, 306
  diagonal shear effects, 230
  nuclear loads
    on above ground structures, 178, 180,
      181
    on buried structures, 184, 185
Arching, soil, 179, 181, 186
Architectural considerations: *see also* Geometry
    of structure
  design approach, comprehensive, 408–410,
      430
    access and approach control, 413–414
    building exterior, 414–416
    building interior, 416
    post-incident conditions, 417
    vital systems, nonstructural, 417
  nuclear loads
    on above ground structures, 178
    on buried structures, 183
  structural element behavior, 197
Area, loaded, 29

Index

Areal weight density, nuclear explosion ejecta, 146
Armor
    conventional explosive device characteristics, 42, 43, 45
    penetration, with conventional devices, 98, 99, 100–101
        comparisons with other materials, 102
        shaped charge devices, 104–105
Armor-piercing (AP) devices
    bombs, 46, 47
    penetration
        armor, 101
        concrete, 90, 93, 94
    types of projectiles, 42–44
Arrival times
    blast design load derivation procedures
        load-time history on buried wall or roof, 187
        pressure-time curve on front wall from external explosions, 187
        pressure-time curve on roof or sidewall, span parallel to shock front explosions, 191
        pressure-time curve on roof or sidewall, span perpendicular to shock front explosions, 190
    blast fragment coupling, 175
    conventional loads on buried structures, 169
    ground shock, 151
Arson, 32
Artificial intelligence, 33
Artillery, 18
    fragment characteristics, 158
    planning and design considerations, 24
    Soviet doctrine, 19–20, 22–23
Assessment, *see* Threat and risk assessment
Asset assessment, 10
Asset definition, 18
Asset rating, 12
ASTM standards, material property specifications, 389
Asymptotic values
    impulsive and quasi-static domains, 426
    P-I diagrams, 328, 330, 331, 335, 336, 337, 338, 340, 341, 342
Attack delay, 412–413
Attack prevention, 412
Attenuation
    ground shock, 151, 152
    transmission through tunnels and ducts and, 323, 324
Attenuation coefficients
    layered systems, 134
    soil properties from explosion tests, 132
Average pressure, loads on above-ground structures, 172
Axial forces
    design approach, comprehensive, load redistribution, 429
    dynamic response and analysis, 250, 357
    flexural behavior, 345
    intermediate approximate methods, 263, 264
    material model testing, 270–271
    P-I diagrams, analytical, 325
    SDOF analysis, 251
    structural element behavior
        cylinders, arches, and domes, 210, 211
        direct shear response, 228
        membrane behavior, 226
        shear resistance, 206
        tensile and compressive members, 206, 207
Azides, 58

B

Background material, *see* Publications, literature and technical references
Back packing, 19
Bacteriological environments, 167
Balanced design, 291
Balance-energy method, *see* Energy balance method, P-I diagrams
Ballistic attacks, planning and design considerations, load determination, 18–19
Ballistic limit velocity, armor penetration, 100–101
Ballistics, threat assessment, 14
Baratol, 154
Barriers, 412, 413
Bars
    connections, joint behavior, 304
    joint strengthening, 232
    principal reinforcement, 207
Basement parking garages, 198
Base motion, dynamic response and analysis, 253
Base shears, blast loading and, 428
Bays, 416
Beam-column interaction equations, 390
Beam geometry, SDOF and P-I computations, 356, 357
Beam regions (B regions), 292, 295

Beam resistance functions, dynamic response and analysis, 270–271
Beams
  comprehensive design approach, structural system and component selection, 431
  concrete
    direct shear response, 351–352
    dynamic resistance function, 346
    dynamic shear force, 352–357
    dynamic structural model, 349
    shearing failure, 325
  connections, *see* Connections and supports
  dynamic response and analysis, 262
    continuous systems, 259–260
    plastic hinge regions, 274–275
    SDOF parameters for fixed beams, 278
    SDOF parameters for fixed boundary beams, 279–280
    SDOF parameters for simply supported beams, 277
    validation procedures, 271–272
    validation with test data, 269–270
  P-I diagrams, 357–362
    dynamic reaction, 340
    energy balance method, 338
  progressive collapse
    GSA guidelines, 386
    sample analyses, 397, 398
    steel frame buildings, 380
  structural element behavior
    diagonal shear effects, 228, 230
    interaction diagram, 204
    natural periods of vibration, 215
    shear resistance, 205–206
Beam-slab systems, structural element behavior, 205
Beam theory, 209, 292
Beam vibration problem, 259
Bending moment
  dynamic response and analysis, 263
  SDOF analysis, 251
Bending shear
  coupled failure modes, 35
  failure modes, 18
Bilinear resistance functions, 346–347
Bishopsgate, 376
Blast curtains, 416
Blast design load derivation procedures, 187–193
  blast load on rear wall, 191–193
  load-time history on buried wall or roof, 187, 188
  time-pressure curve on front wall from external explosion (surface burst), 187–189, 190
  time-pressure curve on roof or sidewall
    span parallel to shock front, 190–191, 192
    span perpendicular to shock front, 190, 191, 192
Blast doors, 310–313, 314
Blast effects/parameters
  design approach, comprehensive, 418
  explosion effects and mitigation, 50–52
  fragment coupling, conventional loads on structures, 175–176, 177
  human body responses to blast loading, 327
  maximizing standoff distance, 6
  mitigation, general principles and overview, 7, 16–17, 24–30
    design and construction considerations, 28–30
    design manuals, 26–28
    technical resources and blast mitigation capabilities, 24
    transitions between response modes, 30
  shock response spectra, 246
  structural element behavior, technical manuals, 196
Blast loading, *see* Loads and loading regimes
Blast pressure, *see also* Pressure
  design approach, comprehensive, 428
  free air equivalent weights, 66
  internal explosions, leakage blast pressure, 77–79
  nuclear and HE explosions, 80
Blast proximity, *see* Distance from explosion
Blast valves, 313–315
Blast wave location ratio, 173
Blast waves
  blast design load derivation procedures, blast load on rear wall, 191
  design approach, comprehensive, 418, 419–420, 421
  loads on above-ground structures, 173
  research needs, 438–439
Blast wave velocity, 191
Bombs
  damage assessment, 32
  explosive processes, devices, and environments, characteristics of conventional weapons, 46–47
  fragment characteristics, 158
  high-explosive, characteristics of, 46
  Soviet doctrine, 23
  threat assessment, 14

Index

types of explosive attack incidents in U.S., 14
Bonds, steel-concrete
  dynamic analysis validation procedures, 270, 271–272
  explosive loads and, 294
  failure of, 303
Boundary conditions
  dynamic response and analysis, 344
    equivalent SDOF system approach, beam response modes, 276
    reactions at, 355, 356
    shear forces, 352
  nuclear explosion-induced shock waves, 121–122, 123
  progressive collapse, steel frame structure analysis, 390–391
  reflection and transmission of one dimensional D-waves between two media, 114–116
  support conditions, 291–292
  variable, 376
Bracing, 420
Breaching
  dynamic response and analysis, 238–239
  material properties and, 237
  planning and design considerations, 28
Break-away elements, 430–431
B (beam) regions, 292, 295
Bridging of structural elements, progressive collapse, 373, 377, 380–381
Brittle failure, progressive collapse of steel frame buildings, 380
Brittleness
  design approach, comprehensive, 425
  dynamic response and analysis, 268
  tension and shear weakness, 237
Buckling
  P-I diagrams, 327
  progressive collapse, 401, 402–403
    ideal connections case, 401–403
    sample analyses, 400
    steel frame structures, 380, 390–391, 392
    truss structures, 379
Buffer zones, 416
Building exteriors, *see* Exteriors, building
Building interiors, *see* Interiors, building
Building materials, comprehensive design approach, 414
Building separation, comprehensive design approach, 408, 409
Buried charges, comprehensive design approach, 421

Buried structures
  anti-scabbing plates, 89
  connections, 306
  geometry of explosion, 106
  ground shock, cratering, and ejecta, HE-induced, 106–141
    application to protective design, 129–135
    computational aspects, 116–117
    cratering, 135–137, 138
    ejecta, 137, 139, 140, 141
    one-dimensional elastic wave propagation, 110
    reflection and transmission of one dimensional D-waves between two media, 114–116
    shock waves in one-dimensional solids, 117–119
    stress wave propagation in soils, 119–125
    three-dimensional stress wave propagation, 106–110
  ground shock, cratering, and ejecta, nuclear device-induced, 141–153
    cratering, 141, 142, 143–145
    ejecta, 145–147, 148–149
    ground shock, 147, 149, 150–153
  loads on
    conventional, 169–170
    nuclear, 179–185
  planning and design considerations, 6–7, 19, 25
  underground storage facilities, 433
Burning/combustion phenomena, 59
Burster slabs, 130

C

C4, 58, 66, 154
Cable and conduit penetrations, 315–317
Cable structure, load redistribution, 429
Caliber, armor penetration, 100–101
Caliber density, and fragmentation penetration, 157–158
Camouflet, 137
Cannon projectiles, 42, 45
Cascading failures, progressive collapse, 401
Casings
  blast fragment coupling of cased charges, 175
  design approach, comprehensive, 418, 419, 421
  and fragmentation, 154, 155, 157–158

Catastrophic failure, *see also* Progressive collapse
  design approach, comprehensive, 412
  membrane behavior, 219
  planning and design considerations, 5
  reinforced floor slabs, 208
Catenary action, 416, 429
Ceilings, 416
Central difference operators, steel frame structure analysis, 394
Chapman-Jouguet (C-J) plane, 61
Characteristic impedance, 113
Charge weight
  design approach, comprehensive, 420
  internal explosions, gas pressure phase, 77
Chemical and biological weapons
  door designs, 311
  general principles and overview, 16, 17
  planning and design considerations, 5, 7
  Soviet doctrine, 19, 20–21, 23
  threat assessment, 11
  threats, 167
Chemical explosions
  accidental, 51
  chemistry and physics of, 54–56
Chemistry, explosions, 53–54, 60–61
Circular natural frequencies
  continuous systems, 261
  dynamic response and analysis, 261, 276
  loading regimes, 330, 350, 351
Civil defense agencies, 405, 406
Cladding systems, 413–414
Classical resistance models, design manual limitations, 25
Clearing time
  loads on above-ground structures, 170–171
  pressure-time curve on front wall from external explosions, 188
Closed-form analytical solutions, 238
  dynamic response and analysis, 262
  P-I diagrams, 331–335, 336, 341, 366
    rectangular load pulse, response to, 331, 333–334
    triangular load pulse, response to, 334–335, 336
Closed loops, connections, 304
Close-in detonation, *see also* Distance from explosion
  blast effects and mitigation, 52
  blast fragment coupling, 176
  design approach, comprehensive, 418
  geometry of charge, 70
  planning and design considerations, 34
    HE device behavior, 16–17

nonspherical charge and complex geometries, need for validation, 19
  pressure, 65
  scaling laws, 67
  structural element behavior, design applications, 234
Cluster bombs, 47
Collaboration, multidisciplinary, 38
  comprehensive design program, 11, 36, 435, 436, 437–438
  facility design, 8
  research needs, 38
  terrorism and insurgency, 24
Collaboration, multinational, 36, 37
Collapse, catastrophic, *see* Catastrophic failure
Collapse, progressive, *see* Progressive collapse
Collapse pressure, structural element behavior, 210
Collapse prevention, comprehensive design approach, 408–409
Column removal, *see* Removal of structural elements
Columns
  connections, *see* Connections and supports
  design approach, comprehensive, 420
    load redistribution, 429
    short standoff bomb blasts, 428
    structural system and component selection, 430, 431
    structural system behavior, 425
  design experiments, 375
  DOD analysis recommendations, 384
  DOD design recommendations, 381–382
  load transfer from slab, 238
  P-I diagrams
    dynamic reaction, 340
    energy balance method, 338
  progressive collapse, 377
    analysis and acceptance criteria, 387
    concrete structures, monolithic, 379
    GSA guidelines, 386
    redesigns of structural elements, 389
    sample analyses, 396, 399, 401
    steel frame buildings, 380
    steel frame structure analysis, 390–391, 392
  structural element behavior
    shear resistance, 206
    shear walls, 212, 213
    tensile and compressive members, 206
Combined effects
  blast fragment coupling, 177
  blasts and fragments, 175–176, 177

Index

Combined size and rate effects, structural element behavior, 230–231
Combustion phenomena and processes
 chemistry and physics of explosions, 53–58
 explosion effects and mitigation, 59–60
Communication systems, 417
Composite doors, 311, 312
Composite materials, penetration characteristics, 101, 102, 166–167
Composite P-I diagrams, 343, 361
Composite systems, 442
Composition C4, 58, 66, 154
Comprehensive design approach, *see* Design approach, comprehensive
Compression
 comprehensive design approach, structural system and component selection, 431
 concrete
  dynamic increase factor, 202
  and flexural resistance, 202–203
  normalized triaxial compression data, 201
  stress-strain curves, 199–200
 concrete penetration, 90–91, 94, 95
 connections, 296, 304
 diagonal, 237
 dynamic response and analysis, 250, 357
  division into layers, 345
  flexural behavior, 346
 joint behavior, 304
 membrane behavior, 218, 219, 221, 222, 224, 225, 226
 natural periods of vibration, 216
 P-I diagrams
  analytical, 325
  SDOF and P-I computations, for reinforced concrete slabs, 364
 progressive collapse, truss structures, 379
 shear wall compressive strength, 213
 size effects and combined size and rate effects, 230, 231
 structural element behavior
  compressive members, 206–207
  cylinders, arches, and domes, 211–212
  direct shear response, 228
  and flexural resistance, 205
  joints, 232
Compression field theory, modified, 206, 270
Compression wave, ground shock, 150
Computational analysis
 advanced, *see* Advanced analysis and computation methods

application examples for SDOF and P-I computational approaches, 357–371; *see also* Pressure-impulse (PI) diagrams
 reinforced concrete beams, 357–362
 reinforced concrete slabs, 363–366, 367–371
design approach, comprehensive, 419
dynamic response, 262–267, 272–289
 advanced approximate methods, 264–265
 applications to support analysis and design, 276–281, 282–287
 approximate procedure for multi-segmental forcing functions, 281, 288–289
 equivalent SDOF system approach, 274–276
 intermediate approximate methods, 263–264
 material models, 265–267
 validation requirements, 267–272, 273
empirical data, 16
ground shock, 151
ground shock, cratering, and ejecta, HE-induced, 116–117
loads on above-ground structures, 172
penetration effects, 90
planning and design considerations, 24
 current state and future needs, 33
 requirements and capabilities, 30
progressive collapse, 376
research needs, 439
shock waves in three-phase media, 64
Computer codes
 design approach, comprehensive, 418
 dynamic response and analysis, validation of, 269–270
 FACEDAP, 327
 loads on above-ground structures, 172
 membrane behavior, 226
 nuclear loads on buried structures, 185
 penetration effects, 90
 planning and design considerations
  empirical data, 16
  nonspherical charge and complex geometries, need for validation, 19
  prediction of load parameters, 16
  single-degree-of-freedom, 28
 pressure-time histories, 421
 progressive collapse
  advanced frame structure analysis, 394
  material properties and structural modeling, 389

research needs, 439
SDOF analysis, DSAS, 362
SHOCK, 319
stress-time histories, 151
structural element behavior, diagonal shear effects, 230
Concentrated mass, SDOF system analysis
    fixed beams, 278
    fixed boundary beams, 279–280
    fixed slabs, 282–283
    simply supported beams, 277
    simply supported slabs, 281
Concrete
    connections, 291–292
        advanced truss analogies, 299, 300, 301
        detailing behavior studies, 296–307
    conventional explosive effects, 43
    cratering, 136, 138
    design approach, comprehensive, 430
        structural system and component selection, 431
        structural system behavior, 424–425
    dynamic response and analysis
        finite element approach, 271–272
        impact loading tests, 269
        intermediate approximate methods, 263
        material models, 266, 267
        validation procedures, 270
    failure modes, 237–238
    ground shock coupling factor as function of scaled depth, 131
    penetration, 87, 91
        aggregate size and, 88
        with conventional (nonnuclear) devices, 87–92, 93, 94, 95
        fragment, 159, 161, 162, 163, 164
    progressive collapse, structure types
        monolithic concrete, 378–379
        precast concrete, 378
    research needs, 440
    structural element behavior
        diagonal shear effects, 228
        direct shear response, 228
        dynamic effects, 202
        material properties, 199–202
        plastic hinge regions, 232–233
        shear resistance, 205–206
        structural design applications, 233
        technical manuals, 197
Concrete, reinforced
    charge weight for bridge pier breach, 139
    comprehensive design approach, structural system and component selection, 431, 432

connections, 293; *see also* Connections and supports
cratering, 136
dynamic analysis approach, 344–346
dynamic resistance function, 346, 347
dynamic response and analysis, 250
equivalent SDOF system approach, 349
failure modes, 237–238
penetration, 90–91, 93, 94
    compound materials, 101, 102
P-I diagrams
    analytical, 325
    beams, 357–362
    multiple failure modes, 341
    slabs, 363–366, 367–371
    threshold points, 339–340, 341
progressive collapse, GSA guidelines, 386
research needs, 440
shearing failure, 325
structural element behavior, 195, 199
    connections and support conditions, 231–232
    cylinders, arches, and domes, 210
    diagonal shear effects, 228, 229
    flexural resistance, 202–203, 204
    joints, 232
    membrane behavior, 219, 222–223, 225
    natural periods of vibration, 216–218
    structural design applications, 234
    technical manuals, 196
    tensile and compressive members, 206–207
Concrete-earth, penetration of, 101
Concrete-steel, dynamic response and analysis
    material models, 267
Concussion grenades, 46
Confined dust explosion, 330
Confinement
    concrete
        analytical P-I diagrams, 325
        and structural element behavior, 200, 201
    dynamic response and analysis, 250
Connections and supports, 291–310
    background, 292, 294–296
    design approach, comprehensive, 420
        short standoff bomb blasts, 428
        structural system and component selection, 431
    detailing behavior studies
        structural concrete, 296–307
        structural steel, 308–309
    DOD design recommendations, 381–382
    doors, 310, 313

Index

dynamic response and analysis, finite element approach, 265
government guideline limitations, 29
progressive collapse, 377, 400–402
    analysis and acceptance criteria, 388
    catenary action, 416
    GSA guidelines, 386
    redesigns of structural elements, 389
    sample analyses, 395, 397–399, 400
    semi-rigid connections, 392–394, 395–397, 398
    steel frame buildings, 380
structural element behavior, 231–233
    membrane behavior, 221–222
    plastic hinge regions, 232–233
    principal reinforcement, 207–208
tying or bridging of structural elements, 380–381
Conservation equations, 61, 62–63, 118, 123, 126–127, 256, 257
Constant pressure detonation state, 62
Constitutive models, dynamic material and, 343–344
Construction
    planning and design considerations, 7, 8, 440
    technical resources and blast mitigation capabilities, 28–30
Consulting firms, 36–37
Contact detonations
    cratering in concrete, 136
    design approach, comprehensive, 420, 424
    failure modes with, 237
Containment conditions, comprehensive design approach, 421
Continuity/continuous structural systems
    comprehensive design approach, 409, 429–430
    dynamic response and analysis, 239, 258–262
        applications to support analysis and design, 276
        equivalent SDOF system approach, 274
        SDOF, 250, 252
    equivalent SDOF system approach, 349
    P-I diagrams, energy balance method, 338, 340
    progressive collapse, 377
        concrete structures, monolithic, 377–378
        GSA guidelines, 386
Continuum mechanics theories, 31
Control of access and approach, 4, 413–414
Control rooms, 432, 433

Conventional devices/weapons/environments
airblast, 68–79
    external explosions, 68–74
    internal explosions, 75–77, 78
    leakage blast pressure, 77–79
explosive processes, characteristics of, 41–49
    bombs, 46–47
    conventional weapons, 41
    conventional weapons summary, 49
    direct and indirect fire weapon projectiles, 42–44
    grenades, 44, 46
    rockets and missiles, 47–48, 49
    small arms and aircraft cannon projectiles, 42, 43, 44, 45
    special-purpose weapons, 48–49
    summary, 49
fragmentation, 153–167
general principles and overview, 16
ground shock, cratering, and ejecta, HE-induced, 106–141
    application to protective design, 129–135
    computational aspects, 116–117
    cratering, 135–137, 138
    ejecta, 137, 139, 140, 141
    HE charges and conventional weapons, 106
    one-dimensional elastic wave propagation, 110
    reflection and transmission of one dimensional D-waves between two media, 114–116
    shock waves in one-dimensional solids, 117–119
    stress wave propagation in soils, 119–125
    three-dimensional stress wave propagation, 106–110
loads on structures, 169–177
    above-ground structures, 170–174
    blast fragment coupling, 175–176, 177
    buried structures, 169–170
    mounded structures, 175
    surface-flush structures, 175
penetration, 83–106
    armor, 98, 99, 100–101
    concrete, 87–92, 93, 94, 95
    other materials, 101, 102
    rock, 92–96, 97
    shaped charges, 103–105
    soil and other granular material, 96, 97, 98, 99

planning and design considerations
  research needs, 33–34
  technical resources and blast mitigation
    capabilities, 25
  Soviet doctrine, 23
Corridors, 416
Cost-benefit analysis, 8–9, 12–13, 434
Coulomb friction model, 266
Coupled equivalent SDOF systems, 251, 252, 349
Coupling
  blast fragment, 175–176, 177
  failure modes, 18, 35
  ground waves, 148
  structural element behavior, size effects and combined size and rate effects, 231
Coupling factor, ground shock, 131, 133
Cracking
  concrete, 87–88
  connections, 296, 299
    flexural reinforcement, 298
    joint behavior, 304
    statically loaded knee joints, 297
Cratering
  conventional environments, HE-induced, 135–137, 138
  failure modes, structural responses, 237
  nuclear and HE explosions, 79, 141–142, 143–145
  structural system behavior, 424–425
Criminal threats, 11, 13
Criteria, *see* Publications, literature and technical references
Critical buckling ratios, rings, 210
Critical structural elements, dynamic response and analysis, 239–240
Cross-sectional area or shape
  blast valves, 314
  comprehensive design approach, structural system and component selection, 431
  connections, 299, 307
  dynamic resistance function, 346, 347
  dynamic response and analysis, cross-sectional equilibrium, 271
  failure modes, cratering and spalling effects, 237
  progressive collapse, steel frame structure analysis, 390–391
  structural element behavior
    direct shear response, 227
    tensile and compressive members, 207
Cross-section analysis, moment-curvature relationship, 344

Crushing
  concrete, 87–88
  impact and penetration effects, 84
Cube root scaling, 67–68, 82
Culverts, reinforced concrete, 225–226
Curtains, blast, 416
Curtain walls, 413
Curvature, structural element behavior, 203–204
Curve fitting, P-I diagrams, 338
Cycling, load, 202
Cylinders, arches, and domes, structural element behavior, 208–212
Cylindrical charges, 34
  design approach, comprehensive, 418
  scaling laws, 67
Cylindrical structures
  nuclear loads on buried structures, 183–184
  soil arching, 186

D

Damage estimation
  guidelines, need for, 441
  P-I diagrams, 365, 370
  practical damage and response limits, 234–236
  rotation and, 233
Damage levels, P-I diagrams, 327
Damage limits
  progressive collapse, Department of Defense guidelines, 384
  structural element behavior, 234–236
Damping
  design approach, comprehensive
    structural member behavior and performance, 423–424
    structural system behavior, 425
  dynamic resistance function, 347
  dynamic response and analysis, 250, 252, 254–255
    finite element approach, 265
    simple SDOF system, 243
  membrane behavior, 224
  P-I diagrams, 330, 331
Damping matrix, 254, 265
Data
  experimental studies, *see* Experimental analysis/data
  risk assessment, 12
Debris, *see* Particles, dust, and debris
Decay
  blast load on rear wall, 193

Index

exponential, dynamic response and analysis, 281, 289
Decision support tools, 10, 32
Defense Nuclear Agency (DNA), 25
Defense Special Weapons Agency (DSWA), 25
Defense Threat Reduction Agency (DTRA), 25
Deflagration, 59
Deflected shape functions, dynamic shear force, 354
Deflections
    dynamic response and analysis
        comparisons of experimental and analytical results, 272, 273
        equivalent SDOF system approach, beam response modes, 275
    SDOF analysis, simple, 242
    structural element behavior
        cylinders, 209, 210
        membrane behavior, 223, 226
        shear walls, 213
Deflections-to-depth ratio, membrane behavior, 226
Deformable projectiles, penetration, 87
Deformation, response time of system, 328
Deformation modes
    dynamic amplification factor, 239
    dynamic response and analysis, 239, 263–264
    progressive collapse, steel frame buildings, 391–392
    structural system behavior, 425
Deformed configuration, equivalent SDOF transformation factors, 350
Degrees of freedom
    multiple, *see* Multi-degree-of-freedom (MDOF) systems
    single, *see* Single degree of freedom (SDOF) systems
    steel frame structure analysis, 394
Demand-capacity ratio, 388
Demonstrations, technological developments, 436
Density
    rock, 97
    and soil/granular material penetration, 96
Department of Defense, *see* U.S. Department of Defense
Department of the Army Tri-Service Manuals, *see* U.S. Department of the Army technical manuals
Depth of burial, buried structures
    nuclear loads, 180
    and soil arching, 186
Depth of burst (DOB)

cratering, 136, 137, 141
ejecta, 146
Depth of structural members, shear resistance, 206
Design
    balanced, 291
    dynamic response and analysis, computational systems, 276–281, 282–287
    general principles and overview
        design and construction considerations, 28–30
        development and implementation of methodology, current state and future needs, 37–39
        manuals, 26–28
        philosophy of planning and design, 5–8
        technical resources and blast mitigation capabilities, 24–30
    ground shock, cratering, and ejecta from HE devices, 129–135
    guidelines for progressive collapse mitigation and prevention
        Department of Defense, 380–384
        GSA, 384–390
    structural element behavior
        applications, 233–234
        membrane behavior, 225–226
    structural element behavior applications, 233–234
Design approach, comprehensive, 405–442
    development and implementation of effective technology, 434–441
    education, training, and technology transfer needs, 437–438
    recommended actions, 435–436
    research and development, recommendations for, 438–441
    load considerations, 417–422
    multi-hazard protective design, 433–434
    need for, 10–11
    other safety considerations, 434
    planning and design assumptions, 410–411
    protection approaches and measures, 407–410
    siting, architectural, and functional considerations, 411–417
        access and approach control, 413–414
        building exterior, 414–416
        building interior, 416
        perimeter line, 413
        post-incident conditions, 417
        vital systems, nonstructural, 417

structural behavior and performance, 422–424
structural system
  behavior of, 424–430
  and component selection, 430–433
Design range, 234
Detailing parameters, 294
  behavior of
    structural concrete, 296–307
    structural steel, 308–309
  design approach, comprehensive, 431, 433
  dynamic response and analysis, finite element approach, 265
  research needs, 440
Detonation processes
  chemistry of explosions, 55
  constant pressure detonation state, 62
  defined, 59
  distances from explosion and dynamic loads, *see* Distance from explosion
  explosion effects and mitigation, 60–64
  shaped charge devices, forcing premature detonation, 105
Diagonal bars, joint strengthening, 232
Diagonal compression
  connections, 296
  failure modes, local, 237
Diagonal cracks, connections, 303
Diagonal element mass matrices, progressive collapse, 394
Diagonal matrix, dynamic response and analysis, 254
Diagonal reinforcement, connections, 298, 307
  joint behavior, 304
  strut and truss mechanisms, 296
Diagonal shear
  dynamic response and analysis, 250, 344
    flexural behavior, 345
    intermediate approximate methods, 264
    material model testing, 270–271
  failure modes, local, 237
  P-I diagrams
    analytical, 325
    composite, 361
    SDOF and P-I computations, 357, 358
  SDOF analysis, 251
  structural element behavior, 205, 228–230
Diagonal tensile strain, connections, 300
Diaphragms, 428
Dilatational wave, 111
Dimensionless loading parameters, P-I diagrams, 330
Dimensionless P-I diagrams, 338

Direct and indirect fire weapon projectiles, 42–44
Direct-induced waves, 148, 149
Direct shear, 357
  dynamic response and analysis, 250, 251, 271, 357
    applications to support analysis and design, 276
    intermediate approximate methods, 264
  failure modes, 238
    coupled, 35
  P-I diagrams, 367, 368
    composite, 361
    dynamic structural model, 351–357
    multiple failure modes, 343, 362
  SDOF analysis
    coupled equivalent systems, 252
    equivalent SDOF system approach, 349
  SDOF and P-I computations, 358–359
    for reinforced concrete slabs, 363, 365
  structural element behavior, 226–228
Direct shear failure
  design approach, comprehensive, 425
  P-I diagrams, 360
Discontinuity
  direct shear response, 226
  D regions, 295
  at slab-wall interface, 238
  stress distribution at, 292
Dispenser bombs, 47
Displacement
  direct shear and flexural response, 365, 367, 369, 371
  P-I diagrams, 331, 368
  SDOF analysis, 240
  SDOF and P-I computations, 363, 367
    direct shear response, 371
    for reinforced concrete slabs, 364, 371
Displacement function, SDOF analysis, 249, 288
Displacement vector, dynamic response and analysis, 252
Dissipation, load, 328
Distance between buildings, 408, 409
Distance from explosion
  blast fragment coupling, 175, 176
  close-in detonation, *see* Close-in detonation
  design approach, comprehensive, 408, 418, 419–420
    short standoff bomb blasts, 428
    structural system behavior, 424
  dynamic loads, comparisons of weapons, 66–67

Index

internal explosions, shock pressure phase, 76–77
nuclear and HE explosions
  dynamic pressure impulse as function of scaled height of burst and scaled ground ranges, 89
  height of burst and ground range for intermediate overpressures, 89
  peak dynamic pressure as function of scaled height of burst and ground ranges, 88
planning and design considerations, 6, 18, 19, 34
  HE device behavior, 16–17
  minimizing, 6
scaling laws, 67
structural element behavior, 198
  frames, 214
  structural design applications, 234
  TN-5-1300 assumptions, 196
zero standoff, *see* Contact detonations
Distributed load, membrane behavior, 224
Distributed systems, 417
Documentation, design, 8
Domes, nuclear loads, on above ground structures, 178
Doors, *see also* Openings and interfaces
blast, 310–313, 314
emergency exits, 317, 318
flying debris minimization, 409
Dowel action, 238, 425
Downing Assessment Task Force recommendations, 35, 37
Drag/drag pressure
conventional loads on above-ground structures, 172, 173
nuclear loads on above ground structures, 177, 178, 180, 181–182
D (discontinuity) regions, 292, 295
Drop panels, 425
Drucker-Prager failure criterion, 266
Drucker-Prager model, 266, 267
DSAS, 269–270, 362
Ductile failure, progressive collapse of steel frame buildings, 380
Ductility
design approach, comprehensive, 433
dynamic response and analysis
  applications to support analysis and design, 276
  multi-segmental forcing functions, 289
  P-I diagrams, 331
  progressive collapse, 377
  structural element behavior, 197

structural system behavior, 424
Ductility factor, 247
Ductility ratios, 197, 289
Ducts
airblast transmission through, 319, 322–324
cable and conduit penetrations, 315–317
Duhamel's integral, 242–247
Duration of impact or load, *see also* Pressure-time history; Time-pressure curves
blast design load derivation procedures
  blast load on rear wall, 193
  load-time history on buried wall or roof, 187
  pressure-time curve on front wall from external explosions, 188
  pressure-time curve on roof or sidewall, span parallel to shock front explosions, 190
  pressure-time curve on roof or sidewall, span perpendicular to shock front explosions, 190
comparison of response times of loading regimes, 329
conventional loads on above-ground structures, 173
design approach, comprehensive, 418, 420, 421
dynamic response and analysis, 239
external explosions, 68
failure modes, 237
impulsive loading regimes, 328
internal explosions, gas pressure phase, 77, 78
loading regime quantification, 330
natural frequency and, 328
penetration process, energy absorption, 84
P-I diagrams, 326
planning and design considerations, 25, 405
structural element behavior, distances from explosion and dynamic loads, 198
Dust, *see* Particles, dust, and debris
D-waves, reflection and transmission between two media, 114–116
Dynamic amplification factor, 239
Dynamic coupling, shock isolation, 319, 320–321
Dynamic direct shear model, 357
Dynamic equilibrium equation, 108
intermediate approximate methods, 263
nuclear loads on buried structures, 183–184
Dynamic equilibrium of forces, dynamic shear force, 353
Dynamic failure, multiple failure modes, 343

Dynamic increase factors (DIFs), steel connections, 309
Dynamic load factor, SDOF analysis, 242
Dynamic loads
  design approach, comprehensive, 422
  P-I diagrams, 330, 339
    comparison of response times of loading regimes, 329
    versus shock spectrum, 327–328
    structural element behavior, 198, 202
    diagonal shear effects, 230
Dynamic models, P-I diagrams, numerical approach, 339
Dynamic pressure, 66–67
  blast design load derivation procedures
    pressure-time curve on front wall from external explosions, 188, 189
    pressure-time curve on roof or sidewall, span parallel to shock front explosions, 191
    pressure-time curve on roof or sidewall, span perpendicular to shock front explosions, 190
  conventional loads on above-ground structures, 173
  external explosions, 72
  loads on above-ground structures, 170, 177
  nuclear and HE explosions, 80, 81, 82
    loads on above ground structures, 177
    peak, 82–83
    peak dynamic pressure at shock front as function of peak overpressure at sea level, 87
Dynamic region approximation, P-I diagrams, energy balance method, 336–338, 339
Dynamic resistance function, 250, 251, 346–347, 348
Dynamic response and analysis, 237–289
  advanced analysis requirements, 238–239
  computational capabilities, validation requirements, 267–272, 273
  computational systems, intermediate and advanced, 262–267
    advanced approximate methods, 264–265
    intermediate approximate methods, 263–264
    material models, 265–267
  computation support for protective analysis and design activities, 272–289
    applications to support analysis and design, 276–281, 282–287
    approximate procedure for multi-segmental forcing functions, 281, 288–289
    equivalent SDOF system approach, 274–276
    SDOF parameters for fixed beams, 278
    SDOF parameters for fixed boundary beams, 279–280
    SDOF parameters for simply supported beams, 277
    SDOF parameters for simply supported slabs, 281
  continuous systems, 258–262
  design approach, comprehensive, 422–423; *see also* Design approach, comprehensive
  modes of failure, 237–238
  multi-degree-of-freedom (MDOF) systems, 251–258
    numerical methods for transient responses analysis, 255–258
    principles of, 251–255
  nuclear loads on buried structures, soil-structure interactions, 179
  P-I diagrams
    closed-form solutions, 331–335, 336
    numerical approach, 339
    transition point searches, 339
  P-I diagrams, dynamic analysis approach, 343–348
    direct shear behavior, 347–348
    dynamic material and constitutive models, 343–344
    dynamic resistance function, 346–347
    flexural behavior, 344–346
  planning and design, 8
  planning and design considerations, 7, 25, 29, 405
  pressure-impulse diagrams, *see* Pressure-impulse (PI) diagrams
  progressive collapse, steel frame structure analysis, 394
  SDOF analysis, simple, 240–251
    advanced approaches, 247–250
    graphical presentations of solutions, 247, 248, 249
    numerical solutions, 247–250
    theoretical solution for, Duhamel's integral, 242–247
  structural element behavior
    material properties, 202, 203
    structural design applications, 233
    technical manuals, 196, 197
Dynamic shear force, 353, 357

Index

Dynamic structural analysis techniques
  design approach, comprehensive, 416
  DOD recommendations, 383–384
Dynamic structural model, P-I diagrams, 348–357
  direct shear response, 351–357
  flexural response, 349–351

E

Earthquakes, *see* Seismic loads/stresses
Education, 437–438, 441
Effective mass, equivalent SDOF system approach, beam response modes, 275
Effective spring constant, SDOF parameters
  fixed beams, 278
  fixed boundary beams, 279–280
  simply supported beams, 277
  simply supported slabs, 281
Ejecta
  conventional environments, HE-induced, 137, 139, 140, 141
  cratering, 135, 136, 142
  from nuclear weapons, 145–147, 148–149
  planning and design considerations, 7
Elastic behavior
  concrete elastic modulus, 199–200
  cylinders, 209
  domes, 212
  response spectra, 244–246
  rings, 210
Elastic isotropic solid, three-dimensional stress wave propagation, 108–109
Elastic models, 250, 265
  dynamic shear force, 354
  loading regime quantification, 329
  membrane behavior, 222, 224
  P-I diagrams
    analytical, 325
    closed-form solutions, 331–335, 336
  SDOF
    energy solutions, 337, 338
    equivalent SDOF transformation factors, transitions between response modes, 350
  SDOF analysis
    for fixed one-way slabs, long edges fixed and short edges simply supported, 286–287
    for fixed one-way slabs, short edges fixed and long edges simply supported, 284–285
    for fixed slabs, 282–283
Elastic-perfectly plastic models
  dynamic response and analysis, 346–347
  membrane behavior, 224
  P-I diagrams, 330, 325
  SDOF, energy solutions, 337, 338
Elastic-plastic models, 239, 250
  design manual limitations, 25
  membrane behavior, 222, 224
  P-I diagrams, 330
  SDOF analysis, 248–250
    fixed one-way slabs, long edges fixed and short edges simply supported, 286–287
    for fixed one-way slabs, short edges fixed and long edges simply supported, 284–285
    for fixed slabs, 282–283
    parameters for fixed beams, 278
    parameters for fixed boundary beams, 279–280
Elastic theory, 212
Elastic wave propagation, one-dimensional, 110
Elastic zone, cratering, 136
Elastoplastic steel, flexural resistance, 202–203
Electromagnetic pulse, 5, 7, 311
Element removal, *see* Removal of structural elements
Element types, finite element approach, 265
Elevator shafts, 417
Emergency exits, 317, 318
Emergency functions, 412–413, 417
Enclosure systems and technologies, 439–440, 442
End rotation, progressive collapse, 397–398
Energy
  blast effects and mitigation, 52
  chemistry and physics of explosions, 53–58
  conservation of, 61, 118, 256, 257, 426
  cube root scaling, 67
  design approach, comprehensive
    sacrificial elements, 430, 432–433
    structural system and component selection, 432
  dynamic response and analysis, 256–257
  fragmentation and, 34
  and penetration, deformation and, 87
  propagation of
    internal explosions, 75
    scaling laws, 67
  shock waves
    reflected, 70
    in three-phase media, 63–64

Energy balance method, P-I diagrams, 331, 335–339, 341–342
  approximating dynamic regions, 336–338, 339
  continuous structural elements, 338, 340
Engineering classification for intact rock, 96
Enhancement factors, 430
Entrance tunnels, 310
Entropy, 53, 55, 63
Envelopes, research needs, 439–440
Environmental effects, research needs, 441
Equations of motion
  continuous systems, 259, 262
  MDOF analysis, 252
  P-I diagrams
    dynamic structural model of direct shear effects, 351
    dynamic structural model of flexural response, 349–350
  SDOF analysis, 241, 247–248
  SDOF systems
    equivalent SDOF system approach, beam response modes, 275–276
    flexural equivalent, 349–350
  shock waves in one-dimensional solids, 118
  three-dimensional stress wave propagation, 108–109
Equilibrium equation
  dynamic shear force, 353
  SDOF
    approximate, 423
    simple, 240
Equipment, 442
  multiple detonation effects, 35
  planning and design considerations, 7, 18, 29–30
    risk assessment, 12
    robustness, 29–30
Equivalent load factor, 172, 173, 425
Equivalent loads
  blast design load derivation procedures, 190
  static, 7
Equivalent pressure, conventional loads on buried structures, 169
Equivalent SDOF systems, 274–276
  coupled, 251, 252, 349
  coupled systems, 349
  dynamic response and analysis, 357
  dynamic structural responses, 349
  P-I diagrams, 339
Equivalent TNT values, *see* TNT equivalents
Equivalent uniform pressure-time history
  above-ground structures, 172
  buried wall or roof, 187, 188

Equivoluminal (shear) waves, 109
Euler-Bernoulli models, 292
Eulerian methods, 117
Evacuation of facility, 6, 412
Exhaust systems, 417
Existing construction
  prevention of collapse, 408–409
  progressive collapse
    Department of Defense guidelines, 380–381
    General Services Administration (GSA) guidelines, 385
  retrofits, *see* Retrofits
Exits, emergency, 317, 318
Exothermic reactions, 53, 59
Experimental analysis/data
  blast fragment coupling, 175
  computer code requirements, 16
  cratering, nuclear explosions, 143
  design approach, comprehensive, 418, 420
  dynamic response and analysis, 344
    material model testing, 267
    numerical method comparisons, 272–274
    validation procedures, 267–272, 273
  joint behavior, 301
  membrane behavior, 223–224
  military focus of previous work, 25–26
  and P-I models, 368, 370
    analytical, 325
    SDOF and P-I computations, for reinforced concrete slabs, 363, 364
  requirements and capabilities, 30–31
  soil properties from explosion tests, 132
Explicit method, steel frame structure analysis, 394
Explosive constants, 154
Explosive devices and explosions, 41–69
  airblast, 68–79
    external explosions, 68–74
    internal explosions, 75–77, 78
    leakage blast pressure, 77–79
  analysis requirements and capabilities, 31
  connections and support conditions, 292
  conventional devices, processes, and environments, 41–49
    bombs, 46–47
    direct and indirect fire weapon projectiles, 42–44
    grenades, 44, 46
    rockets and missiles, 47–48, 49
    small arms and aircraft cannon projectiles, 42, 43, 44, 45
    special-purpose weapons, 48–49

Index

effects and mitigation, 50–64
  blast effects, 50–52
  chemistry of, 53–54
  combustion phenomena and processes, 59–60
  detonation processes and shock waves, 60–64
  physics of, 54–58
  types and properties of explosives, 58–59
loading characteristics, 292, 294
  complexity of, 17–18
membrane behavior, 223–224
nuclear devices, processes, and environments, 41
planning and design considerations, 4, 5–6, 7
  computer code requirements, 19
  load determination, 18–19
  multiple devices, 16
  technical resources and blast mitigation capabilities, 25
  threat assessment, 14
  types of explosive attack incidents in U.S., 14
transitions between response modes, 30
types and properties of explosives, 58–59
Explosive energy constant, Gurney, 156
Explosive facility design, 196
Exponential decay
  dynamic response and analysis, 281, 289
  loading regime quantification, 329
Exponential load, P-I diagrams
  energy balance method, 336–337
  response functions, 332
Exteriors, building
  comprehensive design approach, 414–416
  design approach, comprehensive, 413, 414
  progressive collapse, analysis and acceptance criteria, 387
  research needs, 439–440
  threat assessment, 14
External criteria screening (ECS) techniques, 376
External elements, alternate path analysis, 384
External explosions
  airblast, conventional weapons and high-explosive devices, 68–74
  planning and design considerations, 29
  time-pressure curves, front wall (surface burst), 187–189, 190
External parameters, dynamic analysis approach, 344

F

Fabrics, 442
Facade shielding, 432
FACEDAP (Facility and Component Evaluation and Damage Assessment Program), 327, 426
Face mats, concrete protection, 88
Facility and Component Evaluation and Damage Assessment Program (FACEDAP), 327, 426
Factored shear force, 205
Failure, conditions resulting in failure, 327–328
Failure modes
  coupling, 18, 35
  design approach, comprehensive
    load redistribution, 428–429
    structural system behavior, 424–425
  direct shear response, 351–352
  dynamic response and analysis, 250, 357
    flexural behavior, 345
    material models, 266
  dynamics of, 237–238
  explosive loads and, 294
  membrane behavior, 226
  P-I diagrams, 325, 327
    dynamic reaction, 340
    multiple, 341, 342, 343
    SDOF and P-I computations, 358, 360
  planning and design considerations, 5
  progressive collapse, 373–374
    concrete structures, monolithic, 377–378
    steel frame buildings, 380
    steel frame structure analysis, 390–391
  structural analysis research needs, 35
Failure moment, statically loaded knee joints, 297
Fallback, cratering, 135, 142
Fasteners, *see* Connections and supports
Federal Emergency Management Agency (FEMA) guidelines, 6, 27, 28, 407
FIBCON, 101
Fiber-reinforced concrete, penetration of, 101
Finite element and finite difference methods
  ABAQUS/Explicit, 270, 273
  dynamic response and analysis, 264–265, 271–272
  membrane behavior, 224–225
  planning and design, 8
Finite element model, 271–272
Finite rise time, P-I diagrams, 330
Fire(s), 167
  combustion phenomena and processes, 59

comprehensive design approach, 417
fire and incendiary bombs, 47–48
hazard mitigation, 6
multi-hazard protective design, 433–434
planning and design considerations, 7
secondary threats, 32
types of explosive attack incidents in U.S., 14
Fireball, nuclear devices, 79
Fixed beams, SDOF parameters, 278, 279–280
Fixed slabs, SDOF analysis, 282–283
    one-way with long edges fixed and short edges simply supported, 286–287
    one-way with short edges fixed and long edges simply supported, 284–285
Flat plate slabs, load transfer, 238
Flexural moment capacity, 228–229
Flexural response
    buried structures
        conventional explosive loads, 170
        equivalent uniform loads, 188
    comprehensive design approach, load redistribution, 429
    concrete
        flexural response, face mats and, 88
        reinforcement and, 89
    connections, reinforcement and, 304–305
    diagonal shear effects, 228–229
    dynamic response and analysis, 250, 357
        equivalent SDOF system approach, 274–275
        intermediate approximate methods, 264
        material model testing, 270–271
        SDOF, 251
    equivalent SDOF system approach, 274–275, 349
    failure modes, 238; *see also* Failure modes
        coupling/coupled, 18, 35
        dynamic resistance function, 347
        local, 237–238
        P-I diagrams, dynamic reaction, 340
        P-I diagrams, multiple failure modes, 341, 343, 362
        SDOF and P-I computations, 358
    government guideline limitations, 29
    P-I diagrams, 327, 357, 367, 368
        composite, 361, 362
        multiple failure modes, 341, 343, 362
        SDOF and P-I computations, 360, 361, 363, 364, 371
    P-I diagrams, dynamic structural model, 349–351
        equation of motion, 349–350
        transformation factors, 350–351

progressive collapse
    GSA guidelines, 385
    sample analyses, 395
    SDOF analysis
        assumptions, 325
        coupled equivalent systems, 251, 252
    SDOF and P-I computations, 357, 360, 361
        for reinforced concrete slabs, 363, 364, 371
    structural element behavior, 202–205
        cylinders, arches, and domes, 212
        diagonal shear effects, 228–229, 230
        joints, 232
        plastic hinge regions, 232
        shear resistance, 205, 206
        technical manuals, 197
        tensile and compressive members, 207
Flexure, equivalent SDOF system approach, 349
Floors, 416
    arch joints, 306
    comprehensive design approach, 430–431
    DOD design recommendations, 381
    ground floor analysis and acceptance criteria, 387
    progressive collapse, 402–403
        sample analyses, 396, 397, 398
    reinforced, catastrophic failure, 208
Flow charts, 381
Flow tube, 315, 317
Flow velocity, and dynamic pressure, 80
Focusing, shock waves, 17, 34
Force, P-I diagrams, 327
Forced entry, 14, 18–19, 29
Forced vibration responses, P-I diagrams, 333–334
Force-impulse diagrams, *see* Pressure-impulse (PI) diagrams
Force-impulse term, P-I diagrams, 327
Force redistribution, progressive collapse, 379
Forces, military, 19, 35
Force-time histories
    dynamic response and analysis, 271
    finite element approach, 265
    P-I diagrams, idealized transient loading profile, 326
Force vector, dynamic response and analysis, 252
Forcing functions, 242, 246
    dynamic response and analysis, 276
    dynamic shear force, 352
    equivalent SDOF transformation factors, 351
    multi-segmental, 281, 288–289
    SDOF analysis, 245

Index

Foreign intelligence services, 11, 24
Fortification science, 25, 32–33, 52, 405
Foundation, load transfer to, 427
Fracture-based approaches, dynamic response and analysis, 269
Fracture fatigue failure, 380
Fragility data for equipment, 7, 29–30
Fragmentation, 153–167; *see also* Particles, dust, and debris
    blast fragment coupling, 175–176, 177
    computer code requirements, 19
    conventional loads on mounded structures, 175
    design approach, comprehensive, 418, 419
        flying debris minimization, 409
        research needs, 440
    and loading parameters, 34
    planning and design considerations, 5, 6, 7, 16, 19
    and shock wave energy, 70–71
    structural element behavior, technical manuals, 196, 197
Fragmentation (FRAG) bombs, 46
Fragmentation grenades, 46
Frames, 442
    connections, factors affecting concrete behavior, 300–301
    design approach, comprehensive
        structural system and component selection, 432
        transfer of load, 429–430
    DOD design recommendations, 381–382
    joint size limitations, 232
    progressive collapse, 377–378
        advanced frame structure analysis, 390–404; *see also* Progressive Collapse
        catenary action, 416
        steel frame buildings, 380
    progressive collapse, structure analysis
        computer code requirements, 394
        examples of, 395–404
        semi-rigid connections, 392–394, 395–397, 398
    structural element behavior, 214–215
Free air equivalent weight, 65, 66
Free body diagram
    connections, 303
    deformed slab, 220
    MDOF, 252
Free-field blast wave, 187–189
Free-field pressure
    design approach, comprehensive, 421
    loads on above-ground structures, 170
    load-time history on buried wall or roof, 187
Free-field pressure pulse, 170
Free-field pressure-time variation, 68
Free-vibration response, P-I diagrams, 333–334
Frequency characteristics, 325; *see also* Vibration, natural periods and modes
    dynamic response and analysis, 239, 348
        continuous systems, 261
        SDOF, simple, 241, 242, 243
    membrane behavior, 224
    natural circular frequency, 330, 350, 351
        continuous systems, 261
        dynamic response and analysis, 261, 276
        loading regimes, 330, 350, 351
    natural frequency
        dynamic response and analysis, 254, 276
        loading regimes, 330
        response time of system, 328
        SDOF analysis, simple, 241
        truss structure progressive collapse, 379
    SDOF systems, 348
    structural system behavior, 425
Friction, shock front attenuation, 323
FROG missiles, 48
Fuel-air-munitions, 47, 48–49
Fuel materials, chemistry and physics of explosions, 53–54
Fuel oil devices (ANFO), 31, 58
Fully nonlinear dynamic methods, planning and design, 8
Fundamental frequency, SDOF systems, 348

G

Garages, *see* Parking facilities
Gas constant, 59–60
Gases
    chemistry of explosions, 53, 54
    detonation processes and shock waves, 60–64
    impulsive loading, 418
    internal explosions
        gas pressure phase, 75, 77
        quasistatic phase, 75
    nuclear and HE explosions
        gas density and dynamic pressure, 80
        peak dynamic pressure for ideal gas, 82–83
    shock propagation in, 52
Gas explosions
    design approach, comprehensive, 422
    and progressive collapse, 375

Gasoline, 57
Gauge signals, 175, 176
General principles of protective technology, 1–39
  analysis requirements and capabilities, 30–31
    computational analysis, 30
    experimental analysis, 30–31
  current state and future needs, 31–39
    development and implementation of design methodology, 37–39
    policy and technology needs, 35–37
    research and development, relationships to, 33–35
  load definition from threat and hazard environment, 16–24
    military threat assessment, 19–24
    terrorism and insurgency threat assessment, 24
  methodology, threat, and risk assessment, 8–16
    risk assessment, 11–15
    threat, hazard, and vulnerability assessments, 10–11
  planning and design philosophy, 5–8
  technical resources and blast mitigation capabilities, 24–30
    design and construction considerations, 28–30
    design manuals, 26–28
General purpose (GP) bombs, 46
General resistance function, membrane behavior, 223
General Services Administration (GSA) guidelines, 385–386
  planning and design considerations, 6
  progressive collapse, 384–390
    analysis and acceptance criteria, 387–389
    assessment of existing facilities for potential for progressive collapse, 385
    material properties and structural modeling, 387–389
    mitigation, new facilities, 385–387
    redesigns of structural elements, 389–390
Generators, 417
Geological structure
  conventional loads on buried structures, estimation of, 170
  cratering, 141, 144
  ejecta, 146, 148
  ground shock, surface-burst, 149

Geologic materials, material models, 266, 267
Geometrical load, direct shear response, 226
Geometric model, progressive collapse, 399
Geometry of beam, SDOF and P-I computations, 356, 357
Geometry of casing, and fragmentation, 154, 155, 157–158
Geometry of charge
  cannon projectiles, 45
  computer code requirements, 19
  design approach, comprehensive, 421
  planning and design considerations, 34
Geometry of explosion
  buried structures, 106
  hemispherical surface bursts
    pressure-time curve on front wall from external explosions, 189
    pressure-time curve on roof or sidewall, span perpendicular to shock front explosions, 190
Geometry of explosive/projectile
  blast design load derivation procedures, pressure-time curve on roof or sidewall, span perpendicular to shock front explosions, 190
  design approach, comprehensive, 418
  general principles and overview, 16
    and loading parameters, 17, 34
    research issues, 34
  and penetration
    of concrete, 92
    of rock, 92, 93
    shaped charge devices, 105
    of soil/granular material, 96, 98, 99
  scaling laws, 67
  and shock front geometry, 70
  shock wave parameters for spherical TNT explosions, 71
Geometry of fragments, and penetration, 157–158, 159
Geometry of gas cloud, 422
Geometry of impact, and penetration, 91
Geometry of shock front, 70, 72
Geometry of structure, 412
  design approach, comprehensive, 414, 418, 431
    factors affecting outcomes, 432
    structural system and component selection, 431
  instability, 390
  and nuclear load effects
    on above ground structures, lift and drag coefficients, 177, 178
    on buried structures, 183, 184, 185

Index

soil arching, 186
progressive collapse, steel frame buildings, 391–392
structural element behavior
  cylinders, arches, and domes, 208–212
  membrane behavior, 218–219
German Federal Ministry of Defense, 406
Girders
  progressive collapse, 377
    catenary action, 416
    sample analyses, 395, 396
    transfer, 429–430
Global structural behavior
  design approach, comprehensive, 427
  dynamic load types and, 422
  local response versus, 52
  progressive collapse
    GSA guidelines, 385
    steel frame buildings, 380
Government-academic-industry partnerships, 38, 39, 435
Government resources, *see specific U.S. government agencies*
Grain, and soil/granular material penetration, 96
Graphical presentations, dynamic response and analysis, 247, 248, 249
Gravity loads
  progressive collapse, 398
  tensile and compressive members, 206
Grenades, 25, 44, 46
Grills, door designs, 311
Ground floor, analysis and acceptance criteria, 387
Ground range, *see* Distance from explosion
Ground shock
  blast effects and mitigation, 51–52
  conventional environments, HE-induced, 106–141
    application to protective design, 129–135
    computational aspects, 116–117
    HE charges and conventional weapons, 106
    one-dimensional elastic wave propagation, 110
    reflection and transmission of one dimensional D-waves between two media, 114–116
    shock waves in one-dimensional solids, 117–119
    stress wave propagation in soils, 119–125
    three-dimensional stress wave propagation, 106–110
  conventional loads on surface-flush structures, 175
  coupling factor as function of scaled depth, 131
  design approach, comprehensive, 420–421
  from nuclear weapons, 147, 149, 150–153
  planning and design considerations, 5, 7
  soil properties for calculations, 131
  time of arrival, 128–129
Ground surface, *see* Surface (of ground)
GSA guidelines, *see* General Services Administration (GSA) guidelines
Guidelines, *see* Publications, literature and technical references
Gurney equation, 156

H

H-6, 66, 154
Hardened facilities, 434
  dynamic response and analysis, 239
  subsystem protection, 29
Hardening, structure, 379, 412
Hardness
  and penetration, 87
  terminology, 6
  yielding, hardening effects of, 266
Harmonic loads, simple, 246
Harmonic vibration, SDOF analysis, 242
Hartford Coliseum, 379
Haunch, joint behavior, 304, 305
Hazard assessment, *see* Threat and risk assessment
HBX-1 and HBX-3, 66, 154
Headwaves, 148, 149
Heating, ventilation, and air conditioning (HVAC) systems, 7, 410
Heat of combustion, 55
Heat of detonation, 60–61
Heat of reaction, 55, 65
Heat sensitivity, primary explosives, 58
HE devices, *see* High-explosive (HE) devices
Height of burst (HOB)
  cratering, 137
  nuclear explosions, 141, 143, 145
  nuclear and HE explosions, 79, 80, 82
    dynamic pressure impulse as function of scaled height of burst and scaled ground ranges, 89
    low overpressures, 85
    peak dynamic pressure as function of scaled height of burst and ground ranges, 88

Hemispherical blast pressure, internal explosions, 77
Hemispherical shock waves, 73
 ground surface explosions, 72
 internal explosions, blast environment with, 75
 nuclear and HE explosions, 80
Hemispherical surface bursts
 pressure-time curve on front wall from external explosions, 189
 pressure-time curve on roof or sidewall, span perpendicular to shock front explosions, 190
High-explosive (HE) devices
 airblast, 68–79
  external explosions, 68–74
  internal explosions, 75–77, 78
  leakage blast pressure, 77–79
 anti-tank (HEAT) munitions, 42, 44, 45, 46
  rockets and missiles for launching, 49
 basis for comparison, TNT equivalents, 65
 battlefield support missiles, 48
 blast effects and mitigation, 50
 bombs, 46, 47
 direct and indirect fire weapon projectiles, 42–44
 ground shock, cratering, and ejecta from, 106–141
  application to protective design, 129–135
  computational aspects, 116–117
  cratering, 135–137, 138
  ejecta, 137, 139, 140, 141
  one-dimensional elastic wave propagation, 110
  reflection and transmission of one dimensional D-waves between two media, 114–116
  shock waves in one-dimensional solids, 117–119
  stress wave propagation in soils, 119–125
  three-dimensional stress wave propagation, 106–110
 impulsive loading, 418
 planning and design considerations, 16
  combined/multiple weapons systems, 18
  load parameter estimation, complexity of, 18
  previous testing methods, 25–26
 structural element behavior, frames, 214
 structural failure modes, research needs, 35
High levels of protection (HLOP), 381
High pressure spike, internal explosions, 75

Hinges/hinge regions
 plastic, see Plastic hinge regions
 progressive collapse, 388, 398–399
 three-hinge mechanisms, 406
History, see also Load-time history; Pressure-time history
 dynamic resistance function, 346–347
HLOP (high levels of protection), 381
HMX, 154
Holzer method, 258
Hooke's law, 422–423
Howitzers, 22
HTA-3, 154
Hugoniot equations, 61–62, 118, 119
Human body responses to blast loading, 327, 433
Hybrid finite element and finite difference method, 265
Hydrocode, 419–420
Hyperbolic functions, 330, 336–337, 338, 358, 368
Hyperbolic shapes, 358
Hysteretic loop, dynamic resistance function, 347

I

Ideal gas, 63, 82–83
Idealized transient loading profile, P-I diagrams, 326
Ideal surface peak overpressure curves, 82, 83
IEDs (improvised explosive devices), 31, 41
Impact angle, see Angle of incidence
Impact load transfer, 413–414
Impact parameters
 dynamic response and analysis validation requirements, 270, 271
 and effectiveness of projectile, 43
 ejecta, 141
 penetration process, 84, 91
 P-I diagrams, idealized profile, 326
 planning and design considerations, 24, 25
 risk equation, 11, 12
Impedance, 113
Implementation
 design methodology, 38–39
 technological/design developments, recommended actions, 436
Improvised explosive devices (IEDs), 31, 41
Impulse
 free air equivalent weights, 66
 and peak overpressure, 82

Index

P-I diagrams, *see* Pressure-impulse (PI) diagrams
Impulsive loading
    comparison of response times of loading regimes, 329
    comprehensive design approach, 418
        pressure versus impulse sensitivity, 426
    P-I diagrams versus shock spectrum, 327–328
    pressure versus impulse sensitivity, 426
Incendiary devices, 14, 47–48, 167
Incidence, angle of, *see* Angle of incidence
Incident pressure
    conventional loads on mounded structures, 175
    external explosions, 68, 69
    loads on above-ground structures, 171, 172
    pressure-time curve on front wall from external explosions, 188, 189
Incident wave
    airburst blast environment, 72
    nuclear and HE explosions, 80
    reflection and transmission between two sand materials, 122
Indeterminate structures, 424
Indirect fire weapon projectiles, 42–44
Industrial accidents, 11
Industry, *see* Collaboration, multidisciplinary
Inelastic analysis, progressive collapse of steel frame buildings, 392
Inelastic buckling, progressive collapse of steel frame buildings, 391
Inelastic strain energy, 424
Inertia forces
    and dynamic behavior, 258, 275, 306, 355–356
    P-I diagrams, energy balance method, 335
Inertial loads, seismic loading, 427–428
Insider threats, 11
Inspection of facilities, 8
Insurgency threat assessment, 24
Intake shafts, 417
Intelligence services, 11, 24
Interaction diagram
    flexural resistance, 203–204
    tensile and compressive members, 207
Interactions of forces, *see also* Dynamic response and analysis
    internal, dynamic analysis approach, 343–344
    shock waves
        interference, 64
        internal explosions, 75

Interactions between structural elements or systems, 442
    advanced frame structure analysis, 390
    planning and design considerations, 7
    research needs, 438
    structure-door interactions, 310
Interface, media
    ground shock, surface-burst, 149
    reflection and transmission of one dimensional D-waves between two media, 114–116
    shock waves, 122–123
        nuclear and HE explosions, 123
        reflections from, 133
        without reflected shock, 127
Interface, structural, *see also* Connections and supports; Openings and interfaces
    design approach, comprehensive, structural system behavior, 424–425
    load transfer at, 238
    shear plane, displacement along, 348
    shear transfer at, 357
    static interface shear transfer, 228
Interference, shock wave, 64
Interiors, building
    comprehensive design approach, 416
    connections, 300–301
    progressive collapse, analysis and acceptance criteria, 387
Internal damping, dynamic resistance function, 347
Internal explosions
    airblast, 75–77, 78
    design approach, comprehensive, 421
    multiple detonations, 35
    planning and design considerations, 29
    pressure increases, 319, 322
Internal forces
    dynamic analysis approach, 343–344
    pressure, *see* Pressure, internal
Internal shock and its isolation, 317, 319, 320–321
Internal structure, *see also* Supports/support conditions
    alternate path analysis, 384
    connections, 291–310
        background, 292, 294–296
        structural concrete detailing behavior studies, 296–307
        structural steel detailing behavior studies, 308–309
    internal pressure, 319–322
        airblast transmission through tunnels and ducts, 319, 322–324

increases, 319
internal shock and its isolation, 317, 319, 320–321
openings and interfaces, 310–317
  blast doors, 310–313, 314
  blast valves, 313–315
  cable and conduit penetrations, 315–317
  emergency exits, 317, 318
  entrance tunnels, 310
Internal volume, gas pressure phase of internal explosions, 77
International collaboration, 36, 37, 435
Isobaric reactions, 59
Isochoric reactions, 59
Iso-damage curve, *see* Pressure-impulse (PI) diagrams
Isoentropic process, 63

J

Jet formation, shaped charges, 103
Joints, 296
  connections, 291–292
  limiting conditions, 232
  reinforcement and, 301, 302
  structural system and component selection, 432
Joists, 431
Jones-Wilkins-Lee (JWL) equation, 63

K

Khobar Towers, 10
Kinetic energy, penetration depth and, 84
Knee joints, 296, 301
Knee wall, 413
Knowledge base, 33

L

Lacing reinforcement, 196
Lagrangian method, 117–118
Lamé's constants, 108
LANCE missiles, 48
Landscape, 440
Landscaping, 414
Lap splice, reinforcement, 429
Lateral bracing, 420
Lateral loads, progressive collapse, 404
Lateral strains, longitudinal motions and, 111–112
Lateral torsional buckling, progressive collapse of steel frame buildings, 380
Law enforcement, 24
Layered materials, fragment penetration, 166–167
Layered medium
  ejecta thickness from high yield burst in, 148
  ground shock, surface-burst, 149
  reflections from interfaces, 133
  wave transmission, 116–117
Layered structure, flexural behavior, 345
Layout, comprehensive design approach, 409–410, 416
Lead azide, 58
Leakage blast pressure, 77–79
Length-span ratio, pressure-time curve on roof or sidewall, span perpendicular to shock front explosions, 190
Levels of protection, UFC classification, 381
Lifelines, planning and design considerations, 7
Lift, nuclear loads on above ground structures, 177, 178, 180, 181–182
Light case (LC) bombs, 46
Limit states
  advanced frame structure analysis, 390
  steel frame design, 390
Linear analysis, DOD recommendations, 383–384
Linear elastic analysis, progressive collapse, 387, 404
Linear elastic models, 265
Linear elastic static analysis, progressive collapse, 386, 388
Linearly elastic oscillators, P-I diagrams, 337
Linear SDOF methods, 8
Linear SDOF oscillators, 426
Line elements, finite element approach, 265
Lines, explosive, 67
Liquified natural gas (LNG), 49
Literature and technical references, *see* Publications, literature and technical references
Load, equivalent SDOF transformation factors, 350
Load and resistance factor design (LRFD), 383
Load-bearing members
  comprehensive design approach, 416
  progressive collapse
    DOD analysis recommendations, 384
    DOD design recommendations, 381–382, 383
Load capacity
  design approach, comprehensive, 429
  estimation of, 205

Index

Load cycling, dynamic effects, 202
Load definition
   design approach, comprehensive, 418
   research needs, 438–439
Load deflection
   dynamic resistance function, 347
   and flexural response, 345
   SDOF and P-I computations, 357
   unloading-reloading of load deflection curve, 348
Load-deformation relationship, dynamic response and analysis, 344
Load factor
   dynamic response and analysis
      SDOF analysis for fixed one-way slabs, long edges fixed and short edges simply supported, 286–287
      SDOF analysis for fixed one-way slabs, short edges fixed and long edges simply supported, 284–285
      SDOF analysis for fixed slabs, 282–283
      SDOF parameters for fixed beams, 278
      SDOF parameters for fixed boundary beams, 279–280
      SDOF parameters for simply supported beams, 277
      SDOF parameters for simply supported slabs, 281
   inertia, 355–356
Load function, SDOF and P-I computations, 357, 358
Load history, *see also* Load-time history
   analysis methods, 383
   dynamic resistance function, 346–347
Loading diagram, dynamic response and analysis, 278, 279–280
Loading function
   and connections, 294
   loads on above-ground structures, 172
   P-I diagrams, SDOF system response functions, 332
   SDOF analysis, 281, 288
Load magnification factors, structural element behavior, 202
Load-mass factor
   fixed beams, 278
   fixed boundary beams, 279–280
   fixed one-way slabs
      long edges fixed and short edges simply supported, 286–287
      short edges fixed and long edges simply supported, 284–285
   fixed slabs, 282–283
   simply supported beams, 277
   simply supported slabs, 281
Load paths, DOD design recommendations, 382
Load proportionality factors, 356
Load pulse, 331
Load rate, 330
Load rise time, 330
Loads and loading regimes
   blast design load derivation procedures, 000, 187–193
      blast load on rear wall, 191–193
      load-time history on buried wall or roof, 187, 188
      time-pressure curve on front wall from external explosion (surface burst), 187–189, 190
      time-pressure curve on roof or sidewall, span parallel to shock front, 190–191, 192
      time-pressure curve on roof or sidewall, span perpendicular to shock front, 190, 191, 192
   bridging or tying of structural elements, 373
   connections and support conditions, 292
   design approach, comprehensive, 419–420; *see also* Design approach, comprehensive
      approximate SDOF calculations, 422–423
      load redistribution, 428–429
      reflected pressure, 418
      structural behavior and performance, 417–422
      structural system behavior, 425
   doors, 312
   dynamic response and analysis, 344
      equivalent SDOF system approach, beam response modes, 275, 276
      load-deflection relationship, 250
      load-structure interactions, 239
      material models, 267
      modes of failure, 237–238
      multi-segmental forcing functions, 281, 289
      SDOF parameters for fixed beams, 278
      SDOF parameters for fixed boundary beams, 279–280
      SDOF parameters for simply supported beams, 277
      SDOF parameters for simply supported slabs, 281
      structural analysis methods, 238–240
   dynamic shear force, 355
   fragment, *see* Fragmentation
   fuel-air munitions and, 49

484 Modern Protective Structures

as function of shape, 34
general principles and overview, 16–24
   complexity of, 17–18
   design manual limitations, 25
   internal explosions, 75
   load definition, 34
   mitigation, 4
   natural frequency-duration of forcing or load function relationship, 328
P-I diagrams, 325, 341–342, 344
   idealized transient loading profile, 326
   SDOF and P-I computations, 360
P-I diagrams, characteristics of
   influence of system and loading parameters, 330–331
   loading regimes, 328–330
P-I diagrams, closed form solutions
   rectangular load pulse, 331, 333–334
   triangular load pulse, 334–335, 336
planning and design considerations, 7, 11, 16–24, 31–32, 405
   combined/multiple weapons systems, 18
   load definition, 34
   multiple detonation effects, 35
   prediction of, 16
   previous testing methods, 26
   procedures for determining, 18–19
   research needs, 30, 31, 34
   technical resources and blast mitigation capabilities, 24, 25
progressive collapse
   abnormal loadings, 374–375
   redesigns of structural elements, 389
   steel frame structure analysis, 390–391
redundancies and alternate load paths, 295
research needs, 439
SDOF and P-I computations, 357
structural element behavior
   cylinders, arches, and domes, 210, 211
   direct shear response, 228
   distances from explosion and dynamic loads, 198
   frames, 214
   membrane behavior, 222, 223, 224, 226
   shear walls, 212–213
   structural design applications, 233, 234
   tensile and compressive members, 206–207
unloading-reloading of load deflection curve, 348
Loads on structures, 169–193
   blast design load derivation procedures, 187–193
   blast load on rear wall, 191–193
   load-time history on buried wall or roof, 187, 188
   time-pressure curve on front wall from external explosion (surface burst), 187–189, 190
   time-pressure curve on roof or sidewall, span parallel to shock front, 190–191, 192
   time-pressure curve on roof or sidewall, span perpendicular to shock front, 190, 191, 192
conventional loads, 169–177
   above-ground structures, 170–174
   blast fragment coupling, 175–176, 177
   buried structures, 169–170
   mounded structures, 175
   surface-flush structures, 175
nuclear loads, 177–187
   above-ground structures, 177–178, 179–182
   buried structures, 179–185
   soil arching, 185–187
Load-time history, 18; *see also* Pressure-time history; Time-pressure curves
   buried wall or roof, 187, 188
   and maximum response to impulsive loading, 328, 329
   multiple detonations, 35
   P-I diagrams, 326
Load transfer, *see* Redistribution/transfer of load
Local/localized response
   blast effects and mitigation, 52
   buckling and progressive collapse, 380
   connections, 307
   design approach, comprehensive, 420, 427–428
   load redistribution, 428–429
   structural system behavior, 424
   dynamic load types and, 422
   dynamic response to impact loading, 269
   membrane behavior, 224
   modes of failure, 237–238
   progressive collapse, 373–374
      GSA guidelines, 385
      steel frame buildings, 380
   redundancies and, 295
   separation from structural responses, 237
   structural element behavior, 25
      frames, 215
London bombing incidents, 375, 376
Longitudinal reinforcement, joint behavior, 304
Longitudinal strains, 111
Longitudinal (P) waves, 110

Index

LOP (low levels of protection), 381
Love waves, 110
Low levels of protection (LOP), 381
Lumped mass models, 239, 252, 327

M

Mach front
  airburst blast environment, 72
  triple point, 71–72, 80
Mach reflections
  conventional loads on above-ground structures, 170, 171
  nuclear and HE explosions, 80, 82, 83
    dynamic pressure impulse as function of scaled height of burst and scaled ground ranges, 89
    peak dynamic pressure as function of scaled height of burst and ground ranges, 88
Mach sten, 80
Maintenance and repair, 28
Manuals, *see* Publications, literature and technical references
Masonry, 233, 413, 432
Mass
  comprehensive design approach, 423–424, 425
  conservation of, 61, 118
    nuclear explosion-induced shock waves, 123
    shock waves without reflected shock, 126–127
  dynamic resistance function, 29
  equivalent SDOF transformation factors, 350
  SDOF parameters
    for fixed beams, 278
    for fixed boundary beams, 279–280
    for fixed slabs, 282–283
    for simply supported beams, 277
    for simply supported slabs, 281
  seismic loading, 427–428
Mass density, surface-burst ground shock, 150
Mass factor
  equivalent, 275
  SDOF systems
    fixed beams, 278
    fixed boundary beams, 279–280
    fixed slabs, 282–283
    simply supported beams, 277
    simply supported slabs, 281
  SDOF systems, fixed one-way slabs
    long edges fixed and short edges simply supported, 286–287
    short edges fixed and long edges simply supported, 284–285
Mass matrix, 252
  finite element approach, dynamic response and analysis, 265
  progressive collapse, steel frame structure analysis, 394
Mass-spring-damper systems, SDOF, 422–423
Material failure, progressive collapse of steel frame buildings, 380
Material models
  for computational analysis, 265–267
  connections, steel, 309
  current state and future needs, 32
  design manual limitations, 25
  dynamic response and analysis, 250, 262, 343–344
    validation procedures, 270
    validation requirements, 268
  enhancement factor computation, 430
  membrane behavior, 226
  P-I diagrams, analytical, 325
  structural design applications, 236
Material properties, 442; *see also* Structural elements and structural element behavior
  cratering, nuclear explosions, 141
  explosive loads and, 292, 294
  and failure modes, 237–238; *see also* Failure modes
  impact and penetration effects, 84
  planning and design approach, comprehensive, 432–433
    research needs, 439, 440
    structural system and component selection, 431
  planning and design considerations, 28
    design manual limitations, 25
    research needs, 30–31
  progressive collapse, GSA guidelines, 387–389
  structural element behavior, 198–202
    concrete, 199–202
    dynamic effects, 202, 203
    steel, 198–199
    structural design applications, 233
Mathematical models, direct shear response, 227–228
Matrix methods
  dynamic response and analysis, 253, 254, 265

progressive collapse, steel frame structure analysis, 394
Matrix singularity problems, 394
Mats, concrete protection, 88
Maximum displacement, 325, 327, 347
Maximum peak response, P-I diagrams, 326
Maximum resistance, SDOF analysis
   fixed one-way slabs
      long edges fixed and short edges simply supported, 286–287
      short edges fixed and long edges simply supported, 284–285
   fixed slabs, 282–283
Maximum response parameters
   impulsive loading, load time history-independence, 328, 329
   SDOF systems, 243–244, 348
Maximum stress
   connections, 307
   loads on above-ground structures, 172
MDOF, *see* Multi-degree-of-freedom (MDOF) systems
Mechanical couplers, reinforcement, 207
Mechanical equipment, 29–30; *see also* Equipment
Media
   reflection and transmission of one dimensional D-waves between two media, 114–116
   shock waves, 122–123
   three-phase, shock waves, 63–64
Media-structure interactions
   conventional loads on buried structures, estimation of, 170
   dynamic response and analysis, 250
   planning and design considerations, 19
Medium levels of protection (MLOP), 381
Membrane behavior
   doors, 312
   dynamic response and analysis
      flexural behavior, 346
      intermediate approximate methods, 264
   flexural resistance, 205
   progressive collapse
      concrete structures, monolithic, 378
      sample analyses, 398
   SDOF and P-I computations, reinforced concrete slabs, 363, 364
   structural element behavior, 218–226
Mercury fulminate, 58
Mesh, 269, 271–272
Metal grills, door designs, 311
Metals, impact loading tests, 269
Methodology, 8–16

development and implementation of, current state and future needs, 37–39
   risk assessment, 11–15
   threat, hazard, and vulnerability assessments, 10–11
Military applications, research focus, 25
Military ordnance, 41
Military threat assessment, 11, 19–24
Mindlin plate, 262, 348
Missiles and rockets, 47, 49
   anti-runway bombs, 47
   explosive processes, devices, and environments, characteristics of conventional weapons, 47–48, 49
   fragment characteristics, 158
   penetration
      characteristics of, 86
      shaped charge devices, 104, 105
   projectiles, 43, 44, 45
   Soviet doctrine, 22, 23
Mission requirements, 28
Mitigation, blast, 4
   blast effects and, 50–52
   general principles and overview, 24–30
      design and construction considerations, 28–30
      design manuals, 26–28
   research needs, 435
Mitigation, progressive collapse, 385–387
MLOP (medium levels of protection), 381
Modal equation of motion, 262
Mode shapes, 254, 261–262
Modification factors, dynamic response and analysis, 345
Modulation, nuclear loads on buried structures, 179
Mohr circle, 266
Mohr-Coulomb criterion, 266
Moisture, 441
Moment
   dynamic analysis approach, 343–344
   statically loaded knee joints, 297
   structural element behavior
      cylinders, arches, and domes, 208–209, 210, 212
      diagonal shear effects, 228
      interaction diagram, 203–204
      membrane behavior, 226
      plastic hinge regions, 232
Moment capacity
   diagonal shear effects, 228
   dynamic response and analysis
      SDOF analysis for fixed slabs, 282–283

Index 487

SDOF parameters for fixed boundary beams, 279–280
Moment connections, 395, 398, 399
Moment curvature
  continuous structural elements, 259–260
  dynamic response and analysis, 344–345
  dynamic resistance function, 346, 347
  intermediate approximate methods, 263
  load versus deflection for beam, 360
  SDOF and P-I computations, 357
Moment-rotation relationships, progressive collapse, 392–393, 397–398, 399, 400, 401
Moment-thrust interaction diagrams, 207
Momentum, conservation of, 61, 118, 123, 127, 243
Monolithic concrete, progressive collapse, 378–379
Mortar shells
  explosive characteristics, 43, 44, 45
  fragment characteristics, 158
Mounded structures
  conventional loads on, 175
  nuclear loads on, 185
Mounts, *see* Connections and supports
Multi-axial stress-strain relationships, 262
Multi-degree-of-freedom (MDOF) systems, 339
  dynamic response and analysis, 251–258
    dynamic analysis approach, 343
    numerical methods for transient responses analysis, 255–258
    principles of, 251–255
  planning and design, 8
  progressive collapse of steel frame buildings, 394
Multidisciplinary collaboration, *see* Collaboration, multidisciplinary
Multi-facility conditions, 440–441
Multi-hazard protective design, 433–434, 438–439
Multi-linear signal, 288
Multinational collaboration, *see* Collaboration, multinational
Multiple detonations or multiple weapons system attacks
  equipment response, 35
  interference between shock waves, 64
  planning and design considerations, 17–18, 34
Multiple failure modes, 341, 342, 343, 361
Multi-segmental forcing functions, 281, 288–289
Mushroomed projectiles, 95

N

National academic support consortia (NASC), 436
National centers for protective technology research and development (NCPTR&D), 38, 436
National Research Council (NRC), 3–4, 36, 405
National Security Council, Technical Support Working Group (TSWG), 36
Natural frequency, *see* Frequency characteristics
Natural gas, liquified (LNG), 49
NCPTR&D (national centers for protective technology research and development), 38, 436
Near-surface bursts
  cratering, 143, 145
  nuclear explosions, 79–80, 145
Newmark β Method, 247–249, 255, 350, 351
Newton's second law, 183–184, 243
Nitrates, 53, 58
Nitrocellulose, 58
Nitroglycerine, 54–55, 58, 65
Nitromethane, 58
Nonharmonic loads, P-I diagrams, 326
Non-ideal load pulses, 330
Nonlinear dynamic methods
  DOD recommendations, 383–384
  planning and design, 8
  progressive collapse, steel frame structure analysis, 394
Nonlinear problems
  blast effects and mitigation, 51–52
  comprehensive design approach, 424, 425
  dynamic resistance function, 346–347
  dynamic response and analysis, 250–251, 252, 270–271
    intermediate approximate methods, 263
  membrane behavior, 224
  P-I diagrams, 366, 370
  progressive collapse
    steel frame building connections, 392–393
    truss structures, 379
  SDOF systems, flexural equivalent, 350
  shock waves
    in one-dimensional solids, 117
    in three-phase media, 63–64
  structural system behavior, 425
Nonlinear static analysis, DOD recommendations, 383
Nonperiodic loads, P-I diagrams, 326
Nonspherical charges

computer codes for, 19
design approach, comprehensive, 421
Nonstandard explosive devices, 17
Nonstructural systems, *see* Vital systems/subsystems, nonstructural
Northridge earthquake, 308, 377, 380
Norwegian Defense Construction Service, 24
Nose shape factor, 98, 99
Nuclear devices/weapons/environments
 airblast, 79–83, 84–90
 cube root scaling, 68
 door designs, 311
 explosive processes, devices, and environments, characteristics of., 41, 48
 general principles and overview, 16, 17
 ground shock, cratering, and ejecta from, 141–153
  cratering, 141, 142, 143–145
  ejecta, 145–147, 148–149
  ground shock, 147, 149, 150–153
 loads on structures, 177–187
  above-ground structures, 177–178, 179–182
  buried structures, 179–185
  soil arching, 185–187
 planning and design considerations, 5, 7, 11
  multiple weapons system attacks, 18
  research needs, 33–34
 radiological environments, 167
 shock waves, 120–121
 Soviet doctrine, 19, 20, 21, 23
 structural element behavior, frames, 214
 structural failure modes, structural analysis research needs, 35
Nuclear power plants, 29
Numerical locking, 224
Numerical methods
 dynamic load proportionality factors, 356
 dynamic response and analysis
  single degree of freedom (SDOF), 247–250
  validation requirements, 269
 membrane behavior, 224–225
 P-I diagrams, 339–343, 348–349
  agreement with, 366
  SDOF and P-I computations, 359, 363, 364
 pressure-impulse (PI) diagrams, 341–343
 progressive collapse, moment and shear connections, 399
 shock propagation, 52
 for transient responses analysis, 255–258
N-waves, 64

O

Oblique impact
 effects of, 43
 internal explosions, shock pressure phase, 76–77
Octol 75/25, 66, 154
Oklahoma City attack, 390
One-dimensional media, shock waves, 117–119, 121–122
One-dimensional waves
 ground shock, 151
  elastic wave propagation, 110
  reflection and transmission of one dimensional D-waves between two media, 114–116
 shock waves, 62, 121–123
One-way slabs, shearing failure, 325
One-way spans, multiple failure modes, 343
Openings and interfaces, 310–317
 airblast transmission through tunnels and ducts, 319, 322–324
 blast doors, 310–313, 314
 blast valves, 313–315
 cable and conduit penetrations, 315–317
 comprehensive design approach, 409, 428, 440
 emergency exits, 317, 318
 entrance tunnels, 310
 guidelines, 29, 197
 leakage blast pressure, internal explosions, 79
 planning and design considerations, 7, 29
 structural element behavior, 197, 198
Open loops, connections, 304
Operations, 4, 6, 7, 412–413
Orientation of weapon, and loading parameters, 17
Orthogonality relations, 262
Oscillators
 dynamic response and analysis, 242, 253
 P-I diagrams
  energy balance method, 337
  SDOF and P-I computations, for reinforced concrete slabs, 364
 SDOF
  linear, 426
  simple, 242
Overhead explosions
 buried structures, 106
 conventional loads on buried structures, 169
 nuclear and HE explosions

Index

dynamic pressure impulse as function of scaled height of burst and scaled ground ranges, 89
peak dynamic pressure as function of scaled height of burst and ground ranges, 88
nuclear devices, 79
shock wave parameters for spherical TNT explosions, 71–72
Overpressure, 66–67
  blast design load derivation procedures, blast load on rear wall, 191
  blast valves, 314
  conventional loads on above-ground structures, 172
  ground shock, surface-burst, 149
  non-ideal load pulses, 330
  nuclear and HE explosions, 80, 81, 82
    ejecta, 147
    height of burst and ground range for intermediate overpressures, 84
    height of burst and ground range for low overpressures, 85
    ideal surface peak overpressure curves, 82, 83
    peak dynamic pressure at shock front as function of peak overpressure at sea level, 87
    shock waves, 120–121
    triangular representation of overpressure-time curves, 85
  nuclear loads on above ground structures, 178, 179, 181, 182
  P-I diagrams, idealized transient loading profile, 326
  transmission through tunnels and ducts, 319, 323
Over-strength factors, DOD design recommendations, 381
Oxidation, chemistry of explosions, 53, 54
Oxygen balance of explosive, 54

P

Pancake effect, 198, 208
Panels
  cladding, 413–414
  drop, 425
  energy-absorbing, 430
  progressive collapse of precast concrete structures, 377
  research needs, 442
Parallel programs, 435

Parking facilities
  basement garages, 198
  design approach, comprehensive, 413, 417, 432
  progressive collapse, analysis and acceptance criteria, 387
Particles, dust, and debris
  cratering, 136
  design approach, comprehensive, 409, 418, 419
  ejecta particle size, 140
  maximizing standoff distance, 6
  planning and design considerations, 5, 6
  pressure pulse combined with particle velocity, 52
Particle velocities, 66–67, 68
  ground shock, surface-burst, 150
  shock waves, 122
    burster slab and soil detonations, 130
    nuclear and HE explosions, 121
    numerical evaluation, 126
  and stress, 112–113
Partitions, 416, 440
Partnerships, *see* Collaboration, multidisciplinary
Passive arching, soil, 186
Passive security measures, 412
Peak deflections, P-I diagrams, 362
Peak dynamic pressure
  conventional loads on above-ground structures, 173
  nuclear and HE explosions
    as function of scaled height of burst and ground ranges, 88
    at shock front as function of peak overpressure at sea level, 87
Peak gas pressure, internal explosions, 77
Peak loads
  blast design load derivation procedures, blast load on rear wall, 191
  P-I diagrams, 326
  response spectra, versus P-I diagrams, 327
Peak overpressure, 80, 81, 82, 83, 180
Peak pressure(s)
  distances from explosion and dynamic loads, 198
  dynamic response and analysis, exponential decay after, 281, 289
  external explosions, 68–69
  incident versus dynamic, 74
  loads on above-ground structures, 172
  pressure-time curves
    on front wall from external explosions, 187, 188, 189

on roof or sidewall, span perpendicular to shock front explosions, 190
Peak reflected pressure, 76–77, 171
Peak resistance, dynamic response and analysis, 276
Peak stress, layered systems, 134
Penetration
 conventional environments, 83–106
  armor, 98, 99, 100–101
  concrete, 87–92, 93, 94, 95
  other materials, 101, 102
  rock, 92–96, 97
  shaped charges, 103–105
  soil and other granular material, 96, 97, 98, 99
 design approach, comprehensive, 419
 explosive processes, devices, and environments, characteristics of conventional weapons, 43
 fragment, 158–167
 planning and design considerations, 25, 197
 structural element behavior, 197
Penetration path
 courses of bombs and, 97
 rock penetration calculations, 95
 and soil/granular material penetration, 96
Penetrations, structural, *see* Openings and interfaces
Pentolite, 66, 154
Perfectly elastic system
beam, dynamic shear force, 352
 equivalent SDOF transformation factors, 350
 loading regime quantification, 329
 SDOF
  closed-form solutions to P-I curves from, 331–335, 336
  energy solutions, 337, 338
  response spectra, 327, 328
Perforation
 with fragments, 153
 impact and penetration effects, 84
Performance
 comprehensive design approach, 433
 P-I diagrams, 370
Perimeter line, 4, 413
Perimeter protection, 7, 440
Personnel protection, 6, 7, 12, 442
 comprehensive design approach, 433
 notification of threat situations, 410
 post-incident conditions, 417
PETN, 58, 66
Petrochemical facilities, 27, 197, 407, 432–433

Physical environment effects, research needs, 441
Physical environments, explosions, 4
Physical security
 design approach, comprehensive, 412, 433
 multi-hazard protective design, 433–434
 petrochemical plant, 432–433
 vital systems, nonstructural, 417
 planning and design considerations, 5, 29, 405
 publications, 407
Physics, explosions, 54–58
PI diagrams, *see* Pressure-impulse (PI) diagrams
Piece-wise nonlinear curves, dynamic resistance function, 346, 347
Plane dilatational wave, 111
Planning, *see also* Design approach, comprehensive; General principles of protective technology
 comprehensive design approach, 410–411
 general principles and overview, 5–8
Plastic concrete, penetration of, 101
Plastic deflection, dynamic resistance function, 347
Plastic deformation
 progressive collapse of steel frame buildings, 380
 reinforced concrete slabs, 364
Plastic hinge regions, 232–233
 design approach, comprehensive, 428, 429
 design issues, 307
 dynamic response and analysis, 250
  equivalent SDOF system approach, 274–275
 government guideline limitations, 29
 progressive collapse, GSA guidelines, 386
Plasticity
 P-I diagrams, 330–331
 theory of, 265, 390
Plastic models
 dynamic response and analysis
  SDOF parameters for fixed beams, 278
  SDOF parameters for fixed boundary beams, 279–280
  SDOF parameters for simply supported slabs, 281
  validation requirements, 268
 equivalent SDOF transformation factors, transitions between response modes, 350
 material models, 266
 membrane behavior, 224
 P-I diagrams, 330

Index

Plastic moment capacity, structural element behavior, 205
Plastic waves, in one-dimensional solids, 117
Plastic zone, cratering, 136
Plates
  anti-scabbing, 89
  P-I diagrams
    dynamic reaction, 340
    energy balance method, 338
  steel, penetration, 87
Plate slabs, load transfer, 238
Point explosions, scaling laws, 67–68
Point-to-point progress, P-I curves, 342
Poisson's ratio, 111
Policy, current state and future needs, 33–35
Polymers, 442
Porosity, soil, 128
Post-incident conditions, comprehensive design approach, 417
Power supplies, 7, 417
Precast concrete
  design approach, comprehensive, 428
  progressive collapse, 377, 378
Precision testing, current state and future needs, 32
Pressure, *see also* Loads and loading regimes; Loads on structures
  barrier walls for reduction of, 413
  blast valves, 314
  close-in detonation issues, 65
  comparisons of weapons/explosives, 66–67
  comprehensive design approach
    pressure versus impulse sensitivity, 426
    short standoff bomb blasts, 428
    venting, 430–431, 432
  connections, steel, 309
  dynamic, *see* Dynamic pressure
  explosions
    chemistry and physics of, 53, 55
    combustion phenomena and processes, 59–60
    shock waves, 61–62, 63
  external explosions, 68, 72–73
  free air equivalent weights, 66
  internal, 319–322
    airblast transmission through tunnels and ducts, 319, 322–324
    increases, 319
  internal explosions, 319, 322
    gas pressure phase, 77
    high-pressure spike, 75
    leakage blast pressure, 77–79
  nuclear and HE explosions
    dynamic pressure impulse as function of scaled height of burst and scaled ground ranges, 89
    ideal surface peak overpressure curves, 82
    peak dynamic pressure as function of scaled height of burst and ground ranges, 88
  P-I diagrams, 326, 327
    conditions resulting in failure, 327–328, 339–341
  piping system protection, 315, 317
  P-waves, 64
  shock front travel, 119–120
  shock waves
    attenuation of, 128
    estimation of decay, 129–130
  soil arching, 186
  transmission through tunnels and ducts, 319, 323
Pressure distributions, comprehensive design approach, 418, 421
Pressure enhancement, planning and design considerations, 16
Pressure-impulse (PI) diagrams, 325–371
  analytical solutions, 331–339
  analytical solutions, closed form, 331–335, 336
    rectangular load pulse, response to, 331, 333–334
    triangular load pulse, response to, 334–335, 336
  analytical solutions, energy balance method, 335–339
    approximating dynamic regions, 336–338, 339
    continuous structural elements, 338, 340
  application examples for SDOF and P-I computational approaches, 357–371
    reinforced concrete beams, 357–362
    reinforced concrete slabs, 363–366, 367–371
  characteristics of, 327–331
    influence of system and loading parameters, 330–331
    loading regimes, 328–330
  comparisons of methods and their applications, 341–343
  composite, 343
  comprehensive design approach, 426–427
  current state and future needs, 32
  dynamic analysis approach, 343–348
    direct shear behavior, 347–348

dynamic material and constitutive
models, 343–344
dynamic resistance function, 346–347
flexural behavior, 344–346
dynamic structural model, 348–357
direct shear response, 351–357
flexural response, 349–351
literature and technical references, 51
numerical approach, 339–343
planning and design considerations, 18
Pressure pulse
blast effects and mitigation, 52
combined effects with particle velocities, 52
positive phase, 68–69
pressure versus impulse sensitivity, 426
thermobaric devices, 49
Pressure rays, and loading parameters, 34
Pressure-time history, *see also* Time-pressure
curves
blast fragment coupling, 175
design approach, comprehensive, 421
estimation of duration of reflected pressure,
169, 170
external explosions, 68–70
internal explosions, 76
layered systems, 134
loads on above-ground structures, 172
load-time history on buried wall or roof, 187
nuclear and HE explosions, 82, 85
research needs, 34
triangular representation of overpressure-
time curves, 85
Pressure-volume space, 61, 63
Primary explosives, 58
Private sector, *see* Collaboration,
multidisciplinary
Probabilistic analysis, threat, 11
Progressive collapse, 373–404
abnormal loadings, 374–375
advanced frame structure analysis, 390–404
computer code requirements, 394
examples of, 395–404
semi-rigid connections, 392–394,
395–397, 398
design approach, comprehensive, 416
load redistribution, 429
research needs, 440
structural system and component
selection, 431
DOD guidelines, 380–384
approaches and strategies, 381–384
damage limits, 384
new and existing construction, 380–381
GSA guidelines, 384–390

analysis and acceptance criteria,
387–389
assessment of existing facilities for
potential for progressive collapse,
385
material properties and structural
modeling, 387–389
mitigation, new facilities, 385–387
redesigns of structural elements,
389–390
observations, 375–377
secondary threats, 32
structural element behavior
distances from explosion and dynamic
loads, 198
membrane behavior, 219
structure types, 377–380
monolithic concrete, 378–379
precast concrete, 378
steel frame buildings, 379–380
truss structures, 379
Projectiles
direct and indirect fire weapon projectiles,
42–44
shielding devices, 432
small arms and aircraft cannon, 41, 42, 43,
44, 45
Propagation of energy
internal explosions, 75
scaling laws, 67
Propagation velocity, external explosions, 68
Propane, 56–57, 58
Proportionality factors, 356
Propylene oxide, 48–49
Protected facility components, 7
Protected space, blast pressure leakage into,
78–79
Protection levels, UFC classification, 381
Protective technology, *see* Design approach,
comprehensive; General principles
of protective technology
Proximity of blast, *see* Distance from explosion
Pseudostatic/quasistatic phase of internal
explosions, 75, 77
Publications, literature and technical references,
79, 195, 196–197
blast effects and mitigation, 51–52
connections, 295–296
conventional weapons information, 49
material models, 267
planning and design considerations, 5, 7, 18,
24–30
current state and future needs, 33

Index

design and construction considerations, 28–30
design manuals, 26–27
need for, 10–11
recommendations, 405–406
security needs assessments, 18
structural engineering guidelines, 28–30
technical resources and blast mitigation capabilities, 24
progressive collapse mitigation and prevention, 380–390
Department of Defense, 380–384
GSA, 384–390
structural element behavior, 195, 196–198
ASCE Manual 42, 197–198
Department of the Army technical manuals, 196–197
Department of the Army Tri-Service Manuals, 195, 196–197
Public sector, *see* Collaboration, multidisciplinary; *specific U.S. government agencies*
Pulse duration, conventional loads on buried structures, 169
Punching shear, concrete
progressive collapse, 377–378
reinforcement and, 89
P-waves, 64

Q

Quality assurance (QA) criteria, 7, 8
Quasi-static loads
connections, 296
loading regimes, response time of, 328, 329
P-I diagrams, 330, 331, 426, 427, 428
energy balance method, 335–336, 337
SDOF and P-I computations, 360
versus shock spectrum, 328
SDOF and P-I computations, 358
structural system and component selection, 430–431
Quasistatic/pseudostatic phase of internal explosions, 75, 77

R

Radial bars, 304
Radial hoops, 304–305
Radiation, 5, 6, 7
Radiological environments, 16, 167
Range, *see* Distance from explosion

Rankine-Hugoniot equations, 61–62, 118, 119
Rarefaction, N-waves, 64
Rate effects, structural element behavior, 230–231, 234
Rayleigh damping, 254
Rayleigh line, 62
Rayleigh quotient, 255, 257–258
Rayleigh waves, 110
Reaction heat, 55
Reaction time, comparisons of experimental and analytical results, 272
Reaction zones
chemistry and physics of explosions, 53–54
detonation processes and shock waves, 60–62
Reactive armor devices, 105
Rebounding, dynamic resistance function, 347
Rectangular load pulse
loads on above-ground structures, 170
P-I diagrams, 330
P-I diagrams, closed form solutions, 331, 333–334
P-I diagrams, energy balance method, 338, 339
P-I diagrams, SDOF system
response functions, 332
shock spectrum for, 333
Redesigns of structural elements, 7, 389–390
Redistribution/transfer of load, 424
comprehensive design approach, 413–414, 427, 428–429
redundancy and, 429–430
structural system and component selection, 431
dynamic response and analysis, 238
progressive collapse
redesigns of structural elements, 389
sample analyses, 395
truss structures, 379
structural element behavior, 228
Redundancy
blast resistance, 295
design approach, comprehensive, 409, 424
load redistribution, 428–429
transfer of load, 429–430
planning and design considerations, 6
progressive collapse, 377
GSA guidelines, 386
Reference material, *see* Publications, literature and technical references
Reflected pressure
angle of incidence and, 70
blast design load derivation procedures

494 Modern Protective Structures

load-time history on buried wall or roof, 187
pressure-time curve on front wall from external explosions, 187, 188
conventional loads, 169
design approach, comprehensive, 418
external explosions, 69–70
loads on above-ground structures, 170–171
nuclear and HE explosions, 80
Reflected wave
nuclear and HE explosions, 80
reflection and transmission of one dimensional D-waves between two media, 114–116, 117
spherical shock wave, 71
Reflection, shock waves, 17, 34
ground surface explosions, 72
internal explosions
propagation and interaction, 75
shock pressure phase, 76–77
layered systems, 133, 134
nuclear and HE explosions, 83
dynamic pressure impulse as function of scaled height of burst and scaled ground ranges, 89
height of burst and ground range for intermediate overpressures, 84
loads on above-ground structures, 177
peak dynamic pressure as function of scaled height of burst and ground ranges, 88
numerical evaluation, 125–126
reflection and transmission between two sand materials, 121–125
spherical shock wave, 71
Refracted wave, surface burst, 149
Reinforced concrete, *see* Concrete, reinforced
Reinforcement fracture, flexural behavior, 346
Reinforcement ratio, membrane behavior, 226
Reinforcement techniques and materials
concrete, 87–88
confining stress effects, 200, 201
and cratering, 136
and penetration, 88, 92
and shear effects, 89
connections, 296, 298
joint behavior, 301, 302, 303, 304
design approach, comprehensive, 413–414
interior, 416
structural system and component selection, 431
structural system behavior, 424–425
upward forces and, 428
dynamic response and analysis, 357

finite element approach, 271–272
flexural behavior, 345
intermediate approximate methods, 263
P-I diagrams, 340
SDOF and P-I computations, 357
validation procedures, 271–272
explosive loads and, 294
failure modes, 237–238
planning and design considerations, 28
progressive collapse, 377
concrete structures, monolithic, 378
sample analyses, 398
structural element behavior, 207–208
connections and support conditions, 231–232
joints, 232
membrane behavior, 221–222
principal reinforcement, 207–208
shear resistance, 205
technical manuals, 196
tying or bridging of structural elements, 373
Relief wave, 171
Reloading, dynamic resistance function, 347
Removal of structural elements
alternate path analysis, damage limits, 384
DOD analysis recommendations, 383–384
progressive collapse, 403–404
concrete structures, monolithic, 379
GSA guidelines, 386
material properties and structural modeling, 389
sample analyses, 400
Repair of facilities, 28, 426–427
Rescue, recovery, and evacuation, 6, 412, 417
Research and development
current state and future needs, 33–35, 37–38
development and implementation of effective technology, 436, 438–441
fortification science, 25
literature and technical references, *see* Literature and technical references
National Research Council (NRC) recommendations, 3–4
pressure response, need for experimental studies, 30–31
recommended actions, 435
Reserve capacity, membrane behavior, 221
Resistance
connections and, 291
design approach, comprehensive, 423–424
dynamic response and analysis
applications to support analysis and design, 276

Index

equivalent SDOF system approach,
  beam response modes, 275
  SDOF analysis for fixed slabs, 282–283
  SDOF parameters for simply supported
    beams, 277
equivalent SDOF transformation factors,
    350
load-time history, 328, 329
structural element behavior
  flexural, 202–205
  shear, 205–206
  shear walls, 213
Resistance curve, 250
Resistance factor design, progressive collapse,
    391
Resistance function
  design approach, comprehensive, 430
  dynamic response and analysis, 249, 250,
    288
  membrane behavior, 224
  P-I diagrams, 338, 363, 366
  structural design applications, 233
Resistance model
  connections, 293
  reinforced concrete slabs, 346
Response history functions, P-I diagrams, 331,
    332
Response levels
  P-I diagrams, 328
  planning and design considerations, 28
  pressure response, need for experimental
    studies, 30–31
Response limits, structural element behavior,
    234–236
Response spectrum
  natural frequency-duration of forcing or
    load function relationship, 328
  P-I diagrams, 326
    closed-form solutions, 332
    shock spectrum versus, 327, 328
  SDOF analysis, 245–246
  simple, 243–244
Response time of system
  design approach, comprehensive, 420
  P-I diagrams, 328
Restraint, lateral, 300
Retrofits
  planning and design considerations, 29,
    408–409
    policy and technology needs, 35, 36
    research needs, 38
    windows, 415–416
  progressive collapse

Department of Defense guidelines,
    380–381
  General Services Administration (GSA)
    guidelines, 385
  structural element behavior, 197
Reversed loading
  reinforced floor slabs, 208
  unloading-reloading of load deflection
    curve, 348
Review, security plan, 24
Ricochet, 43
Rigid body motions, nuclear loads on buried
    structures, 179, 183–184
Rigid-body rotations, three-dimensional stress
    wave propagation, 107
Rigid connections, progressive collapse, 398,
    399
Rigidity, and penetration, 87
Rigid plastic responses
  P-I diagrams, 330
  SDOF, energy solutions, 337, 338
Rigid plastic system, equivalent SDOF
    transformation factors, 350
Rings, structural element behavior, 209–210,
    211
Rise time
  blast design load derivation procedures
    load-time history on buried wall or roof,
      187
    pressure-time curve on roof or sidewall,
      span parallel to shock front
      explosions, 190
    pressure-time curve on roof or sidewall,
      span perpendicular to shock front
      explosions, 190
  external explosions, 68
  ground shock, 129
  loads on above-ground structures, 173
  P-I diagrams, 330, 331, 332, 368
  structural element behavior, 234
Risk assessment, *see* Threat and risk assessment
Risk equation, 10
Risk management approach, 9
Robustness, 28, 295, 441, 442
Rock
  cratering, 144
  ejecta, 137, 139, 140, 146
  layered systems, 133
  material models, 267
  penetration, 92–96, 97, 102
Rocket propelled grenades (RPGs), 25, 44
Rockets, *see* Missiles and rockets
Rock quality designation (RQD), 92, 93, 96
Ronan Point, 374–375, 377

Roof
    loads on above-ground structures
        conventional explosives, 172, 174
        nuclear and HE explosions, 178, 179
    loads on buried structures, 169
    load-time history, 187, 188
    membrane behavior, 226
    plastic hinge regions and, 232–233
    principal reinforcement, 207
    progressive collapse, sample analyses, 396, 397
    time-pressure curves
        span parallel to shock front, 190–191, 192
        span perpendicular to shock front, 190, 191, 192
Rotation
    connections, 307
    damage indications, 233, 234–235
    progressive collapse, 397–398, 399, 400, 401
    structural system behavior, 424
Rotational springs, progressive collapse, 393
Rotatory inertia, 262, 263
Rupture zone, cratering, 136

S

Sacrificial elements, 430, 432–433
SAFEGUARD series, 29
Safe standoff criteria development, 327
Sand
    fragment penetration, 164
    shock wave reflection and transmission between two sand materials, 121–122
SAP (semi-armor piercing bombs), 47, 90–91
Scabbing, concrete
    anti-scabbing plates and, 89
    cratering and, 136
    face mats and, 88
    impact and penetration effects, 84
Scab material, concrete penetration, 92
Scab plate, concrete, 92
Scaled distances
    design approach, comprehensive, 418
    dynamic response and analysis, 238–239
Scaled tests, current state and future needs, 31, 32
Scaled time, 327, 328, 330
Scaling laws, 67
SCUD missiles, 48

SDOF, *see* Single degree of freedom (SDOF) systems
Search algorithm, transition points, 339
Secondary explosives, 58
Secondary load carrying mechanisms, progressive collapse, 377–378
Secondary threats, current state and future needs, 32
Second-order analysis, progressive collapse, 391–392
Second-shock phenomenon, nuclear and HE explosions, 80
S/E_c ratio, 226
Security Engineering Manual (U.S. Army TM 5-583), 26
Security systems, 412
Seismic loads/stresses, 308
    blast loads versus, 422, 428
    column removal and, 404
    connections, steel, 308, 309
    design approach, comprehensive, 427–428
    equipment fragility data, 29
    multi-hazard protective design, 433–434
    SDOF analysis, 244, 245–246
    seismic wave arrival order, 110
    shock wave transmission in soils, 128–129
Seismic velocity
    shock waves
        burster slab and soil detonations, 130
        HE-induced, 128
    soil properties from explosion tests, 132
Semi-armor piercing (SAP) bombs, 47, 90, 93, 94
Semi-rigid connections, progressive collapse, 392–394, 395–397, 398
Separation of buildings, 408, 409
Separation of support systems and utilities, 417
Serviceability of facility
    design principles, 4
    specification of criteria, 410–411
Shaped charges
    penetration, with conventional devices, 103–105
    research needs, 34
Shape functions
    dynamic shear force, 352, 354
    P-I diagrams, 342
Shape of load pulse, P-I diagrams, 330
Shear capacity, structural element behavior, 205
Shear cones, doors, 312, 313
Shear connections, progressive collapse, 395, 398, 399
Shear effects
    blast loading and, 428

Index

comprehensive design approach
    structural system and component selection, 431
    structural system behavior, 424–425
concrete, reinforcement and, 89
connections, 296, 307
    reinforcement and, 305
diagonal shear behavior, 347–348
direct shear response, 238
doors, 312, 313
dynamic response and analysis, 250
    applications to support analysis and design, 276
    computational systems, intermediate and advanced, 262
    dynamic resistance function, 348
    flexural behavior, 345
    intermediate approximate methods, 263, 264
    material models, 266, 267
ground shock, surface-burst, 150
P-I diagrams, 367, 368
    composite, 361
    multiple failure modes, 341, 343, 362
P-I diagrams, dynamic structural model of direct shear effects, 351–357
    dynamic shear force, 352–357
    equation of motion, 351
    shear mass, 351–352
planning and design considerations, 28, 29
progressive collapse
    concrete structures, monolithic, 377
    sample analyses, 395
SDOF analysis, 251
SDOF and P-I computations, 357, 358, 364, 371
structural element behavior
    cylinders, arches, and domes, 208–209
    diagonal, 228–230
    direct, 226–228
    direct shear response, 226–228
    joint strengthening, 232
    membrane behavior, 224
    plastic hinge regions, 232
    resistance to shear, 205–206
    size effects and combined size and rate effects, 230
    technical manuals, 196
S-waves, 64
three-dimensional stress wave propagation, 107
Shear failure
    diagonal shear behavior, 347–348

failure modes, 237–238; *see also* Failure modes
    coupling of, 18
P-I diagrams
    dynamic reaction, 340
    SDOF and P-I computations, 360
SDOF assumptions, 325
structural analysis research needs, 35
structural system behavior, 424
Shear reduction factor, flexural behavior, 345
Shear reinforcement, 196
    door designs, 311
    structural system and component selection, 431
Shear strength, soils, 185
Shear walls, structural element behavior, 212–214
Shear waves, 109, 110
Shelters, temporary, 417
Shielding devices, 430, 432, 440
SHOCK, 319
Shock
    equipment fragility data, 29
    internal, 317, 319, 320–321
    maximizing standoff distance, 6
    planning and design considerations
        isolation devices, effectiveness of, 30
        technical resources and blast mitigation capabilities, 24
Shock flow, one-dimensional, 61
Shock front
    blast design load derivation procedures
        blast load on rear wall, 191, 193
        pressure-time curve on roof or sidewall, span perpendicular to shock front explosions, 192
    conventional loads on above-ground structures, 171
    external explosions, 68
    geometry of, 70
    laws of conservation, 61
    Mach front, 71
    nuclear and HE explosions, 80
        loads on above ground structures, 179, 180, 181
    and peak overpressure, 82
Shock front velocity, 121, 128, 171, 180, 181
Shock isolation, 319, 320–321
Shock loads, structural element behavior, 196
Shock phase, internal explosions, 75
Shock pressure, internal explosions, 75, 76–77
Shock reflection, nuclear loads on above ground structures, 177
Shock response spectra, 246

Shock sensitivity, primary explosives, 58
Shock spectrum, 327–328, 330, 331, 332, 333, 334–335
Shock speed, 61
Shock velocity, superseismic conditions, 147
Shock waves
   blast design load derivation procedures
      pressure-time curve on front wall from external explosions, 189
      pressure-time curve on roof or sidewall, span parallel to shock front explosions, 190
      pressure-time curve on roof or sidewall, span perpendicular to shock front explosions, 190
   blast effects and mitigation, 51–52
   conventional explosion loads
      on above-ground structures, 171–172
      on buried structures, estimation of, 170
   explosion effects and mitigation, 60–64
      cube root scaling, 67
      explosive processes, 60–62
   external explosions, 68–74
   focusing, 34
   formation and deterioration, 120
   fuel-air-explosive (FAE) bomb characteristics, 47
   and loading parameters, 34
   nuclear and HE explosions
      airblast, 79–83
      ground shock, 147, 149, 150–153
      loads on above-ground structures, 177, 179
   in one-dimensional solids, 117–119
   planning and design considerations, 7
   reflections and focusing, 17
   reversed loading of reinforced floor slabs, 208
   stress discontinuity at slab-wall interface, 238
   superseismic conditions, 147
   TNT equivalents, 65
Side burst, buried structures, 106
SIFCON, 101
Silver azide, 58
Simulation-based (SB) design technology, 33
Single degree of freedom (SDOF) systems, 348
   advanced models, 433
   approximate, 422–423
   comprehensive design approach, 422–423, 425
   control room evaluation, 433
   dynamic analysis approach, 343
   dynamic response and analysis, 239, 240–251, 270–271
   advanced approaches, 250–251, 252
   applications to support analysis and design, 276
   comparisons of experimental and analytical results, 272, 273
   equivalent approach, 274–276
   equivalent SDOF system approach, 274–276
   graphical presentations of solutions, 247, 248, 249
   numerical solutions, 247–250
   theoretical solution for, Duhamel's integral, 242–247
   validation, 270
   validation procedures, 271
   equivalent SDOF system approach, *see* Equivalent SDOF systems
   load functions, 294
   membrane behavior, 224
   P-I diagram computational approaches, 357–371
      reinforced concrete beams, 357–362
      reinforced concrete slabs, 363–366, 367–371
   P-I diagrams, 325, 370
      closed-form solutions, 331–335, 336
      comparison of closed-form with numerical analysis, 366
      energy balance method, 336–337
   planning and design, 8
   pulse shape effects, elimination of, 330
   response spectra, 327, 328
Sinusoidal load pulse, 330
Siting, architectural, and functional considerations, 6–7, 411–417
   access and approach control, 413–414
   building exterior, 414–416
   building interior, 416
   perimeter line, 413
   post-incident conditions, 417
   vital systems, nonstructural, 417
Size effect law, 230, 267
Size effects and combined size and rate effects, 31, 230–231
Slab resistance models, 218
Slabs
   burster, 130
   comprehensive design approach, 424–425
      structural system and component selection, 431, 432
   DOD design recommendations, 381
   door designs, 311

Index

dynamic response and SDOF analysis, 282–283
   equivalent SDOF system approach, 276
   fixed one-way slabs, long edges fixed and short edges simply supported, 286–287
   fixed one-way slabs, short edges fixed and long edges simply supported, 284–285
   flexural behavior, 346
   simply supported slabs, 281
floor
   catastrophic failure, 208
   structural element behavior, 198
load transfer, 238
membrane behavior, 226
P-I diagrams
   dynamic reaction, 339–340
   multiple failure modes, 343
progressive collapse
   catenary action, 416
   concrete structures, monolithic, 378–379
   sample analyses, 398
shearing failure, 325
stress discontinuity at slab-wall interface, 238
structural element behavior
   interaction diagram, 204
   membrane behavior, 218–219, 220, 222–223
   natural periods of vibration, 215–216
   plastic hinge regions, 232–233
   principal reinforcement, 207–208
   span-to-slab thickness ratio, 223
structural system behavior, 424–425
Slabs, concrete
   beam-slab system flexural resistance, 205
   direct shear response, 347–348
   dynamic response and analysis, 346–347
   estimation of reflected pressure duration, 169
   government guideline limitations, 29
   P-I diagrams, 363–366, 367–371
   resistance model, 345, 346
Slant burst, buried structures, 106
Slenderness ratio, 390–391
Slip effects, 267
Small arms and aircraft cannon projectiles, 42, 43, 44, 45
Snap-through, 346, 379
Soil(s)
   cratering, 138, 144
   ejecta, 140

explosive processes, devices, and environments
   characteristics of conventional weapons, 43
   shock waves, 63
fragment penetration, 164, 167
ground shock coupling factor as function of scaled depth, 131
mounded structure berms, 174
nuclear loads on buried structures, soil-structure interactions, 179, 181–182
peak airblast pressure estimation from underground explosion, 74
penetration, with conventional devices, 96, 97, 98, 99, 102
planning and design considerations
   detonation-soil interaction, 34
   soil-structure interactions, 7, 34
shock waves, 128–129
stress wave propagation, 119–125
superseismic conditions, 147
Soil arching, 179, 181–182, 185–187
Soil loading wave velocity, 169
Soil penetration index, 98
Soviet doctrine, 19–24
Soviet munitions, 43, 44, 48, 49, 158
Spalling, concrete
   comprehensive design approach, 424
   failure modes, structural responses, 237
   fragment penetration and, 162, 163
   front face mats and, 88
   SDOF and P-I computations, 363
Spans
   design approach, comprehensive, load redistribution, 429
   dynamic resistance function, 29
   P-I diagrams, multiple failure modes, 342–343
   progressive collapse, GSA guidelines, 385, 386
   progressive collapse prevention, 416
   structural element behavior, span-to-slab thickness ratio, 223
Spatial distribution of load
   blast fragment coupling, 175
   design approach, comprehensive, 421
Special-purpose weapons, 47, 48–49
Special weapons, Soviet doctrine, 23
Specific impulse, P-I diagrams, 326
Speed, shock, 61
Spherical blast pressure, leakage blast pressure, 77
Spherical charges

design approach, comprehensive, 418
internal explosions, blast environment with, 75
planning and design considerations, 34
scaling laws, 67
shock wave parameters for spherical TNT explosions, 71
technical data, 16
Spherical shock waves, 71
Splices, reinforcement, 207, 429
Spring constant, SDOF system
  fixed beams, 278
  fixed boundary beams, 279–280
  fixed one-way slabs, long edges fixed and short edges simply supported, 286–287
  fixed one-way slabs, short edges fixed and long edges simply supported, 284–285
  fixed slabs, 282–283
  simply supported beams, 277
  simply supported slabs, 281
Springs, progressive collapse of steel frame buildings, 393
Sprinkler systems, 417
Stability
  advanced frame structure analysis, 390
  progressive collapse, 394
Stagnation pressure, 70
Stairwells, 416
Standards, guidelines, manuals, and criteria, *see* Publications, literature and technical references
Standoff distance, *see also* Distance from explosion
  design approach, comprehensive, 408
  penetration depth and, 84
Standoff weapons, threat assessment, 14
State velocities, progressive collapse of steel frame buildings, 394
Static analysis, progressive collapse, 383, 386
Static/dynamic deflection ratio, SDOF analysis, 242
Static interface shear transfer, 228
Static loads
  connections, 296, 297
  dynamic response and analysis
    flexural behavior, 345
    material models, 267
  planning and design considerations, 7
  SDOF and P-I computations, 358
  structural element behavior
    diagonal shear effects, 230
    joints, 232

shear walls, 212–213
Steel
  comprehensive design approach, 430
  structural system behavior, 424–425
  connections, 294
    detailing behavior studies, 306, 308–309
  door designs, 310–311
  dynamic response and analysis, 250
    finite element approach, 271–272
    impact loading tests, 269
    material models, 266
    steel-concrete bonds, 270
    validation procedures, 270
  ejecta, perforation by, 142
  penetration, 91, 100–101
    of armor, 98
    comparisons with other materials, 102
    by fragments, 158–159, 160, 161
    with shaped charge devices, 103
  progressive collapse, 377
    GSA guidelines, 386
    sample analyses, 398
  research needs, 440
  structural element behavior, 195
    beam-slab system flexural resistance, 205
    connections, 233
    dynamic effects, 202
    flexural resistance, 202–203
    material properties, 198–199
    membrane behavior, 222–223
    natural periods of vibration, 216, 217
    structural design applications, 233, 234, 236
    technical manuals, 196, 197
  structural system behavior, 424–425
Steel frame buildings, progressive collapse, 379–380
Steel plates
  and ejecta, 137
  penetration, 87
Steel reinforced concrete, see also Concrete, reinforced; Reinforcement techniques and materials
  failure modes, 237–238
  penetration, 88
Stiffness
  dynamic response and analysis, 252
  element, dynamic amplification factor, 239
  natural, 423–424
    load redistribution, 424, 429
    structural system behavior, 425
  progressive collapse of steel frame buildings, 391–392

Index

soil, shock wave transmission, 130
structural system behavior, 425
support
  finite element approach, 265
  structural element behavior, 231
Stiffness matrix, 394
Stiffness-to-mass ratio (K/M), 243
Stirrups, 196
  connections, 298
  dynamic response and analysis, 271
  and membrane behavior, 222
  progressive collapse, 379
St. Mary Axe, 376
Stodola method, 258
Storage facilities, 433
Strain
  membrane behavior, 224
  reflection and transmission between two sand materials, 122
  three-dimensional stress wave propagation, 107
Strain hardening, truss structures, 379
Strain range
  dynamic reactions of beam, 352
  SDOF analysis for fixed one-way slabs
    long edges fixed and short edges simply supported, 286–287
    short edges fixed and long edges simply supported, 284–285
  SDOF parameters
    for fixed beams, 278
    for fixed boundary beams, 279–280
    for fixed slabs, 282–283
    for simply supported beams, 277
    for simply supported slabs, 281
Strain rate effects
  connections, steel, 309
  dynamic response and analysis, 250, 262
  research needs, 31
  structural element behavior
    size effects and combined size and rate effects, 231
    structural design applications, 236
Strain rate enhancement, 430
Strategy, military threat assessment, 19–21
Streets, 413
Strength
  dynamic resistance function, 29
  rock, 97
  statically loaded knee joints, 297
Strength criterion, size effect law, 230
Strength reduction factors, 202, 381
Stressors, connections, 299, 300, 301

Stress-slip relationship, direct shear response, 228
Stress-strain relationships
  comprehensive design approach, 430
  connections, factors affecting concrete behavior, 300
  dynamic response and analysis, 262, 344
    dynamic analysis approach, 343–344
    validation procedures, 270, 271–272
  explosive loads, 292, 294
  ground shock, 151
  material models, 266
  metals, 199
  stress discontinuity at slab-wall interface, 238
  structural element behavior
    concrete in compression, 199–200, 201
    structural design applications, 233, 236
  three-dimensional stress wave propagation, 107
Stress-time histories, 151; *see also* Time-pressure curves
Stress vectors, material models, 266
Stress waves
  blast effects and mitigation, 51
  ground shock, cratering, and ejecta from HE devices
    one-dimensional, propagation in soils, 119–125
    three-dimensional stress wave propagation, 106–110
  nuclear loads on buried structures, 179
  and shock wave formation, 119
Striker models, 269
Structural analysis
  diagonal shear effects, 230
  planning and design considerations
    loading regimes for, 17–18
    research needs, 35
Structural damping, membrane behavior, 224
Structural detailing, 294
Structural dynamics
  blast effects and mitigation, 51
  natural frequency-duration of forcing or load function relationship, 328
Structural element removal, *see* Removal of structural elements
Structural elements and structural element behavior
  blast effects and mitigation, 52
  combined size and rate effects, 230–231
  comprehensive design approach
    behavior of, 424–430
    component selection, 430–433

compressive members, 206–207
connections and support conditions,
  231–233, 291–310; *see also*
  Connections and supports
cylinders, arches, and domes, 208–212
damage and response limits, 234–236
design
  analysis requirements and capabilities,
    30–31
  applications, 233–234
  comprehensive approach, 422–424; *see
    also* Design approach,
    comprehensive
  comprehensive design approach,
    422–424
  design manual limitations, 25
  experimental studies, 25–26
  failure modes, coupling of, 18
  media-structure interactions, 19
  membrane behavior, 225–226
  pressure versus impulse sensitivity,
    426
  structural engineering guidelines, 28–30
  structural models for prediction of load
    parameters, 16
  technical resources and blast mitigation
    capabilities, 24–30
distances from explosion and, 198
dynamic effects
  dynamic loads, 198
  material properties, 202, 203
dynamic response and analysis, 238–239
  equivalent SDOF system approach,
    274
  failure modes, 237–238
  need for, 239
  validation requirements, 268
failure modes, 237–238; *see also* Dynamic
  response and analysis; Failure
  modes
flexural resistance, 202–205
frames, 214–215
loading regimes
  dynamic loads, 198
  element thickness, estimation of
    reflected pressure duration, 169
loads on, *see* Loads on structures
local response versus, 237
material properties, 198–202
  concrete, 199–202
  dynamic effects, 202, 203
  steel, 198–199
membrane behavior, 218–226
natural periods of vibration, 215–218

nuclear and HE explosions
  loads on buried structures, 179
  pressure-time histories, 82
P-I diagrams
  energy balance method, continuous
    structural elements, 338, 340
  pressure versus impulse sensitivity,
    426
progressive collapse, 377–380; *see also*
  Progressive collapse
  monolithic concrete, 378–379
  precast concrete, 378
  steel frame buildings, 379–380
  truss structures, 379
rate effects, 230–231
reinforcement, 207–208
research needs, 439
resistance
  flexural, 202–205
  shear, 205–206
response limits, 234–236
shear effects
  diagonal, 228–230
  direct, 226–228
  resistance to shear, 205–206
shear walls, 212–214
size effects and combined size and rate
  effects, 230–231
standards, guidelines, manuals, and criteria,
  196–198
  ASCE Manual 42, 197–198
  Department of the Army Tri-Service
    Manuals, 26, 28, 29, 79, 195,
    196–197
  GSA and Department of Defense
    criteria, 198
  GSA guidelines for redesigns to prevent
    progressive collapse, 389–390
structural design applications, 233–234
support conditions, 231–233
tensile members, 206–207
transitions between response modes, 30
tying or bridging, 373, 380–381
vibration, 215–218; *see also* Vibration,
  natural periods and modes
walls, shear, 212–214
Structural failure: *see also* Failure modes
Structural hardening, 412
Structural modeling
  prediction of load parameters, 16
  progressive collapse, GSA guidelines,
    387–389
Structural resistance model, joints, 298

Index

Structural response, approximate SDOF calculations, 422–423
Structured risk analysis, 8–9
Struts, 307
  diagonal strut and truss mechanisms, joint, 296
  strut-and-tie design, 299, 300, 301
Subsystems, nonstructural, *see* Vital systems/subsystems, nonstructural
Supercomputers, shock waves in three-phase media, 64
Superseismic region, 147, 149, 150
Supports/support conditions, 291–310; *see also* Connections and supports
  concrete, reinforcement and, 89
  design approach, comprehensive, 420
    continuity of, 429–430
    research needs, 440
    structural system and component selection, 431, 432
    structural system behavior, 425
  DOD design recommendations, 381–382
  dynamic reactions at, 355, 356
  dynamic response and analysis
    applications to support analysis and design, 276
    SDOF, simple, 240–241
    SDOF analysis for fixed one-way slabs, long edges fixed and short edges simply supported, 286–287
    SDOF analysis for fixed one-way slabs, short edges fixed and long edges simply supported, 284–285
    SDOF analysis for fixed slabs, 282–283
    SDOF parameters for fixed beams, 278
    SDOF parameters for fixed boundary beams, 279–280
    SDOF parameters for simply supported beams, 277
    SDOF parameters for simply supported slabs, 281
  equivalent SDOF transformation factors, 350
  P-I diagrams
    dynamic reaction, 340
    peak reactions, computation of, 362
  planning and design considerations, 7, 29
  progressive collapse
    analysis and acceptance criteria, 387
    catenary action, 416
  shearing failure at, 325
  structural element behavior, 231–233, 294–309

  cylinders, arches, and domes, 212
  membrane behavior, 223
  shear resistance, 206
  structural concrete, 296–307
  structural steel, 308–309
Support systems, nonstructural, 417
Surface (of ground)
  airburst blast environment, 72
  conventional loads on surface-flush structures, 175
  fragmentation and reflection, 34
  near-surface nuclear and HE bursts, 79, 80
  penetration, blast environment with, 73
Surface (of ground) explosions, 72
  blast design load derivation procedures, 189
  blast environment, 73
  cratering, 137
  nuclear and HE explosions
    near-surface bursts, 79–80, 145
    overpressure-time curves, 86
  reflections from ground, 72, 73
Surface (of structure)
  blast loads, 427–428
  nuclear loads on above ground structures, drag coefficients, 177, 178
  time-pressure curves, front wall from surface burst, 187–189, 190
Surface elements, finite element approach, 265
Surface-flush structures
  conventional loads on structures, 175
  nuclear loads on, 182
Surface reflected stresses, layered systems, 134
Surface waves, 110
Survivability, terminology, 6
Swedish Civil Defense Administration, 405, 406
Swiss Civil Defense Administration, 405, 406
Synergy, multi-hazard protective design, 434
System response
  research needs, 38
  SDOF system, 327

T

Tactical rockets and missiles, 47, 49
Tactics, military, 17, 19–21
Tangent squared relationship, hyperbolic, 337
Tank shells, fragment characteristics, 158
Target distance, *see* Distance from explosion
Team approach
  comprehensive design approach, 11

facility design, 8
terrorism and insurgency, 24
Technical data
  blast effects of spherical charges, 16
  experimental studies, *see* Experimental analysis/data
Technical resources, *see* Publications, literature and technical references
Technology needs
  current state, 33–35
  development and implementation of effective technology, *see* Design approach, comprehensive
Technology transfer, 3–4, 437–438, 441
Temperature, 441
  combustion phenomena and processes, 59–60
  exothermic reactions, 53
  progressive collapse of steel frame buildings, 380
  research needs, 439
  thermobaric devices, 49
Temperature-time histories, 11
Temporary shelters, 417
Tensile forces, connections
  factors affecting concrete behavior, 300
  joint behavior, 304
Tensile members, structural element behavior, 206–207
Tensile membrane behavior
  dynamic response and analysis, 346
  flexural resistance, 205
  progressive collapse of monolithic concrete structures, 378
  structural element behavior, 219, 221–222, 225
    stirrups and, 222
    transition into, 222–223
Tensile reinforcement
  connections, 298
  reinforced concrete slabs, 210
Tensile resistance of concrete, steel reinforcement and, 88
Tensile strength, direct shear response, 227
Tension
  principal reinforcement, 208
  structural element behavior
    membrane behavior, 224
    size effects and combined size and rate effects, 230, 231
Terrorism and insurgency threat assessment, 11, 14, 24, 411–412
Tertiary explosives, stability of, 58

Testing
  current state and future needs, 32
  experimental studies, *see* Experimental analysis/data
Textiles, 442
Theoretical moment capacity
  diagonal shear effects, 228
  flexural resistance, 203
Theory
  combined with experiments, 31
  dynamic response and analysis, *see* Dynamic structural model, P-I diagrams
Thermal radiation, 5
  detonation processes and shock waves, 60–62
  planning and design considerations, 7
Thermobaric devices, 49, 60
Thermodynamic expressions
  chemical explosions, 55–56
  combustion phenomena and processes, 59
  shock waves, 62
Thickness, structural element
  cratering and spalling effects, 237
  estimation of reflected pressure duration, 169
  nuclear loads on buried structures, 181
  structural system and component selection, 431
Threat and risk assessment
  comprehensive design program, 438
  general principles and overview, 3, 8–16
  methodology, threat, and risk assessment, 11–15
  military threat, 19–24
  planning and design requirements, 5–6
  risk assessment, 11–15
  technical resources and blast mitigation capabilities, 24–25
  terrorism and insurgency, 24
  threat, hazard, and vulnerability assessments, 10–11
  military threat, 19–24
    tactics and strategy, 19–21
    weapons systems, 21–24
  planning and design considerations, 18
  progressive collapse potential of existing structures, General Services Administration (GSA) guidelines, 385
  research needs, 38
  terrorism and insurgency threat, 24
Three-dimensional stress wave propagation, 106–110

Index

Three-phase media, shock waves, 63–64
Threshold curves, 327, 339, 341, 348, 358, 364, 366, 368
Threshold loading, 363, 364
Threshold points/curves, 339–340, 341, 342, 358, 363, 366
Thrust
  coupled failure modes, 35
  structural element behavior
    cylinders, arches, and domes, 208–209, 210
    interaction diagram, 203–204
Tie forces, *see* Tying/tie forces
Time, scaled, 327, 328, 330
Time(s) of arrival
  ground shock, 128–129
  layered systems, 134, 135
  nuclear and HE explosions, 82
Time axis, P-I diagram idealized transient loading profile, 326
Time-blast phenomena
  internal explosions, 75
  planning and design considerations, 25
Time domain
  progressive collapse of steel frame buildings, 394
  SDOF analysis, 362
Time histories, 367; *see also* Load-time history; Pressure-time history; Time-pressure curves
  dynamic response and analysis, comparisons of experimental and analytical results, 272
  ground shock, 151
  multiple detonations, 35
  nuclear and HE explosions, 88, 89
    buried structures, late time response, 179
    durations for, 83, 90
  temperature-time histories, 151
Time-pressure curves; *see also* Pressure-time history
  blast on front wall from external explosion (surface burst), 187–189, 190
  blast on roof or sidewall
    span parallel to shock front, 190–191, 192
    span perpendicular to shock front, 190, 191, 192
  research needs, 34
Time scales, research needs, 438, 439, 441
Time versus displacement, direct shear and flexural response, 367, 368, 369, 371

Timoshenko beam, 339, 348
  cross-sectional equilibrium, 271
  dynamic response and analysis, 262, 263, 264
    comparisons of experimental and analytical results, 272, 273
  validation, 270
TM (Tri-Service manuals), *see* U.S. Department of the Army technical manuals
TNETB, 66
TNT, 55
  characteristics of, 58
  explosive constant, 154
  free air equivalent weights, 66
  hemispherical surface bursts, 73
  shock wave parameters for spherical TNT explosions, 71
TNT equivalents, 65
  calculation of, 56–58
  grenade characteristics, 46
  internal explosions
    gas pressure phase, 77
    shock pressure phase, 76
  mortar, artillery, and tank round characteristics, 44
Torpex, 154
Torsion
  blast loads and, 428
  progressive collapse of steel frame buildings, 380
  size effects and combined size and rate effects, 230
Training, 437–438, 441
Transfer girders, 429–430
Transfer of load, *see* Redistribution/transfer of load
Transformation factors, dynamic structural model of flexural response, 350–351
Transient effects, dynamic response and analysis, 255
Transient loading, P-I diagrams, 326
Transition point, P-I diagrams, 329, 333, 334, 335
Transition regions/zones, 336, 337, 338
  detonation processes and shock waves, 60–62
  one-dimensional shock flow, 61
  P-I diagrams, 339
    energy balance method, 337, 338
    threshold curve, 339
  progressive collapse, steel frame structure analysis, 391

Transitions
    compression to tensile membrane behavior, 346
    elastic to plastic behavior, 350
    between response modes, 30
Transmitted wave, one-dimensional wave transmission, 114–116, 117, 122
Transportation infrastructure, 440
Trans-seismic ground shock, 148
Transverse loads, 427
Trapdoor mechanism, soil arching, 185
Travel time, soil loading wave velocity and, 169
Tresca yield criterion, 266
Triangular deformation, failure at one support, 351
Triangular load
    conventional loads on buried structures, 170
    duration of effective triangles for dynamic pressure-time curve representation, 90
    dynamic pressure derivation, 72–73
    dynamic response, 330
        SDOF analysis, 288
    internal explosions, 76
    P-I diagrams, 330, 368
    P-I diagrams, closed form solutions, 334–335, 336
    P-I diagrams, energy balance method, 336–337, 338, 339
    P-I diagrams, SDOF system
        response functions, 332
        shock spectrum for, 333–334
    SDOF analysis, 248–250, 281, 288, 289
Triangular overpressure, 121
Triangular representation of overpressure-time curves, 85, 86
Triaxial compression, concrete, 201
Tributary loading, 430
Trigonometric functions, layered systems, 135
Triple point, 71–72, 80
Tritonal, 66, 154
Truss model for joint, 299, 300, 301
Truss structures/systems
    connections, 299, 300, 301
    diagonal strut and truss mechanisms, joint, 296
    progressive collapse, 379
Tumbling, projectile, 95
Tunnels
    airblast transmission through, 319, 322–324
    design approach, comprehensive, 421, 433
    entrance, 310
Two-way slabs, 204, 224
Tying/tie forces
    DOD design recommendations, 381–382
    structural elements, 373, 380–381

U

UFC, see Unified Facilities Criteria (UFC)
Ultimate moment, connections
    joint behavior, 304
    statically loaded knee joints, 297
Uncertainty factor, airslap-induced motion predictions, 151
Unconventional WMDs, 3
Undamped SDOF system
    closed-form solutions, 332
    pulse shape effects, elimination of, 330
Underground explosion
    cratering, 143
    nuclear devices, 80, 143
    peak airblast pressure estimation, 73, 74
Underground facilities, see Buried structures
Underground storage facilities, 433
Uniaxial compressive strength, shear walls, 213
Uniaxial stress-strain curves, concrete in compression, 199–200
Unified Facilities Criteria (UFC), 27, 196, 380, 407–408
Uniformly distributed loads
    design manual limitations, 25
    structural element behavior
        frames, 214
        membrane behavior, 223
Uniform mass, SDOF parameters
    for fixed beams, 278
    for fixed boundary beams, 279–280
    for fixed slabs, 282–283
    for simply supported beams, 277
    for simply supported slabs, 281
United Kingdom, safe standoff criteria development, 327
Unit resisting moments, two-way slabs, 204
Universal gas constant, 59–60
Unloading
    loading-unloading history, dynamic resistance function, 346–347
    unloading-reloading of load deflection curve, 348
Upgrades, planning for, 410
Upward forces
    design approach, comprehensive, 420, 428
    DOD design recommendations, 381

Index

short standoff bomb blasts, 428
U.S. Department of the Army technical
    manuals, 26
  standards, guidelines, manuals, and criteria,
    28, 29, 79, 195, 196–197
  structural element behavior, 195,
    196–197
  TM 5-1300, 26, 29, 79, 195, 196–197, 199,
    205, 233, 234, 235, 405, 433
    connections, 296, 307, 309
    connections, steel, 309
    door design, 310
    internal explosions, 319
    structural system and component
      selection, 431, 432
  TM 5-583, 26, 28, 29
  TM 5-855-1, 27, 41, 195, 406, 418
U.S. Department of Energy Manual 33, 407
USCNS recommendations, 36
U.S. Department of Defense, 405, 406
  experimental studies, military focus of, 25
  Explosive Safety Board (DDESB), 26
  minimum standards, 411
  mission of, 4
  publications, 407–408
  structural element behavior, 198
  weapons systems tests and data, 18–19
U.S. Department of Defense guidelines, 6
  progressive collapse, 380–384
    approaches and strategies, 381–384
    damage limits, 384
    new and existing construction, 380–381
  Unified Facilities Criteria (UFC), 27, 196,
    380, 407–408
U.S. Department of Energy Manual 33, 27
U.S. Department of Homeland Security,
    437
U.S. government agency publications, design
    standards and guidelines, 6,
    196–197; *see also specific*
    departments
  design approach, comprehensive, 405–407
  Naval Facilities Command (NAV-FAC), 26
  policy and technology needs, 35, 36
  progressive collapse mitigation and
    prevention
    Department of Defense, 380–384
    GSA, 384–390
  security engineering, 29
  structural engineering, 28–30
Utilities, 417, 432, 433

V

Validation
  dynamic response and analysis
    computational systems, 267–272,
      273
    technological and design developments, 8,
      38–39, 436
Valves, blast, 313–315
Variable boundary conditions (VBC), 376
Variable strain rate, 250
Vehicle bombs, 14, 198
Velocity
  dynamic response and analysis, 252
  SDOF analysis, 240, 331
Ventilation, 417
Venting, 430–431, 432
Verification of design, 8
Vertical failure plane, stress discontinuity at
    slab-wall interface, 238
Vertical loads, nuclear loads on buried
    structures, 184
Very low levels of protection (VLOP), 381
Vessel explosion, 330
Vibration, natural periods and modes, 215–218,
    235–236; *see also* Frequency
    characteristics
  dynamic response and analysis, 239, 253,
    254
    continuous systems, 261
    SDOF, simple, 242
  P-I diagrams, 326
  SDOF system response spectra, 327
  structural dynamics, natural frequency-
    duration of forcing or load
    function relationship, 328
  structural element behavior, 215–218,
    235–236
Virtual reality (VR) technology, 33
Virtual truss model, joint, 299, 300, 301
Viscous damping, P-I diagrams, 331
Vital systems/subsystems, nonstructural
  comprehensive design approach, 410,
    412–413, 417, 432
  planning and design considerations, 6, 29
  protected facility components, 7
VLOP (very low levels of protection), 381
Voids, soil properties from explosion tests, 132
Volume
  finite element approach, 265
  gas pressure phase, internal explosions, 77
  pressure-volume space, 61, 63
Volume detonation state, 62
Von Neuman spike, 61

Vulnerability assessment, 10–11, 12, 426; *see also* Threat and risk assessment

W

Waffle slabs, 432
Wall removal, alternate path analysis, 384
Walls, 442
    barrier, 413
    buried, 187, 188
    blast design load derivation procedures, 187, 188
    blast load on, 191–193
    comprehensive design approach, 413
        progressive collapse prevention, 416
        shielding devices, 432
        structural system and component selection, 430, 431
connections, *see* Connections and supports
    DOD design recommendations, 381–382
    duct, 323
    internal explosions, shock reflections from, 75
    loads on above-ground structures, 179
        conventional explosives, 170–171, 172, 173, 174
        nuclear and HE explosions, 178, 179
    nuclear explosion-induced shock waves, 121–122
    perimeter line, 413
    P-I diagrams, multiple failure modes, 343
    progressive collapse
        mitigation/prevention of, 416
        precast concrete structures, 377
    shear, 212–214
    structural element behavior, 212
        plastic hinge regions and, 232–233
        tensile and compressive members, 206
    structural/load discontinuity, 238
    time-pressure curves
        external surface burst, 187–189, 190
        span parallel to shock front, 190–191, 192
        span perpendicular to shock front, 190, 191, 192
Water content
    and cratering, 149
    and soil/granular material penetration, 96
    soil properties for calculations, 131
    soil properties from explosion tests, 132
Waterways Experiment Station (WES), 25
Wave equations, three-dimensional stress wave propagation, 109
Wave propagation
    explosive processes, 60–62
    ground shock, cratering, and ejecta from HE devices
        elastic waves, 1D, 110
        reflection and transmission of one dimensional D-waves between two media, 114–116
        shock waves in one-dimensional solids, 117–119
        stress waves, one-dimensional, in soils, 119–125
        stress waves, three-dimensional, 106–110
    internal explosions, shock pressure phase, 76–77
Wave propagation equation, 110
Wave velocity, soil loading, 169
Weapons
    conventional, *see* Conventional devices/weapons/environments
    nuclear, *see* Nuclear devices/weapons/environments
    threat assessment, 14
    understanding capabilities of, 17
Weapons of mass destruction, 440
    Soviet doctrine, 20–21
    unconventional, 3
Weapons systems
    combined, 18
    military threat assessment, 21–24
    technological and design developments, 434
Welds
    concrete anti-scabbing plates and, 89
    progressive collapse, 377
Wind, explosion dynamic pressure, 72–73
Wind loads, 422, 433–434
Windows: *see also* Openings and interfaces
    design approach, comprehensive, 414–416, 433
    flying debris minimization, 409
Winkler-Bach correction factor, 209
Wiring, faulty, 29–30
Wood penetration characteristics, 102, 162–164, 165, 166
World Trade Center, 10, 36, 390

Y

Yield behavior
    dynamic resistance function, 347
    dynamic response and analysis, 345
    material models, 266

Index

progressive collapse, sample analyses, 400
Yield line/yield strength, 25
    flexural behavior, 346
    membrane behavior, 220, 225
    structural design applications, 234
    structural system and component selection, 431
    tensile and compressive members, 207
    two-way slabs, 204

Young's modulus, 398

Z

Zero initial conditions, SDOF system, 331
Zero rise time, SDOF system, 330–331, 332
Zero standoff distance, *see* Contact detonations